T0203686

MICROBIOMES AND THEIR FUNCTIONS

MICROBIOMES AND THEIR FUNCTIONS

Why Organisms Need Microbes

Vasu D. Appanna

CRC Press
Taylor & Francis Group
Boca Raton London New York

CRC Press is an imprint of the
Taylor & Francis Group, an **informa** business

First edition published 2023
by CRC Press
6000 Broken Sound Parkway NW, Suite 300, Boca Raton, FL 33487-2742

and by CRC Press
4 Park Square, Milton Park, Abingdon, Oxon, OX14 4RN

CRC Press is an imprint of Taylor & Francis Group, LLC

© 2023 Taylor & Francis Group, LLC

Reasonable efforts have been made to publish reliable data and information, but the author and publisher cannot assume responsibility for the validity of all materials or the consequences of their use. The authors and publishers have attempted to trace the copyright holders of all material reproduced in this publication and apologize to copyright holders if permission to publish in this form has not been obtained. If any copyright material has not been acknowledged please write and let us know so we may rectify in any future reprint.

Except as permitted under U.S. Copyright Law, no part of this book may be reprinted, reproduced, transmitted, or utilized in any form by any electronic, mechanical, or other means, now known or hereafter invented, including photocopying, microfilming, and recording, or in any information storage or retrieval system, without written permission from the publishers.

For permission to photocopy or use material electronically from this work, access www.copyright.com or contact the Copyright Clearance Center, Inc. (CCC), 222 Rosewood Drive, Danvers, MA 01923, 978-750-8400. For works that are not available on CCC please contact mpkbookspermissions@tandf.co.uk

Trademark notice: Product or corporate names may be trademarks or registered trademarks and are used only for identification and explanation without intent to infringe.

Library of Congress Cataloging-in-Publication Data
Names: Appanna, Vasu D., author.
Title: Microbiomes and their functions : why organisms need microbes / Vasu D. Appanna.
Description: First edition. | Boca Raton, FL : CRC Press, 2023. | Includes bibliographical references and index.
Identifiers: LCCN 2022027034 (print) | LCCN 2022027035 (ebook) | ISBN 9780367763336 (hbk) | ISBN 9780367749897 (pbk) | ISBN 9781003166481 (ebk)
Subjects: LCSH: Microbial ecology. | Microorganisms.
Classification: LCC QR100 .A67 2023 (print) | LCC QR100 (ebook) | DDC 579/.17–dc23/eng/20220831
LC record available at https://lccn.loc.gov/2022027034
LC ebook record available at https://lccn.loc.gov/2022027035

ISBN: 978-0-367-76333-6 (hbk)
ISBN: 978-0-367-74989-7 (pbk)
ISBN: 978-1-003-16648-1 (ebk)

DOI: 10.1201/9781003166481

Typeset in Utopia
by codeMantra

CONTENTS

PREFACE

Although the microbial world is invisible to the naked eye, its influence on the visible world is omnipresent. Whether it is a simple green alga, a colourful mushroom, a delicious truffle, a somersaulting hydra, a menacing cloud of locusts, a mouth-brooding fish, a blooming orchid, a majestic male peacock, a gigantic blue whale or a human centenarian, microbes are the invisible partners that make these organisms the way they are. This book is a molecular exploration of the functions microbes perform for their hosts and explains why organisms need their microbes. These microbial communities referred to as the microbiomes are part and parcel of most if not all multicellular organisms. The microbiome comprises mostly viruses, bacteria, fungi and algae residing within and on living organisms. This book essentially deals with the roles bacteria play in the lives of their hosts and consists of nine chapters.

Chapter 1 provides a glimpse on how life emerged on planet Earth. The physical, chemical and geological events punctuated by happenstance that transpired during 1 billion years following the Big Bang nearly 4.5 billion years ago were the main drivers leading to the creation of a prebiotic environment. This milieu rich in simple organic and inorganic molecules fuelled the appearance of complex moieties like polysaccharides, polypeptides, lipids and nucleic acids culminating into proto-cells. These proto-cells with metabolic, replicative and asymmetric attributes gave rise to the last universal common ancestor from which emerged bacteria and archaea. These two prokaryotic systems are the building blocks of eukaryotes with the modified bacteria becoming the mitochondrion and the chloroplast. Further evolution of eukaryotes into multicellular organisms led to specific functions being performed by specialized cells, tissues and organs. This multicellularity with assigned components dedicated to defined functions is prompted by microbial input. Microbes became integral constituents of most organisms. Thus, microbes and their hosts constitute the holobiont, the organism we see.

The subsequent chapters elaborate on the functions microbes perform for their hosts and illustrate why organisms are dependent on their microbes. Each chapter deals with different organisms from animal and plant kingdoms ranging from algae to mammals. The chapter begins with a brief description of the anatomy and physiology of each organism, its ecological and economic significance. Then, the microbial constituents, their assembly and their roles in the day-to-day activities of the host are described. The conclusion section in each chapter explores how microbiome engineering can contribute to the betterment of society in agriculture, climate change, health and commercial output. Chapter 2 explains why algae cannot thrive without their microbial partners. The algal microbes that are housed intra- and extracellularly supply nutrients, fix N_2, promote anti-oxidative defence and act as surveillance agents. The importance of algae as part of other holobionts such as lichens, corals, sea anemones, salamanders and plants is explained. In sea anemones, they control sexual behaviours, while in salamanders these photosynthetic machines contribute to the development of embryos by modulating O_2 gradient. Chapter 3 takes a look at the molecular mechanisms bacteria invoke to support fungal life. They are endo- and ecto-symbionts that contribute to the range of functions required for the survival of fungi. They participate in the reproductive process and supply essential nutrients like tryptophan and vitamin B_{12}. They are parts of the defence and signalling networks. For instance, rhizoxin, a phytotoxin, is jointly elaborated by the fungus and its bacteria in order to attack plants. Fungus also constitutes an important part of numerous organisms including humans, plants, insects, birds and fishes. While in humans fungi can block the synthesis of nitric oxide, a pivotal intracellular messenger, in insects, fishes and birds, they help metabolize lignin.

Chapter 4 elaborates on functions executed by bacteria in sedentary aquatic organisms. The inability for these organisms to move freely has compelled them to enlist the assistance of microbial partners to produce antibiotics, to synthesize biominerals and to supply Pi and S. Corals depend on both bacteria and algae to propel their development and thwart pathogenic invasion. Hydra relies on their microbes for development, communication, agelessness and defence. Sponges are able to live near hydrothermal vents and on ocean surface due to the bacteria they select. In the former habitat, they possess methanogenic bacteria, whereas in the latter ecosystem, photosynthetic ones are preferred. The significance of bacteria in the lives of insects is explored in Chapter 5. Some of these microbes are lodged in specialized organs termed the bacteriocytes, while others reside within different organs and on the exoskeleton. They fortify the immune system, aid in digestion, guide the moulting process, impart colour, ensure sexual success, enhance adaptability and modulate social behaviour. For instance, the solitary and communal lifestyle of locusts is responsive to microbial cues. They are also sexual manipulators whereby they can tilt the balance to a female population. Insect societies are organized by chemical prompts generated by bacteria. The ability of insects to feed on blood, nectar, fruits, wood and plant toxins is dependent on the bacteria they harbour.

Chapter 6 is dedicated to microbiomes in fishes and the functions they fulfil for their hosts. These water dwellers possess a wide range of microbes on their skin aimed at dissuading pathogens with the toxins they produce and thwart predators with alarm signals. The production of the deadly tetrodotoxin by the puffer fish will be impossible without the supply of arginine from their resident microbes. Wood-eating Amazonian cat fish devoid of microbes will not exist without their lignin-degrading bacteria localized in the digestive tract. Fish living in deep ocean will be unable to survive without their light-emitting bacteria. The microbes within the swim bladder not only control the gas content of this buoyancy organ, but they also ensure its proper functioning. Chapter 7 examines the avian microbiome. The uropygial gland is home to numerous bacteria involved in generating volatile organic compounds responsible for kin recognition, pathogen defence and nest identification. The digestive tract is replete with bacteria aimed at satisfying the diverse nutritional habits that birds engage in. The blood-sucking vampire finch, wood-boring woodpeckers, herbivorous hoatzin and carnivorous vultures depend on their microbes to extract nutrients from diverse food sources. The ability of birds to migrate and adapt to different ecosystems is dependent on the nature of the microbiomes. Furthermore, seasonal changes and development are associated with marked shift in the microbial profile birds harbour. Feathers that are essential in avian locomotion are maintained by microbial residents. The colours associated with the plumage are also manipulated by microbial cues. In fact, the ornamental feathers numerous birds, including the majestic male peacock, display are important secondary sexual characters that ensure reproductive success. Microbial input is also critical in proper breeding and rearing of offspring. Hence, microbial imprint permeates all aspects of avian life.

In Chapter 8, the link between mammals and their microbiomes is emphasized. Following the description of how microbial communities are assembled from conception to adulthood, the contribution of microbes to the welfare of their hosts is explained. The intimate connection of the microbes to the digestion tract, their influence on the immune system and the impact on the ageing process are explored. The nutritional habits of the armadillo, the baleen whale and the panda are dictated by their microbes. The relationship between diseases such as obesity and cardiovascular illnesses is described. The odours generated by microbes are important components of the communication networks in numerous mammals. The role of mucins and milk in recruiting microbes is illustrated. Hibernating mammals rely on their microbes to optimize their metabolic output. Chapter 9 is dedicated to the functions of the microbiome in plants. Since these complex organisms are rooted in their habitats, they are reliant on microbes to accomplish various tasks. Plants secrete different chemicals to construct the proper microbiome. For instance, monosaccharides help select *Bacillus subtilis*, while amino acids attract N_2 fixers. Even flower development and fruit ripening are responsive to microbial cues. Organisms are the way they are as a consequence of their microbes. These microbial components are integrated from the beginning in most multicellular organisms as very flexible components in an effort most probably to obviate the need for further 'dead-end' cellular specialization. Just imagine how many specialized cells a human or plant will need to have without the presence of the malleable and highly adaptable microbes. Remember cellular specialization optimizes a given functionality but is limited in its ability to adapt. The partnership between microbes and their hosts affords numerous advantages. Hence, this strategy has been adopted throughout the flora and fauna in our ecosystem.

This book draws from data generated from the burgeoning discipline of microbiome. Although more functions that microbes accomplish for their hosts will be uncovered, this book provides a comprehensive understanding as to why organisms from disparate phyla need their microbes. And this relationship has been set from the beginning of life on planet Earth. I am indebted to my wife Dr. Sharina Appanna for meticulously reading and commenting on the text. I thank Alex MacLean for converting my scribbles into vivid captivating illustrations reflective of the words expressed in the book. I am also forever grateful to my all students for having sharpen my ability to convey complex scientific concepts in a simple and enticing format that I am quite confident is evident in this book.

Vasu D. Appanna, PhD
Sudbury, Stoney Creek
Ontario, Canada

AUTHOR

Dr. Vasu D. Appanna has been studying and teaching the molecular intricacies that govern life for more than 30 years in Canada and world-wide. He has published extensively on microbial systems surviving in stressed environments and the very successful and widely distributed book – *Human Microbes-The Power Within - Health, Healing and Beyond.* In excess of 150 highly qualified personnel have been trained and mentored in his laboratory. He is currently a Professor of Biochemistry at Laurentian University, Sudbury, Ontario, where he has also served as Department Chair and Dean of Faculty.

THE GENESIS OF LIFE – HOW IT ALL STARTED?

1

Contents

Keywords

- Prebiotic Molecules
- LUCA
- Microbes
- Multicellularity
- Apoorganisms
- Holoorganisms

DOI: 10.1201/9781003166481-1

1.1 The Beginning: Transformation of Chemical Elements into Prebiotic Molecules

How life started on planet Earth? This question has dogged humanity since the dawn of civilization. There is no simple answer as the exact conditions that prevailed when Earth came in existence around 4.5 billion years ago are not easy to replicate. The physical and chemical environments that existed at the beginning have evolved. Although the birth of other planets in distant galaxies and the meteorites that sometimes pass near Earth reveal the characteristics of the chemical constituents inherent in these galactic masses, the precise physico-chemical and geological make-up at the very beginning of the earthly journey still needs to be fully understood. As it is nearly impossible to go back to those prebiotic situations, one must look at present-day organisms and decipher the common molecular link amongst all living creatures.

To appreciate how life began, one needs to recognize the commonality of the molecules and the processes that are operative in most living entities on this planet. For instance, what is the common thread that links diverse organisms like an elephant, a giant sequoia tree, a human, a squid, a coral, a bacterium and a virus? – they are constituted of the same basic molecules and have very similar biochemical processes. They are all made up of proteins, nucleic acids, carbohydrates, lipids, water and some mineral elements. These proteins comprise essentially 21 amino acids, while the nucleic acids are composed of six bases that find themselves in virtually all organisms (**Figure 1.1**). They use the same chemical currency as the source of energy, namely ATP (adenosine triphosphate), and if any living organism is synthesizing this molecule, substrate-level phosphorylation (SLP), photophosphorylation (PP) and oxidative phosphorylation (OP) are the only processes earmarked for this task. If a living entity is breathing, it has to process its reduced food source via the tricarboxylic acid (TCA) cycle and OP to generate ATP. The former produces the reducing factors with the concomitant evolution of CO_2, while the latter utilizes oxygen to transform the reducing factors into a chemical potential that propels the synthesis of ATP (**Figure 1.2**).

Once the common building blocks in all organisms roaming, flying and swimming on the planet have been established, then it is important to chart an itinerary how these chemical ingredients were assembled together into living entities. The current evidence and hypotheses point to the formation of planet Earth approximately 4.8 billion years ago, while the life did not manifest itself 800 million years later. It is clear that the geological events that existed and evolved during this time were pivotal. Hence, to fully begin to understand the origin of life, a multi-disciplinary approach involving geology, chemistry, biology and physics is essential. In the absence of data related to the geophysical events that led to the formation of planet Earth, it is logical to conclude that the long lag phase before the emergence of life allowed appropriate geological and chemical phenomena to occur that resulted in the genesis of the precursor molecules found in living organisms we see today. The solidifications, vaporizations, condensations, drastic temperature fluctuations and cosmic interactions may have provided the precise physical conditions for the proper chemical reactions to manifest. These organic molecules coupled with the inorganic moieties may have been the trigger behind the biomolecules we see operating in living organisms. The biochemical processes mediated by these molecules will again adapt to the physical and chemical environments prevalent during the geological periods when different forms of life emerged. For instance, even though bacteria can be traced back to 3.6 million years ago, plants did not emerge until 700 million years later. The initial life forms were anaerobic as oxygen (O_2)-containing atmosphere was evident only around 2 billion years ago. Humans did not start roaming this planet until approximately 2.8 million years ago. Hence, it is clear that the biological events mirrored in all organisms have been shaped by the variety of chemical, climatic and geological fluxes that transpired on planet Earth at different epochs (**Figure 1.3**).

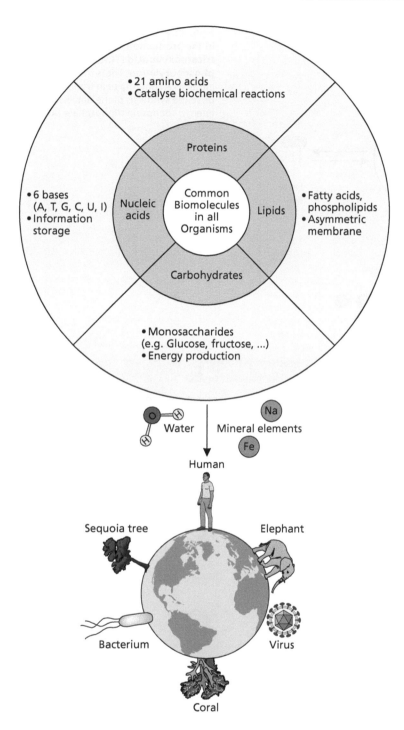

Figure 1.1. Common molecules found in all living organisms and their main functions. A virus, a human, an elephant and a sequoia tree are made of similar building blocks. Monosaccharides, amino acids, fatty acids, nucleic acids and mineral elements constitute the primary compounds that permeate all life forms.

1.2 The Chemical Precursors – The Chemistry of Inanimate Matter

In order to have an appreciation of how life began on Earth, it is important to understand the conditions that helped catalyse the creation of chemicals that became part of all living organisms. As mentioned before, all living organisms have the same building blocks albeit in varying amounts and mostly participate in similar biological functions. Proteins are the workhorse as they are involved in performing the day-to-day activity in a cell, while the deoxyribonucleic acid (DNA) is the guardian of the genetic information each organism is bestowed with. This information is transmitted essentially via the ribonucleic acid (RNA) that translates the message into proteins. Carbohydrates are de facto the main provider of

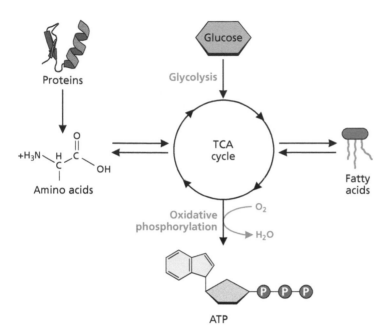

Figure 1.2. Biochemical pathways involved in the production of energy glycolysis, tricarboxylic acid (TCA) cycle and oxidative phosphorylation. These metabolic networks either as an entirety or in truncated versions drive the synthesis of the universal chemical energy, adenosine triphosphate (ATP) in all organisms.

BOX 1.1 ENERGY PRODUCTION IN LIVING ORGANISMS: ATP (ADENOSINE TRIPHOSPHATE)

The universal energy currency in all living systems is ATP, a chain of phosphate anchored to adenosine. The chemical energy stored in the terminal phosphate bond is readily transportable and utilized by virtually all cells. The only route to this high-energy phosphate is via the addition of phosphate to ADP (adenosine diphosphate), a process mediated essentially by three biochemical processes referred to as SLP, OP and PP. During SLP, high-energy compounds produced as a result of cellular metabolism are utilized to phosphorylate ADP. For instance, the 1,3-diphosphoglycerate and phosphoenol pyruvate obtained following glycolysis are converted into ATP. The energy sequestered in reduced nutrient like D-glucose is metabolized during glycolysis and TCA to liberate nicotinamide adenine dinucleotide NADH and flavin adenine dinucleotide $FADH_2$. These electron-rich moieties are oxidized by O_2 via OP with the aid of the electron transport chain ETC to generate a proton motive force (PMF) that drives the addition of phosphate to ADP. PP captures the reductive power of solar energy to drive the cleavage of H_2O during photolysis to form a PMF and O_2. The

former is subsequently channelled into the synthesis of ATP. Both OP and PP invoke the participation of ATP synthase to harness the PMF into the addition of phosphate to ADP. Regardless of the kingdom an organism belongs to, at least one of these biochemical manipulations is responsible for the synthesis of ATP, a moiety that powers the vast majority of biochemical activities (**Box Figure 1**).

Box Figure 1. The ATP-synthesizing machineries in living systems.

energy through their catabolism. Lipids, on the other hand, have taken the role of the well-guarded permeable cellular frontier dedicated to the compartmentalization of biological activities. This intricate chemical landscape then begs the question how all these started? The answer may lie in the chemical changes and the geological events that transpired from the existence of the planet Earth, 4.5 billion years ago and 3.6 billion years since the first life form sprang.

Simple molecules such as H_2, CN, CO, CO_2, NH_3 and H_2O can be traced back to the very beginning when Earth was formed. The presence of these compounds coupled with the transformative power of physical phenomena that were persistent in the nascent Earth may have played a pivotal role in generating moieties that may have propelled the synthesis of biomolecules currently occurring in all living organisms. The constant

Geological period	Million years ago
Man	2.5
Dinosaurs	145
Reptiles	252
Plants	320
Fishes	416
Straight cephalopods	488
Trilobites	542
Bacteria	3,500
Big Bang	4,500

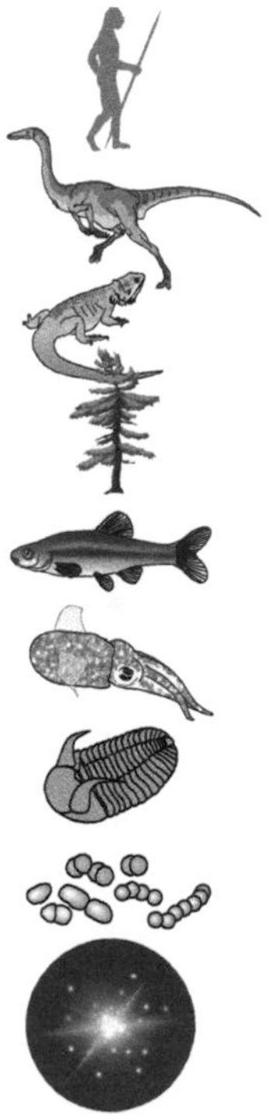

Figure 1.3. **The emergence of disparate living organisms at different epochs during the evolution of planet Earth.** Life emerged around 1 billion years following the genesis of planet Earth with microbial systems occurring first and humans appearing only 2.5 million years ago.

temperature fluctuations, the drastic climatic changes and the ensuing evaporations (this event has the power to concentrate reactants) may have provided an added stimulus to extend the range of chemicals that nature has at its disposal to orchestrate the precise machinery conducive to create living and replicative entities. The cycle of hydration and dehydration coupled with the dynamic range of soluble inorganic elements may have been other catalysing factors. Fluids released from the hydrothermal vents rich in reactive molecular species may have precipitated the production of the precursors of the chemical moieties that constitute present-day living organisms. The volcanic eruption at the Kilauea Mountain, Hawaii in 2018 provides a vivid glimpse of the chemical concoction that can emerge out of such an event. Laze that is laden with hydrochloric acid, sulphur dioxide and other constituents is a fertile environment for a variety of chemical reactions to occur. These ensuing intermediates subjected to elevated temperatures would be ripe for proper chemistry to act and create products that would otherwise not see the light of day.

Although such geological and climatic occurrences magnified by numerous folds that were common on the young planet Earth cannot be readily re-enacted in the laboratory, data obtained by reacting molecules like H_2, CH_4, NH_3 and H_2O in early Earth conditions have revealed the formation of amino acids and nucleic acids. In a seminal experiment attempted by Miller and Urey in 1952, an electric discharge was able to catalyse the formation of amino acids from the chemical constituents predicted to be present in early Earth. Alanine, aspartate and glycine were detected as were evident carboxylic and hydroxyl groups. For instance, the following equation illustrates how the precursors of complex biomolecules might have originated (**Figure 1.4**):

Cyanide (CN) and urea may provide the ingredients to produce amino acids and peptides in the presence of metal catalytic agents such as Ni, Fe

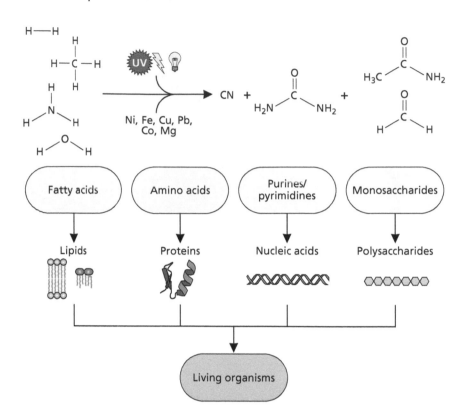

Figure 1.4. **Transformation of simple molecules into prebiotic and biotic chemicals.** Simple molecules like H_2, H_2O and CH_4 with aid of such catalysts as light, heat and metals can readily generate cyanide, a moiety that can give rise to nitrogen compounds. The formation of urea, acetaldehyde and formaldehyde can trigger the synthesis of precursor organic molecules (monosaccharides, amino acids, fatty acids and nucleic acids), the building blocks of proteins, lipids, carbohydrates and nucleotides.

BOX 1.2 PREBIOTIC CHEMICAL PRECURSORS TO BIOMOLECULES

Simple molecules such as H_2, CO, CO_2, H_2O and N_2 lend themselves to a variety of chemical transformations in environments with extreme temperatures, hydrothermal vents, UV radiation and varying pHs. These reactions are also influenced by the presence of metals and minerals. Formate, acetate, pyruvate, urea and CN derivatives that may result from these simple chemical entities can fuel the synthesis of organic molecules, the precursors of biopolymers that constitute the building blocks of all life forms on planet Earth. For instance, the Fischer–Tropsch reaction involving the interaction between CO/H_2 and metal catalyst at high temperatures can afford long-chain hydrocarbon the building unit of lipids with amphipathic characteristics. Aldehydes can result into monosaccharides via the formose transformation, while urea and CN derivatives can provide a route to amino acids, purines and pyrimidines. When subjected to high temperatures, methane and N_2 are known to yield the amino acids glycine and alanine. These initial ingredients that were eventually concocted into molecules of life might have also been delivered to planet Earth via celestial activities like meteorites. Their further processing aided by geological events punctuated by wetting and drying in an enclosed space rich in metals like Mg may provide the ideal environment to form oligomers of carbohydrates, amino acids, lipids and nucleic acids. Specificity dictated by stability would give rise to enantiomers with L-configuration in proteins and D-conformer in polysaccharides. Proteins comprising L and D amino acids are not known to form well-defined tertiary and quaternary structures. These chemical reactions yielding a plethora of molecular entities that were prevalent following the birth of planet Earth most likely may have set the stage for the emergence of life (**Box Figure 2**).

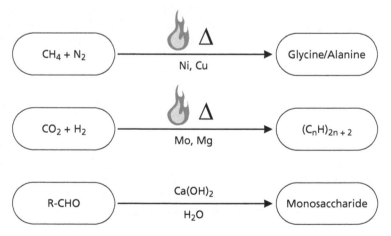

Box Figure 2. **The synthesis of glycine, alanine, alkanes and monosaccharides from elemental precursors.**

or Cu. Cyanohydrins and cyanoacetylene can lend themselves to be the transformed into such heterocycles as purines and pyrimidines, the building blocks of nucleic acids from which are derived RNA and DNA, respectively. The formose reaction mediates the synthesis of monosaccharides with the participation of formaldehyde (HCHO) and catalysts such as Pb and $Ca(OH)_2$. Molybdate and Mg are known to stabilize pentoses and hexoses. The Fischer–Tropsch reaction can produce fatty acids when CO and H_2 react with metal catalyst at high temperature. These would be the starting material to construct phospholipids, cholesterol and terpenes. Free radicals generated by metals such as Fe^{3+} would be a potential initiator of the polymerization process that would help in the formation of the complex polymers that are part of all living organisms.

Even though simulation of the primitive Earth conditions in laboratories has yielded many of these compounds, the possibility that these chemical moieties might have been delivered by meteorite striking the Earth cannot be ruled. Indeed, the Murchison Meteorite did reveal the presence of numerous organic moieties including aromatic compounds and amino acids. The inherent technical difficulties in creating the original Earth conditions have not dampened the search of the geochemical reactions responsible for these starting molecules. Fossilized organic matter has provided clues to some of these primordial chemicals. The changing Earth's climate that has been characterized by the dramatic increase and decrease in temperature may have been an important trigger aimed at the condensation and polymerization of these monomers. This process may also have been aided by the sequestration of these reactants in an enclosed environment like a vesicle where intimate interactions due to close proximity are favoured.

1.3 From Prebiotic Molecules to Polymers and Protocells

The assemblage of complex molecules that were eventually incorporated in all living systems necessitated their interaction in a confined space in the presence of a variety of catalysts and chemical modulators. Clay and vesicle composed of amphiphilic moieties are two possible structural templates where this process can be materialized. Alkyl phosphates, alkyl sulphates and polyprenyl derivatives may have the ideal functionalities to form the non-covalent ensembles such as micelles and supra-structures where the prebiotic molecules may be trapped. Lipid bilayer is a universal constituent of all organisms. In the presence of nucleophiles, electrophiles and redox metals like Fe and Cu that can readily accept or donate an electron, numerous chemical processes can be promoted. The non-equilibrium phase favoured by this arrangement would select for the most thermodynamically stable compounds. The influence of temperature, humidity, UV light, visible light and pH may have further shaped the chemical products that manifest themselves in all organisms on Earth. Carbohydrates, lipids, proteins, RNA and DNA with all their specificity and selectivity may have been acquired when they were assembled randomly in an enclosed space like clay or lipid layer. These molecules are flexible in their activity as they adapt to the environment they are in. For instance, organisms growing in a phosphate-deficient medium tend to substitute their phospholipid bilayer with amino lipids, while substituting the bases in the DNA leads to the synthesis of proteins with disparate amino acids.

1.3.1 The origin of metabolic processes and other ingredients of living systems

The ability of micelle or vesicle to form large congregation and split into smaller entities may have led to the emergence of complex chemical behaviours resulting in self-assembly, self-replication and acquisition of specific functions. Autocatalytic networks and proto-metabolic cycles may have then followed. One possible autocatalytic feature may have led to the addition of formaldehyde to glycoaldehyde culminating in the synthesis of glyceraldehyde and other higher carbon products. A reductive TCA cycle may have originated in this fashion devoid of any enzymatic input. Glycolysis, a key metabolic process aimed at extracting energy from glucose, a product that can be obtained by the formose process, is another example of a primordial network. These lipogenic aggregates may then be assembled into protocellular compartments where select complex molecules are located and are engaged in some unique behaviour dedicated to metabolism or autocatalysis or precursor functions that have become part of living systems. In order for protocells to be the building blocks for a living system, these supra-structures must integrate in a platform with the proper connectivity to self-maintain, self-replicate and self-reproduce with a high degree of fidelity. To execute these tasks, the following conditions have to be fulfilled:

1. A selective permeable barrier with transport processes to maintain a non-equilibrium system must be in force.

2. A metabolic machinery aimed at gathering nutrients and extracting energy must be operative.

3. An information-rich apparatus (genetic) to store and transmit information must be deployed.

These functional entities may have evolved at the same or different periods depending on the conditions that existed on Earth at a given time before the first living organism emerged.

1.4 Components of Protocells

The formation of an asymmetric barrier with a high degree of plasticity is central to this process. Fatty acids, isoprenoids and other molecules possessing a polar head and a hydrophobic tail can aggregate non-covalently with the aid of weak interactions like hydrophobic, van der Waals' or electrostatic force. The flexibility of these delicate contacts would enable a variety of supra-structures depending on the concentration of the amphiphiles, pH and temperature. In living systems, the bilayer configuration is the most predominant one. The incorporation of glycerol, phospho- and glyco-lipids may have added more selectively not just in the uptake and extrusion of chemicals but may have also helped trap proteins and biomolecules with dedicated functions. These compartments may also be constituted with a variety of chemicals with the goal of generating energy. Iron–sulphur moieties may fuel the formation of energy that would drive the synthesis of organic molecules that may become building units of macromolecules or a source of energy. RNA with the ability to store information and catalyse its own replication may have been the critical biomolecule via which life may have originated. Unlike DNA that can store information but does not have catalytic capability and proteins that have the power to facilitate chemical reactions with no information-storing aptitude, RNA can accomplish both these tasks. Hence, this molecule may have been the initial propeller driving the original life form. Indeed, retroviruses like the SARS-CoV-2, agent of the current global pandemic, are organisms with their genetic information residing in RNA moieties. The functional plasticity of ribozyme with its ability to manipulate chemical bonds and store information may have been the initial trigger in the transmittance of information. DNA as a molecule may have evolved later to become the ultimate custodian of cellular information. The elimination of oxygen in the ribose leading to the formation of deoxyribose may provide the stability feature needed in an information-storing chemical like the DNA.

In fact, deoxyribose devoid of the oxygen in the 2-position is numerous fold more stable than ribose. Furthermore, virtually no organism can metabolize deoxyribose. This highly resistant nature of DNA may have contributed to making this biomolecule the preferred custodian of cellular information in almost all organisms.

The dearth of data related to the prebiotic era is a major impediment to precisely pinpoint how life originated. Hence, this hypothesis is rich on speculation but poor on evidence. The emergence of synthetic biology and the ability to create non-existing natural biological information is providing support to the notion as to how an abiotic system can culminate into a biotic entity by a variety of stratagems and happenstance. Dedicated changes in the bases in nucleic acids and amino acids in proteins can result in the introduction of some very selective functions. Thus, it may be within the realm of possibilities that once these three disparate entities encompassing an energy-generating feature, an information-transmitting attribute and a flexible boundary derived from geochemical and physical processes were housed together to create a biotic system. The propensity of the malleable amphipathic structures to fuse into a common space may have triggered the birth of the first living system with the capacity to propagate in a sustainable and adaptable manner. This momentous occurrence may have occurred in hydrothermal vents or other extreme environments where extant biochemical processes are abundant. The high temperature and elevated pressure associated with these environments coupled with the profusion of organic and inorganic matter is an ideal breeding ground for abiotic, biotic and any activity in between spring. These events may have kick-started life and may have led to the emergence of protocell, a supra-molecular structure between inanimate and animate entities (**Figure 1.5**).

1.5 The Last Universal Common Ancestor

The protocell with its metabolic system and its polynucleotide assembled in an enclosed vesicle either separately or together had the necessary molecular components to transition to an animate organization capable of perpetuating itself. This body of molecules resulted in the first living cell referred to as the last universal common ancestor (LUCA) of all living organisms. This living system with the ability to replicate may have appeared around 3.6 billion years ago and most likely was a thermophile thriving in an alkaline environment and bestowed with genes for reverse gyrase, an enzyme involved in modulating DNA structures. Microfossils found in Strelley Pool, Australia, point to the emergence of archaea and bacteria about 3.5 Byr ago, while eukaryotes made their appearance 800 Myr later. LUCA operated in the world of genes and proteins, an environment where all known living organisms function. Phylogenetic trees constructed from these genetic data indicate that numerous genes from LUCA are common to all three domains of life. In fact, an analysis of the genetic information from 2000 bacteria that lived when LUCA was alive revealed that 355 genes were common and may have been acquired from LUCA. The genes uncovered so far reveal an interesting biochemical portrait of LUCA. The production of energy mediated by a proton gradient appears to be a feature shared by all life forms. It is likely that LUCA utilized the gradient generated by Na^+ (sea water) and vent water to create the energy-yielding charge separation. The presence of acetyl CoA synthetase would suggest SLP as an important route to obtain ATP mediated by the trapping of the energy-rich thioester, a process that may be aided by carbon monoxide dehydrogenase. This enzymatic duo would allow the anaerobic growth promoted by H_2/CO_2. Fe-S cluster proteins and O_2-sensitive enzymes coupled with the ability to consume CO_2 and H_2 are tell-tale signs of an anaerobic lifestyle. The harnessing of energy via acetyl phosphate and Na^+/H^+ gradient mediated by a rudimentary ATP synthase would provide the

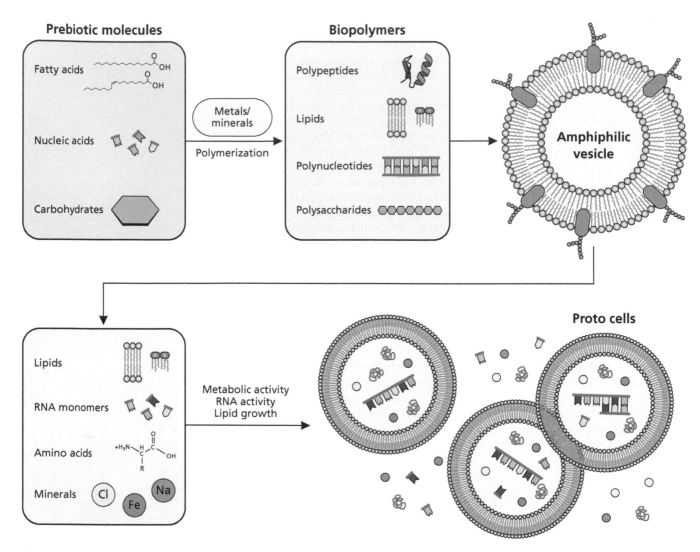

Figure 1.5. From prebiotic molecules to polymers to protocells, precursors to living cells. The prebiotic chemicals gave rise to polysaccharides, polypeptides, lipids and polynucleotides. Lipids with their amphiphilic features became the structural enclosures where chemical reactions can be segregated resulting in metabolic activities. The polyribonucleotides (RNAs) provided the molecular entities to store and transmit information. The protocells with their asymmetrical features, metabolic attributes and information-storing capacity laid the foundation for living cells.

energy needed to sustain life. The presence of reverse gyrase, an enzyme common in thermophiles, supports the notion that life may have unfolded in a hot environment. High salt concentration, presence of minerals and ability to fix N_2 and CO_2 and utilize H_2 as a reductive force may provide the right chemistry to spur a sustainable biotic entity. LUCA had all three ingredients namely an asymmetrical enclosed space, metabolic network and a replicative information apparatus to perpetuate itself. Although this current state of knowledge on a hard-to-locate entity may be ephemeral with the advent of new analytical tools and the discovery of more ancient fossils, it is clear that there is an intimate link between LUCA and all organisms roaming this planet (**Figure 1.6**).

1.6 The Prokaryotes

1.6.1 Archaea: microbes prolific in extreme environments

Prokaryotes are unicellular organisms devoid of any stable membrane-bound organelle including the nucleus. Archaea and bacteria that constitute the bulk of prokaryotes on this planet evolved from LUCA. The domain

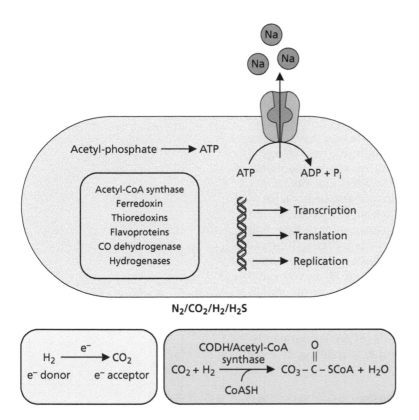

Figure 1.6. Depiction of some biochemical interactions associated with the LUCA. LUCA is the cellular beginning that led to the genesis of prokaryotes. It possessed metabolic networks to synthesize ATP from such gases as H_2, N_2, CO_2 and H_2S. The hydrogenases and dehydrogenases coupled with a variety of other redox proteins generate a Na^+ gradient mediating the production of energy. Acetyl phosphate may provide a ready source of ATP via substrate-level phosphorylation (CODH: carbon monoxide dehydrogenase; ferredoxins: iron redox proteins, thioredoxins: sulphur redox proteins, Na^+ pump to generate energy; CoASH: coenzyme A).

Archaea, a term derived from the Greek word archaio, means original or ancient. They thrive in environments reminiscent of those conditions that prevailed when life first started. Although archaea were first discovered in extreme surroundings such as hydrothermal vents and saline pools, they are known to proliferate in numerous diverse habitats including the human body. Hence, they are no longer considered extremophilic oddities but are indeed part of the global microbial life that all organisms have come to be dependent on. Genomic studies have revealed their uniqueness as approximately 15% of the genes are archaea-specific as no bacterial or eukaryotic homologues have been uncovered. They possess biochemical attributes that are absent in other domains of life. Ether-linked lipids that constitute the membrane barrier are only found in Archaea as phospholipids tend to dominate the other life domains. The membranes are decorated with elongated lipids that can have up to 50 carbons, a structural feat known to bestow unique properties. This feature in tandem with the inclusion of cyclopentane allows these microbes to maintain their fluidity irrespective of the extreme temperatures they are confronted with. They tend to possess unique chemical constituents that are absent or rare in other domains of life. Pyrrolysine is an amino acid almost exclusively confined to the world of archaea and is a key component of proteins involved in methanogenesis, a biochemical network orchestrating the transformation of CO_2 into CH_4, a metabolic pathway inherited from LUCA.

Archaea also harbours a variety of metabolic machineries that set members of this domain of life quite apart from either bacteria or eukaryotes. Glycolysis is a universal metabolic network that enables living organisms to process glucose in such a way as to maximize ATP production and generate metabolites for the TCA cycle and other biosynthetic pathways. This process is usually driven by ATP. The expenditure of two ATP liberates four ATP with a net gain of two ATP. In archaea, ADP is utilized as the initiator of the biochemical reactions resulting in a better energy yield. Enzymes like hexokinase and phospho-fructose kinase are powered by ADP instead of ATP, the usual energy provider in bacteria or eukaryotes. They also partake in the glycosylation of protein, a characteristic limited to eukaryote but not to bacteria. Furthermore, Archaea can transform CO_2 to CH_4 as strict anaerobiosis is required for this biological process to occur. These two gases were prevalent in relatively high amounts in ancient Earth's atmosphere, and most of the other organisms living currently on this planet seem to have lost this characteristic. In fact, the enzyme like methyl coenzyme M reductase with its coenzyme F430 that is critical for methanogenesis is relatively limited to archaea. Hence, this biochemical attribute appears to be a vestige of the ancient world. Nitrogen fixation and ammonia oxidation are two biological processes encountered in Archaea. The former utilizes similar mediators like bacteria to effect the conversion of N_2 into NH_3. While molybdenum is a cofactor of choice of the enzyme nitrogenase in archaea, bacteria seem to recruit the assistance of a variety of metals like iron, molybdenum and vanadium to drive the reduction of N_2. It is also important to note that archaea have also been found in numerous organisms including humans where they perform numerous functions for their hosts. For instance, they play an important role in mitigating the build-up of H_2 in the large intestine, a gaseous component released by anaerobic fermentation promoted by other microbial partners. The transformation of H_2 into CH_4 by such methanogens as *Methanobrevabacter smithii* helps promote improve the energy budget. Small-chain fatty acids (SCFAs) generated by the archaea residing in the colon also fulfil other physiological roles critical to the well-being of the hosts.

1.6.2 Bacteria – the highly adaptive prokaryotes

Bacteria are the other prokaryotes whose ancestry may be traced back to LUCA. They have distinctive features that differentiate them from their archaea counterparts. For instance, the bacterial cell wall is composed of peptidoglycan a feature absent in the archaeal domain of life. They also possess phospholipids and lipopolysaccharides as cellular barriers. The peptidoglycan content is the basis of the Gram stain classification that allows bacteria to be grouped as Gram positive or Gram negative. Bacteria come in various morphologies ranging from spherical to spiral forms. Unlike the archaea and the other domain of life the eukaryote, they do not contain histones, proteins known to be associated with nucleic acids. They are distributed in most if not all ecological niches on this planet ranging from aerobic, anaerobic and any milieu in between. They are known to adapt to virtually any lifestyle including the chemolithotropy whereby energy is derived from inorganic compounds such as hydrogen, nitrite, sulphide, ammonia and Fe. They can acquire their energy by SLP, OP and PP. They encompass a wide variety of metabolic networks including glycolysis, TCA cycle, pentose phosphate pathway and β-oxidation that enable them to adapt to almost any living conditions. It is hypothesized that at

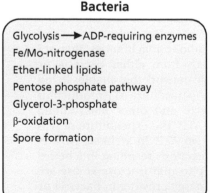

Figure 1.7. Biological characteristics of prokaryotes: archaea and bacteria. Although archaea and bacteria are descendants of LUCA, they have numerous characteristics that are organism specific. For instance, archaea possess ether-linked lipids and introns while bacteria have ester-linked lipids and pentose phosphate pathway. Some of these unique features from both the prokaryotes are prevalent in eukaryotes. For instance, numerous aspects of the genetic machinery of archaea are found in higher organisms.

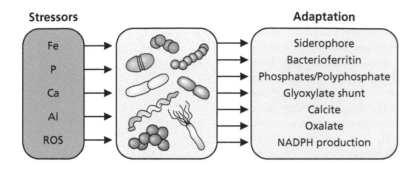

Figure 1.8. Bacteria have tremendous ability to adapt. Owing to the non-specialized nature of bacteria, they can readily adapt to a variety of stresses. A diverse range of metabolites are utilized to combat these challenges. If there is a dearth of Fe, bacteria produce iron scavengers known as siderophores, molecules with very high affinity for this trivalent nutrient. They can reconfigure their metabolic networks to produce chelators like oxalate that can immobilize disparate metal toxicants. Ca is trapped as calcite, while NADPH is a powerful anti-oxidant involved in maintaining proper cellular redox potential.

any given time a bacterium utilizes only 5% of its genetic information to live and is ready to deploy the remaining part to adapt to fluctuating conditions in its habitat.

This arrangement provides the bacterium with a wide-ranging molecular toolbox to accommodate any environmental conditions that may prevail unlike the specialized cells limited in their ability to respond to changing situations. For instance, a lack of Fe can bring the demise of most aerobic organisms. Humans will die of anaemia but not bacteria. They have evolved to secrete siderophores, nano-soldiers aimed at scavenging Fe. These biomolecules have a very high affinity for Fe and help fulfil the need for this micro-nutrient. On the other hand, if iron becomes excessive iron overload can be a problem. Bacteria elaborates bacterioferritin where this mineral is safely sequestered. The metabolic flexibility afforded by the vast amount of under-utilized genetic information allows bacteria to proliferate under disparate nutritional regimes. While many organisms would not survive a constant supply of fatty acids or acetate, bacteria can readily proliferate when challenged by these nutrients on an ongoing basis. Instead of invoking the TCA cycle to metabolize these nutrients, bacteria switch to the glyoxylate shunt whereby the acetyl CoA is fixed into malate and not converted into CO_2, an end product of the TCA cycle. Malate can then be transformed into glucose and ribose, moieties known to propel cellular growth. This versatility has also enabled bacteria to accomplish a variety of other tasks that are essential for the sustainability of the ecosystem, an event crucial for the survival of planet Earth. They contribute to the carbon cycle and nitrogen cycle and play a key role in the mineralization of all if not most wastes. Whenever any living system dies, its constituents are laboriously consumed by a variety of natural recyclers. However, the final process of generating CO_2, N_2 and H_2O is the realm of the bacteria as they possess the molecular machineries to do the terminal dismantling (Figure 1.8).

1.7 Eukaryote – The Initiator of Specialized Life

The term eukaryote is derived from the Greek 'eu' meaning true and 'karyon' meaning nut as it is an organism with a true nucleus. It has an enclosed compartment where the genetic information is stored and also possesses a variety of distinct intracellular arrangements where specific biological functions for the proper operation of the cell are executed. Eukaryotes can be unicellular and multicellular organisms where the division of labour within the cells is on display and intricately controlled. In this fully compartmentalized intracellular space, biomolecules are trafficked in a deliberate fashion and are not subject to random diffusion. The nucleus, the largest organelle, is involved in the transmission of genetic information where replication and transcription occur. While the mitochondrion is the home of the energy-producing machinery of a eukaryotic organism, the membranous enclosure referred to as the endoplasmic reticulum performs a wide range of functions including the modification of proteins. The lysosome is the acidic organelle responsible for the degradation of proteins tagged to be recycled. The peroxisome houses numerous enzymes with a mandate to detoxify the harmful reactive oxygen species. This domain of life with an intricate division of functions did not occur until 1.5 Byr later following the emergence of the prokaryote. In the absence of a definitive intermediate cellular organization resembling a eukaryote, various hypotheses have been advanced to explain how the eukaryote may have come into existence. The recent discovery of fossilized matter has uncovered an organism referred to as Lokiarchaeorta (Loki) that appears to have numerous features common with present-day eukaryote.

1.8 The Genesis of Eukaryote

Hence, eukaryotes are sisters of archaea with a large portion of genetic information that can be traced to bacteria. Although molecular details of eukaryogenesis still need further elucidation, it is becoming increasingly clear that both archaea and bacteria contributed to the formation of eukaryotes. This archaea–bacteria symbiosis may have been the beginning of the present multicellular organisms we see roaming on this planet. The genes dedicated to information processing arose from archaea, while bacteria provided the genetic framework for operational functions like metabolism. The Loki seems to have the information necessary for the ubiquitin system, phagocytic capability, small Ras-like GTPases and multiple pathways for gene expression that are all eukaryotic signature proteins. The mitochondrial tRNA has common features with bacteria, a feature that supports the view that this organelle is bacterial in origin. The actin/tubulin duo of the cytoskeleton, nuclear pore, spliceosome and proteasome can be traced to the archaea. In fact, the recently discovered Asgardarchaeota may be the nearest prokaryotic relative of eukaryotes. It appears that the symbiotic interaction between the archaea and the bacteria may have been the trigger that has culminated in the distinctive entity referred to as eukaryote.

The outside-in model posits that endosymbiotic or phagocytic events may have allowed the engulfing of the mitochondrion (a bacterium) and the nucleus by an archaea host. The inside-out postulation invokes the participation of membrane-bound blebs originating from extracellular protrusions that may fuse with an archaeon to generate a eukaryote. This random partnership between the bacteria and the archaea may have occurred over an extended period to eventually coalesce into a highly organized biological system. Although the discovery of the Loki is providing some clues to this eukaryotic conundrum, these rational models are speculative at this juncture. The *Paramecium* and *Euglena* are two examples of unicellular eukaryotes. The genus *Euglena* is a single-cell flagellate found in both fresh and salt water. Since it possesses chloroplasts and a red eye spot composed of carotenoid pigments, this eukaryote can perform photosynthesis and synthesizes its own food, the paramylon a form of starch. When deprived of light, the *Euglena* can sustain itself over a long period of time. The species *Euglena gracilis* is considered to be half animal and half plant. The *Paramecium*, on the other hand, feeds on bacteria and algae and uses the movement of its cilia to gather nutrients. It resembles the sole of a shoe and has a size that can range from 50 to 300 μm. The trichocysts anchored at the plasma membrane are deployed in a spear-like shape in response to temperature fluctuation or chemical shock. These unicellular eukaryotes that have all their biological functions fulfilled by a singular cell will later evolve into multicellular organisms where diverse cells are dedicated to disparate functions, all contributing to the common goal of the organism. For instance, the astrocyte is a very specialized cerebral cell that contributes to the proper functioning of the brain and the host where this organ is lodged (**Figures 1.9 and 1.10**).

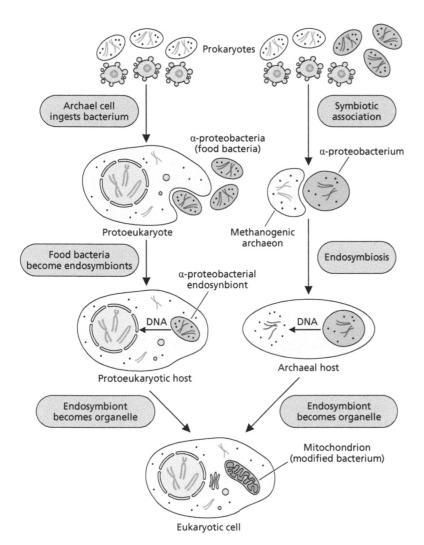

Figure 1.9. Genesis of eukaryotes. Eukaryotes originated following intimate interactions amongst bacteria and archaea as eukaryotic cells are endowed with numerous features derived from these prokaryotes. The interplay between these cells with one engulfing the other and/or membrane blebs from one being entrapped by the other over millions of years has led to the emergence of eukaryotic organisms. The mitochondrion and chloroplast are specialized organelles that can be traced to their bacterial origin. Within eukaryotes, mitochondria are the main energy generators while chloroplasts can convert solar energy into ATP.

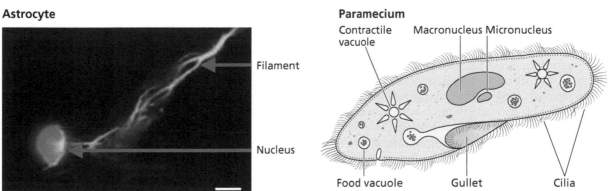

Figure 1.10. An astrocyte, a specialized cell and a paramecium, a unicellular organism. Astrocyte is a major component of the brain and is involved in numerous functions that enable this organ to function effectively (blue: nucleus stained with Hoechst 33528; green: actin filament visualized with phalloidin, a unique characteristic of this brain cell; adapted from Lemire et al. *J. Neuroscience Res.* 87, 1474–1483, 2009). Paramecium is a free-living organism with all the cellular machineries to lead an independent life. Cilia are involved in locomotion and the gullet is a collector of foods.

1.9 The Emergence of Multicellular Organisms

1.9.1 The prokaryotic adventure into multicellularity

Prokaryotes and single-cell eukaryotes are the most abundant organisms on this planet as this lifestyle offers some unique features conducive to its prolific exuberance and survivability. Prokaryotes, bacteria in particular,

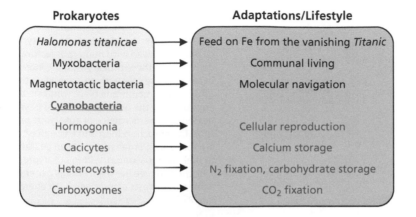

Prokaryotes

Halomonas titanicae
Myxobacteria
Magnetotactic bacteria
Cyanobacteria
Hormogonia
Cacicytes
Heterocysts
Carboxysomes

Adaptations/Lifestyle

Feed on Fe from the vanishing *Titanic*
Communal living
Molecular navigation

Cellular reproduction
Calcium storage
N_2 fixation, carbohydrate storage
CO_2 fixation

Figure 1.11. Prokaryotes living a multicellular lifestyle. Although prokaryotes are singular cells living relatively independently, some of them can lead a multicellular lifestyle in order to fulfil some select functions. For instance, *Myxobacteria* feed in the group, a process that maximizes their food intake, while *Halomonas titanicae* assembles in biofilms to devour Fe within the *Titanic*. Cyanobacteria, the photosynthetic bacteria, assume specialized cellular entities in order to store Ca, fix N_2, convert CO_2 into saccharides and reproduce.

can be found virtually in all ecological niches due to their ability to adapt to fluctuating surroundings. The high volume-to-size ratio of these microscopic organisms provides relatively easy access to nutrients, a feature that fits in perfectly with their solitary feeding habit. This hermit-like existence with seemingly no commitment and coordination with other similar cellular entities necessitates the limited genetic information required to go about the daily business. These single-cell microbes are primarily concerned with their own nutrition and transfer of genetic information mostly through cellular division whereby the parental DNA is replicated to give rise to two daughter cells. This event occurs as the cell grows to a reasonable size that still permits the easy intake of nutrients. With a very restricted ability to move, prokaryotes are literally compelled to maintain a volume-to-size ratio conducive to their feeding requirement. If this physical characteristic is too cumbersome, the prokaryote splits into two. Despite the inherent benefits bestowed by a solitary way of life, there are some examples whereby these unicellular organisms band together to experience life as a collective group (**Figure 1.11**).

1.10 Biofilm – Home for Prokaryotic Group Living

The microbial mat where numerous independent cells share an existence as a community is a common occurrence in nature. These cellular assemblies are made up of biofilms that consist mostly of polysaccharides, proteins and DNA. The cellular aggregation propelled by biofilms appears to be well coordinated and is triggered in response to a variety of environmental cues. Increase in temperature, resistance to antibiotics and lack of nutrients are some of the factors that compel some microbes to live together albeit transiently. The elimination of these conditions usually tends to disband the aggregation. The presence of freeloaders not contributing to the common good of the community usually leads to the demise of this communal assemblage. Infectious microbes are notoriously known to invoke this strategy to inflict on their hosts. They utilize these sticky moieties to attach to the surface of the tissue where they tend to cause damage. Dental plaque and cystic fibrosis are two medical conditions initiated by biofilm formation. These biofilm-housed microbial assemblies are resistant to antibiotic and usually not readily accessible to the defence mechanisms orchestrated by the host immune system. However, the microbes can share information by the quorum sensing molecules and exchange DNA in order to proliferate in a controlled fashion. These biofilm-encapsulated microbes are also not easily displaced and they can cling to virtually any surface even in the open sea where the water current can be intense. For instance, shipwrecks lying at the bottom of the oceans are known to attract bacteria that elaborate extensive biofilms. These microbial communities work collectively to feed on these metal structures and contribute to the biogeochemical cycle of various chemical elements. In fact, the imminent disappearance of the *Titanic* from the bottom of the ocean is being partly attributed to these biofilms housing billions of bacteria known as *Halomonas titanicae* (**Figure 1.12**).

1.11 Bacteria Feeding Together

Myxobacteria are also known to engage in communal living in an effort to feed in the group. They move collectively and acquire their foods cooperatively. They utilize coordinated movements forming a dynamic assemblage of microbes that go about 'hunting' their prey. A dearth of nutrients is known to trigger this momentary multicellular organization characterized by the secretion of exopolysaccharides and is referred to as the swarms. These biopolymers form trails at the leading edge of the temporary cellular arrangement that at the root of this cooperative microbial behaviour aimed solely at acquiring nutrients. When the swarms come in contact with a prey, some cells target the prey colony for lysis. These microbial aggregates then secrete a pool of hydrolytic enzymes and metabolites in order to optimize the intake of the degraded biomass. This short-lived multicellular organization aimed at the unique attribute of nourishing where cooperative motion, collaborative hunting and collective feeding are on display is a precursor to the multicellularity that evolved later on this planet. One of the key elements of multicellular organisms is the presence of specialized cells dedicated to the common good of all the community of cells constituting the living system.

1.12 Manifestation of Specialized Cells in Prokaryotes

1.12.1 Magnetotactic multicellular bacteria

Some bacteria are known to possess tailor-made cells and organelles that are designed to perform specific functions aimed at enabling these prokaryotes to thrive in their ecological niche. Magnetotactic bacteria utilize miniature compass needles that allow them to navigate their environment and move along geomagnetic fields. These magnetic crystals known as magnetite (Fe_3O_4) and greigite (Fe_3S_4) are encased in lipid organelle referred to as the magnetosomes. The magnetostatic molecules can be located in a single bacterium or in a multicellular arrangement. The latter can be either spherical or ellipsoidal. The spherical microbial assembly varies in width ranging from 3 to 12 μm where 10–40 cells are organized while the ellipsoidal multicellular organization can have up to 100 cells that are 2–23 μm long and a width of 7–17 μm. These multicellular magnetic bacteria are widespread in the marine environment and play a pivotal role in the cycling of iron. The formation of these biominerals is intricately modulated and is dependent on oxygen gradient. Iron uptake, crystal nucleation and vesicle formation are tightly coordinated. The lipid envelope that is rich in glycolipids, phospholipids and neutral lipids provides the proper chemical backdrop for the emergence of these crystals following the uptake of Fe. In oxygen-deficient surroundings, the sulphur derivative is more common than the magnetite (**Figure 1.13a**).

1.13 Cyanobacteria with Photosynthetic and Nitrogen-Fixing Cells

Cyanobacteria are oxygen-evolving microbes that have contributed to the oxidative environment we are currently living in. These prokaryotes are known to lead a unicellular and multicellular existence and may have evolved approximately 2.5 Byr ago with their photosynthetic capability, an

Figure 1.12. The *Titanic* being devoured by *Halomonas titanicae*. These bacteria are attached to the hulk and are consuming Fe voraciously. Image courtesy of Lori Johnston, *RMS Titanic* Expedition 2003, NOAA-OE.

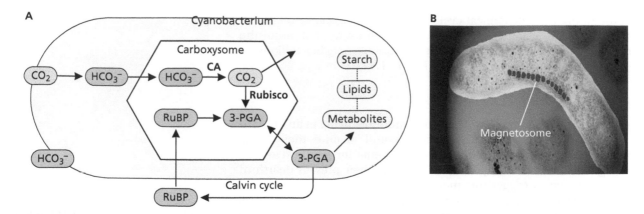

Figure 1.13. Prokaryotes with specialized cellular systems. A: Magnetotactic bacteria have specialized structures referred to as magnetosomes that allow the microbes to navigate the Earth's magnetic field. They are rich in magnetic Fe minerals. The synthesis of these biominerals is intricately modulated and is dependent on numerous factors. B: Cyanobacterium known for their photosynthetic attribute do also possess specialized cellular structures dedicated to specific functions. Carboxysomes are involved in conversion of CO_2 into carbohydrates, a process mediated by the enzymes ribulose 1,5-biphosphate carboxylase/oxygenase (Rubisco) and carbonic anhydrase (CA).

event that has triggered the appearance of aerobic organisms. The unicellular cyanobacteria and their multicellular assemblies are prolific and are found in virtually all ecological habitats including marine, freshwater and terrestrial in most temperature zones. They play an important role in carbon and nitrogen cycling. Cyanobacteria are known to grow as filaments where more than 100 cells can exist as a multicellular organization. Here, different types of cells contribute to the well-being of whole aggregate by performing specific tasks. This multicellular arrangement may provide a defensive shield, helps in the acquisition of nutrients, allows information sharing amongst the cells, limits interactions with freeloaders or non-cooperative cells with a unicellular mindset and maximizes the benefits derived from the division of labour. To attain a harmony amongst a community of cells striving for the common good, three main conditions have to be fulfilled:

1. Cell-to-cell adhesion has to be properly programmed and executed so that each cell is aware of its boundary.

2. Intracellular communication needs to be effectively relayed and responded to.

3. A degree of terminal differentiation should lead to specialized activity that contributes to the health of all the components of the multicellular entity.

The filamentous multicellular cyanobacteria essentially exhibit two cell types, the vegetative and the heterocyst. The former converts CO_2 into saccharides while the latter fixes N_2. However, other cell types do occur in response to a variety of factors. The akinetes are specialized cyanobacterial cells composed of a thick-walled envelope that are non-motile and resistant to desiccation. These are usually dormant, a condition precipitated by environmental conditions such as temperature fluctuation and lack of nutrients. Cells known as hormongonia manifest themselves during cellular reproduction and calcicytes are calcium accumulating cells laden with Ca-ATPases that enable some cyanobacteria to live in a calcium carbonate-rich environment.

The vegetative cells are the main component of the multicellular filamentous cyanobacteria as they are involved in the capture of sunlight and the subsequent conversion of H_2O and CO_2 into saccharide-rich derivatives. These cells house all the photosynthetic pigments within intracellular membranes, the thylakoids. The outer membrane of the consecutive vegetative cells is continuous along the filament, thus providing a favourable arrangement for intracellular communication. The metabolic machinery involving ribulose 1,5 biphosphate carboxylase/oxygenase (Rubisco) and other enzymes are stationed in the carboxysomes where CO_2 and H_2O are transformed into carbohydrates that are stored as glycogen granules. During the differentiation of vegetative cells into heterocysts, an event triggered by a dearth of nitrogen, both the carbohydrate inclusions and the carboxysomes are lost. The rearrangement of the thylakoid membranes paves the way for the emergence of the heterocyst with the initial appearance of the honeycomb, an array of highly contorted membranes. The main task of the heterocyst is to fix nitrogen into ammonia, a process that necessitates an anoxic environment as the key enzyme nitrogenase is oxygen sensitive. To create such an environment, the heterocyst has two layers made up of glycolipids and polysaccharides outside the outer membrane in an effort to limit the permeability of gases. The accumulation of α-ketoglutarate in response to nitrogen deprivation appears to be pivotal for the differentiation of the mother vegetative cell into a specialized cell dedicated to the fixing of this essential ingredient. Numerous transcription factors are recruited in this process that facilitates the transformation of photoheterotrophic cell into one with all the machinery responsible for the conversion of N_2 into NH_3 and eventually glutamine. For this to occur, a well-defined signalling system is operative in order to enable the exchange of metabolites between these two disparate cells. Once, the protection of the main proponent, the nitrogenase via the generation of an oxic surrounding is secured, it is essential that the NH_3 produced is transformed. To execute this function glutamine synthetase comes into action. This enzyme requires substrates such as glutamate and nucleotides that are provided by the vegetative cell and in turn receives the glutamine where it is converted into glutamate by glutamate synthase to be then utilized for the synthesis of N-rich metabolites critical for the well-being of the multicellular filamentous cyanobacteria (**Figure 1.12b**).

These examples illustrate how bacteria manage to evolve specialized cell or organelle in order to survive in environments that are not suited to their unicellular lifestyle. This phenomenon necessitates intricate coordination and communication amongst the cells that are forging this relationship. Whether it is to acquire nutrients, fend against predators or navigate the geomagnetic field, these multicellular assemblies transient or long lasting are beneficial to all the stakeholders. In the case of the filamentous cyanobacteria, even the molecular cohesiveness is on display. In this instance, the multicellularity that is exhibited in response to nitrogen deprivation requires an intimate partnership between the two cell types. The heterocyst produces NH_3 and glutamine while the vegetative cell synthesizes glutamate from glutamine and distributes it throughout the organism including to the heterocyst for further processing into glutamine. Although cells with a specialized function(s) may have originated in prokaryotes, their manifestations are more pronounced in eukaryotes, a situation that has given rise to multicellular organisms with a fascinating array of attributes (specialized cells/tissues/organs) responsive to the ecosystem they are living in (Box 1.3).

1.14 The Emergence of Multicellularity in Eukaryotes

1.14.1 More the merrier: togetherness breeds strength

Although multicellularity may have started in prokaryotes, this process got fine - tuned in eukaryotes and gave rise to all the multicellular organisms we see roaming the planet. The unicellular lifestyle has numerous shortcomings that are resolved in multicellular systems. Uptake of nutrients becomes an impediment when a unicellular organism grows large. The formation of a specialized cell or organ like the mouth helps rectify this situation. Mobility, defence, removal of

Cyanobacteria can fix both CO_2 and N_2. While the photosynthetic process liberates O_2 with the concomitant formation of ATP, the conversion of N_2 into NH_3 necessitates a reductive environment. To overcome this dilemma, some unicellular cyanobacteria perform photosynthesis during the day and resort to their N_2 fixing attribute at night in an O_2 devoid surrounding, an event dictated by the circadian clock. Filamentous cyanobacterium like *Anabaena* spp. tends to transform its vegetative cell into a specialized cell termed the heterocyst whose primarily function is to provide nitrogen nutrients to the organism. These diazotrophic cells are larger and more round than the cells they originate from. They have diminished pigmentation and are encased in thicker cell envelopes, structural features designed to create an O_2-limited space conducive to the nitrogen-fixing machinery. Lack of nitrogen and accumulation of α-ketoglutarate trigger the reprogramming of the vegetative cells that differentiate into the heterocysts. The O_2-producing photosystem II is dismantled, and thick envelope layers composed of glycolipids and polysaccharides create an environment for O_2-sensitive nitrogenase to operate. This genesis of the heterocyst occurs within a day and with a continuous periplasm with the adjoining vegetative cell, an intracellular communication network is established whereby the latter provides reductants and carbon nutrients while the former supplies fixed nitrogen. The introduction of sufficient nitrogen is known to arrest this multicellular arrangement. Hence, the need for nitrogen nutrients compels the microbe to generate a unique cell structure in order to survive and reveals how multicellularity may have emerged (**Box Figure 3**).

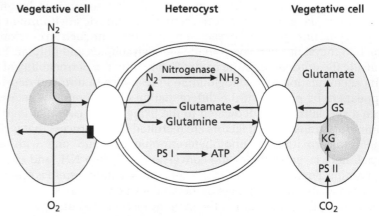

Box Figure 3. **N_2 fixation resulting in the synthesis of NH_3, a moiety stored as glutamine.**

wastes and vulnerability to the environmental changes are real issues that unicellular cell has to contend with while a multicellular organization is more apt to deal with. The banding together of cells with specific tasks to perform for the general well-being of the multicellular organism ushers in new possibilities like combating predators and adapting to changing habitats. To achieve such a feat, the cells need to adhere to each other in a controlled fashion but not encroach on each other territory, i.e. has a clear line of demarcation otherwise such invasion can lead to abnormal situation like cancers. Remember during the development of the liver, the biggest organ inside the human body, the contact with the diaphragm signals the end of the territorial expansion of the hepatic cells. There must be a communication network that receives signals and transduces these into activities conducive for the proper functioning and proliferation of the multicellular entity. Furthermore, some specialized cells must be differentiated terminally in order to perform assigned duties in the complex cellular arrangement. These attributes must come together for multicellular systems to exist.

1.15 The volvocine green algae: a window into eukaryotic multicellularity

This green alga is an excellent model to understand how multicellular organisms came into being. The volvocine family consists of both unicellular and multicellular organisms that exist even till today. *Chlamydomonas reinharditii* is the unicellular type, while *Volvox carteri* has an existence as a multicellular entity. The former possesses two flagella that help in its locomotion in wet environments and act as sensory transducers. However, during reproduction or dividing phase, these flagella are resorbed and migrate to nucleus to be utilized in DNA replication. During the reproductive process, this single-celled organism is non-motile as the flagella moonlight in multiple cellular fission whereby multiple daughter cells are generated. In this instance, two physiological functions mobility/sensing and reproduction are segregated, i.e. they cannot be executed simultaneously; it is more or less the mouth performing the two tasks of chewing food and listening. One of these activities has to wait for the other one to be accomplished before being executed. *Chlamydomonas* is an obligate unicellular organism where the flagella perform two disparate functions albeit at different times. In *V. carteri* that has a multicellular lifestyle, this issue has been resolved as these two functions are discharged by two different cell types. The somatic cells that number around 2,000 closely resemble the unicellular *Chlamydomonas* and are small with two flagella tucked into the gelatinous outer cellular matrix. These have as their mission to propel the organism in wet surroundings and sense light for photosynthetic activity. However, these cells cannot divide, and this function is assigned to the specialized cell type referred to as the gonidium. The gonidia are large and non-motile mostly dedicated to reproduction and propagation of the species. Approximately, 16 of these gonidia undergo a series of embryonic divisions in order to produce the cells for the next generation of volvocine. This division of labour was the beginning of what eventually led to the emergence of specialized tissues, organs and multicellular organisms. In this example, the terminally differentiated somatic cell provided most of the day-to-day functional entity of the organism, while the stem-like gonidia fulfilled the reproductive task. The *Volvox* algae form ball-like structures with cells ranging in number from 500 to 60,000, and all lead to a multicellular existence (**Figure 1.14**).

1.16 Choanoflagellates – a bacteria-mediated multicellularity

Choanoflagellates are microbial eukaryotes that are considered to be the closest living relatives of animals. They can lead to a unicellular or multicellular existence. They are oval-shaped blobs with a single tail-like actin-rich flagellum utilized for locomotion. These organisms feed on

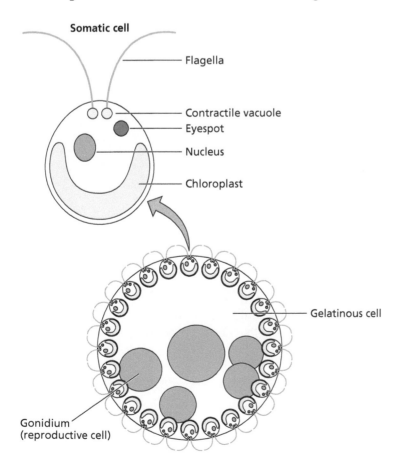

Figure 1.14. **The emergence of multicellularity.** To perform disparate functions separately, organisms evolved cells with specific functions. The volvocine green alga *Volvox carteri* is a multicellular organism with two different cell types, the somatic cell and the reproductive cell. The somatic cell is endowed with two flagella that are involved in numerous activities including movement. The reproductive cell termed the gonidium participates in the reproduction and propagation of the species. This alga forms a ball-like structure and comprises up to 6,000 cells.

bacteria, an activity aided by the flagellum in both trapping and ingesting the food. The single-celled *Salpingoeca rosetta*, common in coastal areas, can develop into a multicellular assembly. This arrangement originates from a single cell and ends up with multiple cells after repeated division. These are referred to as rosettes that stably adhere to each other, an event triggered by a variety of environmental cues. This multicellular morphology generates increased water flux and is more effective in capturing bacteria compared to the unicellular organism. *Algoriphagus machipongonensis* is one of the bacterial species *S. rosetta* nourishes on. It appears that a bacterial component referred to as the rosetta-inducing factor-1 is the catalyst that mediates the multicellular arrangement. This development programme is orchestrated by a lipid extract belonging to the sulphonolipid family that can be isolated from the bacterium. This bacterial–host interaction that results in morphological changes is akin to the influence of exopolysaccharides in the formation of nitrogen-fixing nodules in the encounter between *Rhizobium* and leguminous plants. A similar phenomenon has been observed whereby bacterial signals are pivotal in the morphogenesis of animal gut or the light organ maturation in Hawaiian bobtail squid. Lipopolysaccharide from the resident microbe *Vibrio fischeri* is one of the inducing morphogens enabling the genesis of the light organ. The same marine microbe has been shown to secrete a protein termed EroS that orchestrates a reproductive behaviour in the unicellular choanoflagellates and enables the exchange of DNA, resulting in genetically distinctive offspring. The protein EroS is an enzyme that targets chondroitin sulphate, a component of the cell wall. This interaction softens the outer barrier and allows the cells to fuse and undergo a reproductive cycle. These environmental signals, many of which are microbial in origin, reveal a central role that microbes play in the evolution of multicellular organisms and their specialized components. Choanoflagellates also possess genes for tyrosine kinases, a key mediator of intracellular communication. Proteins such as cadherins and c-lectins that help cells stick together are part of the molecular machinery these organisms have in order to ease their transition to a multicellular assemblage (**Figure 1.15**).

Hence, it is clear that the advantages afforded by multicellularity led to the evolution of multicellular organisms, a process that is closely modulated by environmental conditions and other organisms like bacteria in their midst. The presence of prokaryotes like *V. fischeri* and *A. machipongonensis* has contributed to multicellular organization. The presence of the exocellular biomolecules either aimed to provide the prokaryote nutrients or signal the presence of food like in the case of choanoflagellate may have resulted in cellular complexity in order to ensure survivability. Numerous organisms like sponges and corals start their life as larvae in wet surroundings that are saturated with bacteria. It is quite within the realm of possibilities that these prokaryotes have a wealth of molecular cues responsible for the further development of these larvae into full-fledged adults. Ecological habitats modified by climatic fluctuations, physical-built areas and human activities are also important contributors to the physiological and morphological transformation in numerous organisms. For instance, frequent hurricanes in the Atlantic Ocean have resulted in lizards having more grip to cling to a structure in order to survive and swallows develop short wings in order to navigate the built urban landscape. The poaching habit of humans has forced female rhinoceros to lose their tusk-producing genes, while exposed to constant human fishing activity, some variety of fish has a smaller mouth. Multicellularity that precedes specialization of cells, tissues and eventually morphological manifestation is dictated by numerous factors including the bacteria-laden chemical signals. Hence, microbes have an important role in guiding the development of all multicellular organisms (Box 1.4).

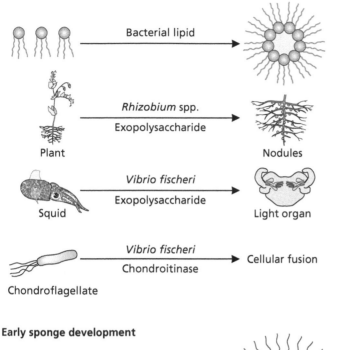

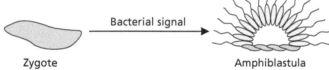

Early sponge development

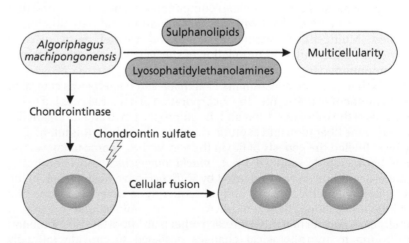

Figure 1.15. Bacteria trigger morphological changes in numerous organisms. Bacteria are important constituents of multicellular organisms and participate in numerous biological activities including the initiation of morphological differentiation. The rosetta-inducing factor, a sulfonolipid derived from bacteria, promotes multicellular arrangement in choanoflagellate, Salpingoeca rosetta; bacterial exopolysaccharides, on the other hand, aid in nodulating plant roots and inducing the genesis of light organs in squids. Cellular fusion and reproduction are mediated by the microbial enzyme chondroitinase in choanoflagellates. Development of sponges is also orchestrated bacterial signals.

BOX 1.4 MICROBIAL CUES TRIGGER MULTICELLULARITY IN *SALPINGOECA ROSETTA*

Choanoflagellates are single-celled organisms that are one of the closest relatives of animals. Each cell has a rounded body attached to a flagellum. This structural arrangement with a tail-like structure and head resembles the sperm. These organisms feed on bacteria by phagocytosis, a process supported by the flagellum. In the species known as *S. rosetta*, the development of a rosette-shaped multicellular arrangement occurs when a single cell undergoes repeated division creating numerous adherent cells. This multicellular configuration helps in the feeding and reproductive habit. This cellular organization is initiated by chemical signals released by the bacterium *A. machipongonensis*. Sulfonolipids and lyosophatidylethanolamines acting as activators and synergistic enhancers trigger the morphological switch geared towards multicellularity. The rosette-inducing activity is apparently thwarted by a rosette inhibitory feature exhibited by a structurally different sulfonolipid. This saturated lipid antagonizes the cellular transformation. Furthermore, a chondroitinase referred to by EroS released by marine microbes to access nutrients is known to promote sexual reproduction. In this instance, the microbial enzyme targets chondroitin sulphate, a polymer constituting the extracellular matrix of the choanoflagellate. The softening of this layer allows two cells to fuse and undergo sexual reproduction. Lipopolysaccharides and tracheal cytotoxin produced by *Vibrio* spp. induces the development of light organ where the bacteria are housed in order to enable the host to hunt effectively. Microbial fingerprint is evident in the development and survival of all multicellular organisms (**Box Figure 4**).

Box Figure 4. Bacterial sulfonolipids and chondroitinase promoting multicellular development.

1.17 Multicellularity: An Irreversible Process

As multicellular organisms confer numerous benefits compared to an existence as a single cell, it became critical that multicellularity stayed a permanent phenomenon. Reversibility leads to the instability of a multicellular organism and eventually to its demise. For instance, some prokaryotes converge in order to overcome challenges such as lack of nutrients, increase in temperature and limited oxygen. *Pseudomonas fluorescens* elaborates saccharide-rich environment where its colonies assemble together to better access oxygen. However, since these cells have their own individuality, some of them do not feel obligated to contribute to this arrangement. They benefit from this process but do not feel compelled to participate in the maintenance of the biofilm. They do not expend any capital to synthesize the precursors necessary to generate the exopolymer. These 'free loaders' eventually lead to the extinction of the multicellular arrangement. Hence, a sense of obligation to multicellularity has to be of paramount importance. The cells that constitute the multicellular arrangement need to be mutually reliant on each other or depend on attributes they do not possess in order to survive. Solitary traits technically become disastrous while living in groups offer numerous benefits. For instance, the human body cannot survive as discrete and separate units. It is an impossible proposition for a head or an arm to live without the benefit of the whole body machinery. Thus, a ratchet-like system has to be operative where each component requires the activity of the other components in order to survive and proliferate. In organisms that have evolved as multicellular entities, the solitary existence of a cell, tissue or organ is not tolerated.

1.18 How Microbes Became an Integral Component of Multicellular Organisms

Multicellular organisms are made up of numerous terminally differentiated cells or organs that are limited to the specific function they perform. They do also possess numerous stem cells that can be programmed to be specialized on demand or when the differentiated cells are not performing well. Furthermore, this cellular assembly needs to be accompanied by other components including microbes if the well-being of the whole organism is to be maintained. These intimate companions of the multicellular organisms are an integral part of the living system. Without microbial constituents, these multicellular organizations cannot survive. In fact, all multicellular organisms evolved in surroundings replete with microorganisms and their molecular components. Hence, it is natural that these living systems evolved with the incorporation and participation of microbes. Multicellular organism cannot be viewed as a discrete autonomous entity but as a compilation of specialized cells, microbes and their components (**Figure 1.16a**).

The exchange of the constituents of archaea and bacteria was central to the development of eukaryote. The incorporation of bacterial cells led to the biogenesis of the mitochondrion and the chloroplast, two crucial organelles involved in the liberation and capture of energy. Most likely, this interaction may have fuelled the genesis of life in the way we have come to know it on planet Earth. Lipids secreted by *A. machipongonensis* and chondroitin sulphate-degrading enzyme released by *V. fischeri* have been pivotal in the development of multicellularity and subsequently complex organisms. The symbiotic association between microbes and multicellular organisms whereby each partner benefits from each other is an important phenomenon contributing to morphological changes designed to provide mutually profitable characteristics. The nodule that houses *Rhizobium* spp. and the light organ that shelters *Vibrio* spp. help in acquiring nitrogen, an important nutrient for leguminous plants and in emitting light, an essential hunting

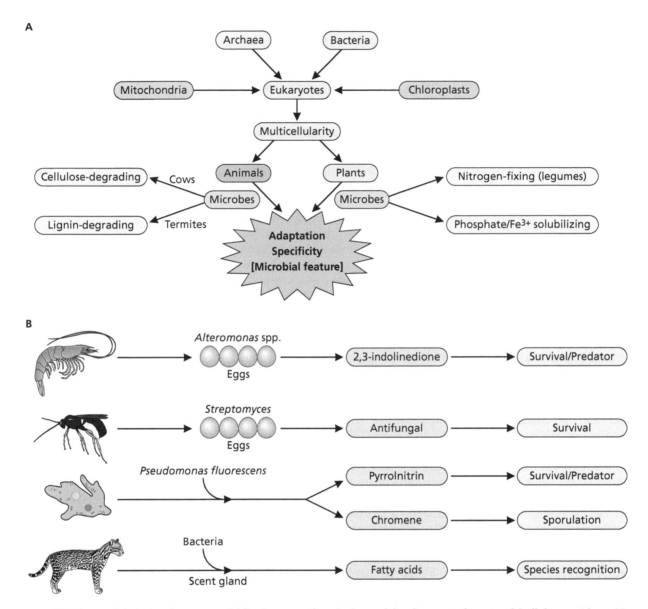

Figure 1.16. **Microbes and their signals are essential for the proper functioning and development of most multicellular organisms.** (a) Microbes are important constituents of multicellular organisms. Although archaea and bacteria supplied the necessary ingredients for the genesis of eukaryotes, these prokaryotic cells are integral components of all higher organisms. Even the mitochondrion and the chloroplast can be traced to their bacterial origin and are organelles that no multicellular organism can survive without. Microbes in multicellular organisms help perform essential functions that the hosts are dependent on and also allow for enhanced adaptive flexibility. For instance, cows will be unable to digest cellulose and termites incapable to metabolize lignin without their bacteria. Plants need their microbes to fix N_2 and solubilize essential nutrient like Fe. (b) Microbial signals play a vital role in the development of multicellular organisms. The development of numerous higher organisms depends on their microbial cells. Insects spray their eggs with antibiotic- and antifungal-producing bacteria in order to increase the survivability of their offspring. Bacteria housed in the scent glands of animals generate volatile fatty acids responsible for recognition of kin and the mating partner.

tool for squids respectively. This close link between microbes and multicellular organisms manifests itself in virtually all spheres of biological processes. Intracellular cohesion of sponges depends on their resident bacteria, while in the human gut, the presence of microbes transforms the indigestible carbohydrates into value-added products that play important roles in human physiology. Enzymes and cofactors of microbial origin are important for the efficiency of the digestive and absorption processes in the human gut. This ability of bacteria to transform complex nutrients may be the driving force behind the evolution of the dual stomach that ruminants possess. One component of this organ is the home of microbes where they are busy converting cellulose into energy-rich products that the host utilizes. Cellulose, a polymer that contains glycosidic bonds made up of β-glucose, is hydrolysed by cellulase, an enzyme only a few organisms have in their genetic repertoire. The microbial systems housed in the ruminants use this glycosidase to liberate glucose that is eventually converted into short-chain fatty acids, an excellent source of energy. Termites are insects that consume wood as their staple diet. They would not be able to adopt this eating habit if it was not for the microbes within them. These bacteria do not only possess cellulase but also have ligninase, an enzyme degrading lignin, a polymer of phenol. The latter is relatively limited to a few organisms on planet Earth.

1.18.1 Molecular conversation between microbes and their hosts

This microbial influence can also be seen during the developmental stages in numerous multicellular systems. Without such input, the propagation of the respective lineages will be in jeopardy. Many multicellular organisms will not be able to survive if it was not for their association with microbes. The ability of a species of the bacterium *Alteromonas* to synthesize the metabolite 2,3-indolinedione on the surface of the embryo is key to the survival of some shrimps. Without this chemical, the embryo would become a quick meal for the fungus *Lagenidium callinectes*. Species of *Wolbachia* inhabiting insects like *Asobara tabida* are crucial for the propagation of these organisms as the microbes are essential for egg maturation. The *trans*-ovarian transmission of bacterial partners is a clear indication of the essentiality of microbial input in multicellular organisms. A wide variety of benthic organisms receive signals from marine microbes that promote the development of their larvae. The European beewolf wasp uses its antenna to spread *Streptomyces* spp. on the brood cells as these microbes fend against any fungal invasion. 'Farming' amoebae are unicellular organisms that harbour microbes as food. In *Dictyostelium discoideum*, the two strains of *P. fluorescens* live to accomplish disparate tasks. While one of them is a food source, the other produces two metabolites, pyrrolnitrin and a chromene derivative. The former has anti-fungal properties, while the latter is known to enhance spore formation of the host and inhibit sporulation in the 'non-farmers' amoebae. The mating or reproductive role of bacteria is also evident in other organisms where scent generated in form of volatile fatty acids acts as a potent signal. In hyaenas, bacteria lodged in the scent glands help generate species-specific fatty acids, while in fruit flies, the cuticular hydrocarbon profiles are indicative of the bacterial symbionts residing within (**Figure 1.16b**).

The integrity of microbial cells as part of the multicellular make-up of all complex organisms is vividly displayed in the communication networks that the constituent cells respond to. In humans when the level of glucose in the blood is relatively high, insulin, a hormone secreted by the pancreas, informs this situation to various organs including the liver and then initiates a programme to trap glucose as glycogen and lipids. The reverse is true. When the blood glucose concentration is low, glucagon orchestrates the release of glucose. Again other organs like the liver and muscles spring to action. Signals generated by the microbial residents of multicellular organisms are decoded by different cellular components. The response elicited contributes to the well-being of the hosts. Peptidoglycan, a biochemical of microbial origin, is known to interact with the pattern recognition receptors and promote developmental processes in various organisms including humans. Polysaccharides produced by *Bacteroides fragilis* induce the immune cells to suppress inflammation in the gut. This microbial signal even allows some species to seek and settle in the proper ecosystem. The marine bacterium *Vibrio anguilarum* and the green seaweed of the species *Enteromorpha* are involved in such a signalling operation. The *N*-acyl homoserine lactones secreted by the microbe are recognized by zoospores and enable them to develop and establish in a select habitat. Bacterial communities in floral nectar play pivotal in modulating how pollinators interact with plants. The metabolites generated in these nectars are important cues on the mode of pollination. Hence, communication networks triggered by bacteria within these multicellular organisms do not only contribute to the proper functioning of the host but are an obligatory requirement for survival. These organisms devoid of their microbial collaborators termed 'apo-organisms' will not be able to have the lifestyle they are accustomed to.

The integration of the microbes and hosts is a properly vetted undertaking that has yet to be fully delineated. This partnership is intricately controlled and highly selective. The exopolysaccharides produced by *Rhizobia* have codes that are recognized by leguminous plants. Following the decoding of these encryptions, the plants allow the microbes to establish residence in a bacterium–plant-specific manner. In this instance, the nature of the monosaccharide constituents and the substituent groups they are decorated with appear to be critical in the selection process as is the ability of the host to respond to the signal. Microbes residing in the intestine are selected on the basis of mucin secreted by the human gut. These glycoproteins that serve as an intestinal barrier are also a source of food for a select group of microbes that are chosen to reside in this part of the human body. This type of programmed molecular matchmaking is initiated at the very onset of human life. Post-embryonic development is mediated by the enzymes like fucose transferases that help decorate the mammalian intestine with fucose, a molecular magnet for microbes like *Bacteroides thetaiotomicron*, an effector of human health. The content of milk is designed to favour the colonization of infants' intestines with the desired bacteria. For instance, the oligosaccharides in the milk that babies are fed by their mothers are not intended for recipients as they do not possess the biomolecular machinery to metabolize these moieties. They are ideal nutrients for the bacteria that are privileged to establish residence within these toddlers and play an integral role in their development as does say the brain with the secretion of the growth hormone. Faecal pellets laden with bacterial symbionts are deposited adjacent to secreted eggs in order to enable larvae to access these critical developmental agents in numerous organisms. This proximity favours the interaction between these select microbes with the hosts. The mode of transmission and the molecular communication ensure that the proper microbial population interacts and resides with and within the multicellular living systems. This high degree of microbial and host fidelity is essential if precise functional properties pivotal for the development and survival of these organisms are to be conferred (**Figure 1.17**).

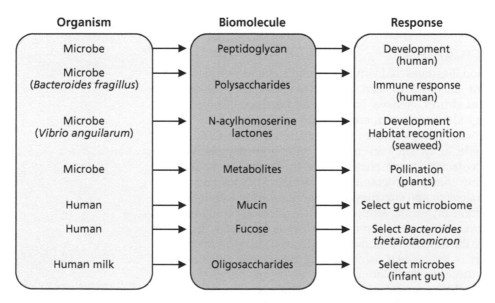

Figure 1.17. The molecular cross-talk between bacteria and multicellular organisms. Microbes produce polymers that mediate development and fortify the immune system in humans. In seaweeds, microbial lactones help in identifying the proper habitats while monosaccharides produced by the resident microbes in plants promote the pollination process. Mucin and fucose are important microbial selectors, and oligosaccharides in milk help nurture beneficial bacteria in infant's gut.

1.19 Intimate Relationship Cemented by Gene Transfer

The molecular language between the microbes and the hosts that ensures the 'completeness' of multicellular organisms is further fortified by the genetic exchange operative in these systems. Mobile genetic elements like transposons, plasmids and phages play a crucial role in shaping the multicellular genome and in imparting physiological functions deemed necessary to inhabit a specific ecosystem. For instance, the ability of humans to live on different diets reflective of the environments they are surrounded by is to a large extent a result of this genetic transfer. Seaweed is an important nutritional component in many countries including Japan. The ability of Japanese people to enjoy this delicacy can be attributed to the horizontal gene transfer between microbes residing in these algae and those in the gut of the Japanese population. The seaweeds are made of unique galactans consisting of L-galactose 6 sulphate. These complex carbohydrates are degraded by a group of carbohydrate-active enzymes (CAzymes referred to as porphyranases produced by marine bacteria like *Zobellia galactanivorans* residing in these seaweeds. Upon consumption of the delicacies, the genetic information is transmitted to the microbial dwellers within the intestine. Individuals who are not used to consuming these sea foods do not have these enzymes. The indigestion one encounters when visiting a different country can be traced to this phenomenon. In fact, almost 60% of human genes have originated from microbial organisms. *Methanobrevibacter smithii*, the most abundant archaeon in the human intestine, has acquired 15% of its genetic material from the bacterial genome in order to survive and contribute to the communal lifestyle in the intestine. Metal transporters and glycosyltransferases are bacterial in origin. The transfer of genetic information is also prevalent amongst microbes within other multicellular organisms. Genes from *Wolbachia*, an endosymbiont in various insects, have been shown to be located in the X chromosome. The genetic information for the biosynthesis of carotenoids in aphids is of fungal origin and is known to give these insects red or green colouration. Multicellular organisms are linked to their resident microbes, even though their genetic information and this exchange that is constantly occurring is a key component of the ability of the hosts to perform critical functions but also allow them to occupy diverse ecosystems.

1.20 Multicellular Organism Is Not an Apoorganism But Is a Holoorganism

The foregoing section illustrates that virtually all multicellular living systems need their resident microbes to lead an effective life. These microbes composed mostly of viruses, archaea, bacteria and fungi are involved in a variety of physiological functions that are part and parcel of the daily life of multicellular organisms. Many enzymes need cofactors in order to do their catalytic function efficiently. They are referred to as holoenzymes. When deprived of their cofactors, these enzymes are ineffective in mediating biochemical reactions they are programmed to execute. They are known as apoenzymes. In analogy to apo enzymes that cannot function without their cofactors, multicellular organisms stripped of their microbial components termed apoorganisms are unable to survive in their ecological niche. It is like putting someone on life support if the brain or the heart is removed. Microbes are an integral part of the cellular assembly that constitutes multicellular organisms. Hence, these multicellular organisms are indeed holoorganisms that need their microbial components to thrive. These are discrete living entities within the ecosystems made up of metaorganisms. The metaorganisms depend on each other and help maintain the ecological homeostasis where the holoorganisms, unicellular prokaryotes and eukaryotes inhabit. The unicellular organisms may live independently while the holoorganisms need their microbial partners. However, the biological activities performed herein are pivotal for a healthy habitat needed to support the metaorganisms operating in that environment. In the holoorganisms, the microbial components are mostly an integral part of the multicellular organisms, while the free-living organisms in the metaorganisms are in close proximity and interact via the activities they execute and/or the chemicals they exude.

The microbial partners in the holoorganisms are essential biochemical participants in the lifestyle of the multicellular assembly. How a human gut can manage without its microbiome with its tremendous attribute of possessing enzymes that aid in digestion? The anatomical feature of this body part has been guided by these microbes and evolved accordingly. They also produced a variety of biochemicals like SCFA, bile hydrolase and tryptophan derivatives. While butyrate, an SCFA, is a morphogen and a source of ATP, bile hydrolase modulates the levels of bile acids in the blood, a feature that has a direct impact on fatty acid metabolism. Tryptophan-derived metabolites are well known for their influence on the neuronal system. How the skin, mouth and lungs can function without the infusion of microbial signals and chemicals? Just imagine the fate of corals that are dependent on their algal partners. The latter provide nutrients obtained from photosynthesis while the corals supply a healthy dose of amino acids to keep the multicellular organization thriving. The apoorganism that is devoid of algae leads to coral bleaching and to eventual death. The host-associated microbiome is uniquely qualified to perform specific tasks that are pivotal for survival. This characteristic ensures that the sustainability of the holoarganisms is preserved and that the apoorganism cannot exist without its microbial constituents in a specific habitat.

Adaptability is another feature that is bestowed by the bacteria residing within the holoorganisms. These programmed and mutually beneficial assembly of microbial and multicellular entities as a compound organism is central to the ability of holoorganisms to occupy different ecological niches. This phenotypic plasticity is again a result of microbial input. The metabolic and adaptive capacity encoded by the host is often limited to a narrow range of environmental conditions. The ability of numerous species of the same animals and plants to live in disparate ecosystems is a product of the microbial ingenuity granted to the holoorganisms. The genetic adaptation in the host is slow and can take several generations before a phenotypic modification can occur. However, the genetic adaptation in prokaryotes is fast where a generation elapses in minutes as opposed to years in humans. This rapid genetic rearrangement triggered by the environment and other epigenetic forces orchestrates swift and expeditious adjustments in the holoorganisms' physiology. The injection of porphyranase gene by the gut microbes of the Japanese population is such an example as is the fact that similar species like cows are found around the globe in different climatic zones due to the microbes they harbour.

Thus, it is quite evident that microbes within and on apoorganisms render these entities into holoorganisms which can then live as metaorganisms in a specific ecosystem. This organization is beneficial to all the participants and enables the ecological habitat to evolve accordingly. Hence microbes, essential constituents of multicellular organisms, do not only perform essential tasks but are pivotal to the phenotypic plasticity of the hosts. The metabolic capacity is enhanced as is the ability of the multicellular systems to extend their geographical range. The ensuing chapters will look at the multicellular organizations in various living phyla and discuss the roles microbes play to ensure the ability of the host to survive and thrive. The biomolecular networks that the microbiomes elaborate will be the key focus and the functions ensuing from these molecular modules will be explained (**Figure 1.18**).

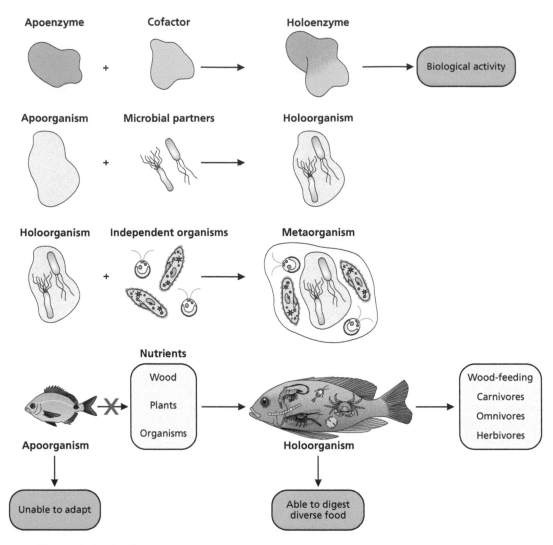

Figure 1.18. The intimate relationship linking bacteria and multicellular organisms – its similarity with enzymes and their cofactors. An apoenzyme (non-functional) becomes a holoenzyme (functional) when associated with its cofactor. In a similar manner, an apoorganism is an organism devoid of microbes and is ineffective. When constituted with all its microbes, the organism is referred to as a holoorganism and is fully operational. These multicellular organisms with their bacteria live in association of other independent organisms in an ecosystem termed a metaorganism. Without their microbes, multicellular organisms like fish and plants cannot exist.

1.21 Conclusions

All multicellular organisms have evolved from single-cell organisms, possess microbial partners and are referred to as holoorganisms or holobionts. A holoorganism without its resident microbes termed an apoorganism is unable to function properly as is an ineffective apoenzyme lacking its cofactors. Microbes are not only important constituents of most organisms, but they have also shaped the genesis of eukaryotes and subsequently contributed to the establishment of multicellular organisms where microbial input is evident in the anatomy, physiology and biochemical processes. Indeed, microbes are an integral component of multicellular organisms and perform a plethora of functions essential for the survival and proliferation of the hosts. The prebiotic molecules were transformed to biomolecules that were assembled as self-sustaining molecular entities and eventually culminated in the emergence of microbes. These microbes became the building blocks for multicellular life where they were incorporated as critical partners enhancing the adaptability and the survivability of the host. The subsequent chapters explore the numerous functions executed by microbes that almost all multicellular organisms have come to depend on (**Figure 1.19**).

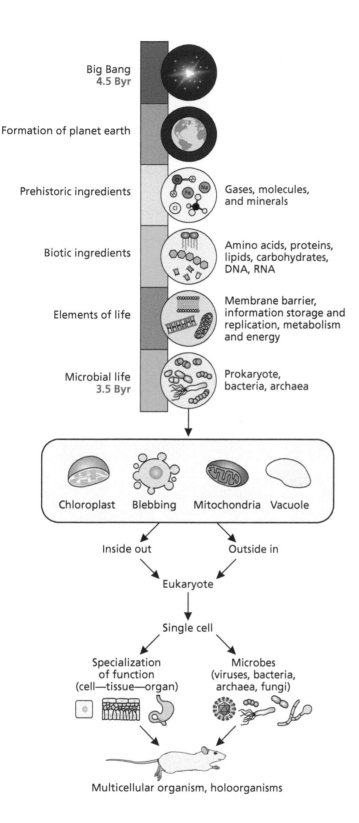

Figure 1.19. A schematic depiction of the emergence of life and multicellular organisms. Prebiotic and biotic molecules led to the emergence of last universal common ancestor (LUCA). The latter laid the foundation for prokaryotes that resulted in the genesis of eukaryotes. These organisms acquired microbes to fuel their development into multicellular organisms where an intricate bond between the host and resident bacteria emerged. These integral microbial components living within and on multicellular organisms are the microbiome.

SUGGESTED READINGS

Appanna, V., et.al. (2016). Phospho-transfer networks and ATP homeostasis in response to an ineffective electron transport chain in *Pseudomonas fluorescens*. *Archives of Biochemistry and Biophysics, 606*, 26–33. doi:10.1016/j.abb.2016.07.011.

Baum, D. A., & Baum, B. (2014). An inside-out origin for the eukaryotic cell. *BMC Biology, 12*(1). doi:10.1186/s12915-014-0076-2.

Ben-Jacob, E., et al. (2016). Multispecies Swarms of social microorganisms as moving ecosystems. *Trends in Microbiology, 24*(4), 257–269. doi:10.1016/j.tim.2015.12.008.

Fiore, M., & Strazewski, P. (2016). Prebiotic lipidic amphiphiles and condensing agents on the Early Earth. *Life, 6*(2), 17. doi:10.3390/life6020017.

Foster, K. R., et al. (2017). The evolution of the host microbiome as an ecosystem on a leash. *Nature, 548*(7665), 43–51. doi:10.1038/nature23292.

Herron, M. D. (2016). Origins of multicellular complexity: Volvox and the volvocine algae. *Molecular Ecology, 25*(6), 1213–1223. doi:10.1111/mec.13551.

Hug, L. A., et al. (2016). A new view of the tree of life. *Nature Microbiology, 1*(5). doi:10.1038/nmicrobiol.2016.48.

Kaiser, D. (2001). Building a multicellular organism. *Annual Review of Genetics, 35*(1), 103–123. doi:10.1146/annurev.genet.35.102401.090145.

Knoll, A. H., & Nowak, M. A. (2017). The timetable of evolution. *Science Advances, 3*(5). doi:10.1126/sciadv.1603076.

Kumar, K., et al. (2010). Cyanobacterial Heterocysts. *Cold Spring Harbor Perspectives in Biology, 2*(4). doi:10.1101/cshperspect.a000315.

Lim, S. J., & Bordenstein, S. R. (2019). An introduction to phylosymbiosis. doi:10.7287/peerj.preprints.27879v1.

Lyons, N. A., & Kolter, R. (2015). On the evolution of bacterial multicellularity. *Current Opinion in Microbiology, 24*, 21–28. doi:10.1016/j.mib.2014.12.007.

Mayer, C., et al. (2017). Selection of prebiotic molecules in amphiphilic environments. *Life, 7*(1), 3. doi:10.3390/life7010003.

Mcfall-Ngai, M., et al. (2013). Animals in a bacterial world, a new imperative for the life sciences. *Proceedings of the National Academy of Sciences, 110*(9), 3229–3236. doi:10.1073/pnas.1218525110.

Muñoz-Dorado, J., et al. (2016). Myxobacteria: Moving, Killing, Feeding, and Surviving Together. *Frontiers in Microbiology, 26*(7), 1–18. doi:10.3389/fmicb.2016.00781.

Peacock, K. A. (2011). Symbiosis in ecology and evolution. *Philosophy of Ecology*, 219–250. doi:10.1016/b978-0-444-51673-2.50009-1.

Queller, D. C., & Strassmann, J. E. (2016). Problems of multi-species organisms: Endosymbionts to holobionts. *Biology & Philosophy, 31*(6), 855–873. doi:10.1007/s10539-016-9547-x.

Ruiz-Mirazo, K., et al. (2013). Prebiotic systems chemistry: New perspectives for the origins of life. *Chemical Reviews, 114*(1), 285–366. doi:10.1021/cr2004844.

Woznica, A., et al. (2016). Bacterial lipids activate, synergize, and inhibit a developmental switch in choanoflagellates. *Proceedings of the National Academy of Sciences, 113*(28), 7894–7899. doi:10.1073/pnas.1605015113.

Woznica, A., et al. (2017). Mating in the closest living relatives of animals is induced by a bacterial chondroitinase. *Cell, 170*(6). doi:10.1016/j.cell.2017.08.005.

Yan, L., et al. (2012). Magnetotactic bacteria, magnetosomes and their application. *Microbiological Research, 167*(9), 507–519. doi:10.1016/j.micres.2012.04.002.

ALGAE AND THEIR MICROBIOMES

2

Contents

Keywords

- Algae
- Microbiome
- Climate Change
- Quorum Sensing
- Siderophore
- Sexual Behaviour
- Sloth

2.1 Algae and Their Occurrence

The term alga is derived from Latin and means seaweed. These organisms were first differentiated from the plant world by Jussieu in 1789. They are autotrophic organisms that are found in almost all environments ranging from aquatic habitats to glaciers. Those that are planktonic reside on the surface of water while the benthic species live at the bottom of water bodies. The cryophilic algae elect to be housed on snow and the epiphytic species have their residence on the surface of plants and animals. Hydra and sponges harbour the endophytic varieties. They can also occur in hot springs and in deserts. Numerous algae are also known to have symbiotic relationships with fungi and corals. They can be unicellular to multicellular with no roots or leaves, stems and well-defined vascular tissues. They exhibit a variety of sizes ranging from a minuscule 1 μm as in the case of diatoms (microalgae; *Micromonas pusilla*, the smallest algae) to the giant expanse of 60 m of macroalgae like the kelp. The latter resembles submerged forests in the oceans. They can exhibit a variety of colours including green, red and brown. Algae display a plethora of shapes and can be spiral, flagellate, spherical, filamentous and tubular. It is estimated that there may be between 30,000 to a million algal species (**Figure 2.1**).

DOI: 10.1201/9781003166481-2

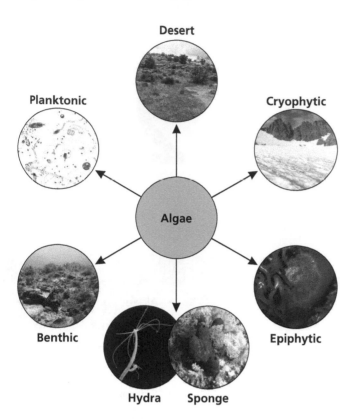

Figure 2.1. Algae and their habitats. Algae are widespread in nature and are found in such disparate ecosystems as in sandy and snowy regions. They are also known to live with other multicellular organisms like hydra and sponge.

2.2 Evolution of Algae – The Microbial Connection

There is a general understanding that algae originated via the endosymbiosis of cyanobacteria, prokaryotes known to perform photosynthesis in eukaryotic cells. The former was engulfed by the latter and led to the genesis of the chloroplast, an organelle dedicated to utilizing the energy of the sun for the conversion of CO_2 and H_2O into carbohydrates. The single-celled algae started to acquire specialized functions mediated by differentiated cells, an event that ushered the emergence of multicellular algae. Recent evidence points to archaea as the host recipient of the cyanobacteria. *Chlamydomonas* and *Volvox* are the closest relatives resulting from this endosymbiotic interaction. Bacterial endosymbiont has also been reported in the multicellular *Pleodorina japonica*. The subsequent conversion of a major portion of the cell wall into the extracellular matrix provided further incentive towards the march of algae to multicellularity and to acquire the numerous advantages afforded by this lifestyle including the attainment of a larger biomass. The influence of microbes on algae is not only limited to cyanobacteria or the initial host but also is reflected in the wide array of microbes that reside within and perform functions that are critical for the survival and growth of the algae. These associations can be symbiotic where both partners benefit. In some cases, commensalism may be the preferred relationship. In this instance, the association is advantageous to only one partner but brings no harm to the other (**Figure 2.2**).

2.3 The Ultrastructure of Algae and Biological Functions

The algal thallus is the main component of this organism and it houses all the constituents necessary to execute the variety of biological functions required for growth. This structure can be motile, sedentary, singular or multicellular in nature. In the multicellular organization, the cells are held together by a common gelatinous matrix. The flagellate forms are motile

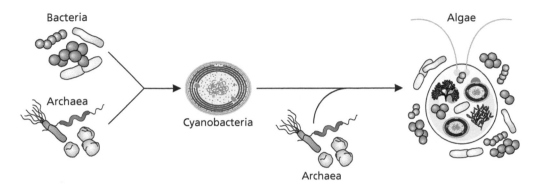

Figure 2.2. The microbial connection and algal evolution. Bacteria and archaea are the main prokaryotes resulting in the genesis of algae. Cyanobacteria with their photosynthetic attribute were central in the emergence of algae. Algae also host a range of bacteria and archaea either on the surface or/and intracellularly.

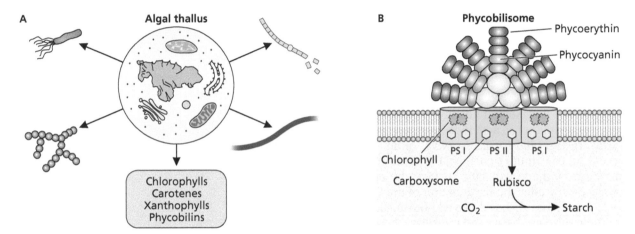

Figure 2.3. Algal structure and photosynthetic pigments. Algae have different forms and some of them exhibit a siphon-like structure known as thallus. They possess a variety of light-absorbing bodies such as the phycobilisomes. The carboxysome is involved in fixing CO_2 and is rich in Rubisco (ribulose 1,5 biphosphate carboxylase/oxygenase). PSI and PSI – photosystems I and II.

while those that have a coccoid shape are sessile. Filaments are usually generated by cellular division and the resulting daughter cells are held together by the end of their cell wall. Branched structures like in *Cladophora* and unbranched organizations as observed in *Spirogyra* can also occur. Some multicellular thallus has a siphonous form as exhibited in *Codium*.

Despite the diverse organizational structures of the different algal species, they all invariably possess the necessary biochemical machineries to capture light and convert CO_2 into carbohydrate. These light-absorbing molecules consist of chlorophylls, carotenes, xanthophylls and phycobilins and allow these organisms to maximize the utilization of light energy. They also give some algae their characteristic colours and enable them to proliferate in a variety of ecological systems. For instance, fucoxanthin is responsible for the brown colouration in brown algae while red algae acquire its colour from phycoerythrin. These pigments are located in the chloroplast and phycobilisome. The latter contains accessory pigments such as phycocyanin and carotene, moieties responsible for extending the light-harvesting capacity and helping combat oxidative stress. The chloroplast that houses the main light transducing molecules, the chlorophyll also possesses granular bodies termed carboxysomes. These defined structures contain the enzyme ribulose 1,5 biphosphate carboxylase/oxygenase (Rubisco) involved in orchestrating the fixation of CO_2 into starch. Although this polysaccharide is one of the main food reserves in algae, they do also stock calorie-rich oils and other derivatives of fatty acids (**Figure 2.3**).

The algal cell wall has as its primary constituents the fibrillar cellulose, mannans and xylans. This barrier is also composed of amorphous alginic acid and galactans that act as a matrix where the fibrillar moieties are embedded. While the fibrillar cellulose is a derivative of β-glucose, the amorphous alginic acid is a polymer composed of uronic acids. These chemical components are an important source of nutrients for the microbiome residing within the algae and may also be playing an important role in selecting the proper microbes to interact with the host. Algae can reproduce both asexually and sexually. In the asexual process, vegetative methods and spore formation processes are invoked. In the vegetative reproduction, the thallus becomes specialized and is detached from the parent to grow as an independent offspring. Budding and fragmentation are commonly observed. Akinetes are one of the numerous specialized spore-containing cells that enable algae to proliferate even in adverse situations. These are thick-walled spores with ample food reserves to survive unfavourable conditions and germinate. During sexual reproduction, two different gametes fuse to form a zygote that then becomes the source of genetically diverse offspring.

2.4 Why Algae Are Critical to the Ecosystem and for Our Survival?

The presence of a diverse and healthy population of algae is crucial to the well-being of the planet as activities executed by these micro- and macro-organisms are essential for the survival of all multicellular species. For instance, algae contribute to nearly 50% of the atmospheric oxygen owing to their photosynthetic capacity. In fact, the emergence of aerobic life forms started only after the oxygenation of the Earth, a process mediated by the cyanobacteria. They are also responsible for the nitrogen content in the air. Their nitrogen-fixing (nif) attribute is pivotal in modulating the levels of this inert gas that keeps our oxidative environment from burning. CO_2 is another gas that is maintained at an appropriate concentration by algae. The capture of CO_2 during photosynthesis and its subsequent transformation into glucose-rich nutrients enable these organisms not only to nourish themselves but also to keep CO_2 level in check. This greenhouse gas is an important effector of climate change and the rising global temperature.

Algae are also either directly or indirectly an important source of food for all marine organisms. They are part of the initial nutrient chain that fuels life in all aquatic environments. They secrete a variety of metabolites that are readily consumed by other organisms and help create ecological communities where diverse life forms thrive. They also harbour energy-rich polysaccharides and oils that heterotrophs utilize to proliferate. The forest-like canopy created by macroalgae on the water surface or submerged in oceans also provides an ideal habitat for other organisms to shelter from predatory extinction. Algae are known to contribute to the biogeochemical cycles of silicon (Si), phosphorus (P) and sulphur (S). Some algae prefer to use silicic acid as a building block for their cell wall instead of the energy-expensive organic precursors like cellulose. The biogenic silica is deposited in a controlled manner to form intricate glass barriers referred to as frustules. The algae with their frustules are then deposited within the sediments in the water bodies from where the Si may re-enter the biological landscape. P mainly as phosphate incorporated in algae is the conduit via which this mineral is cycled in the environment. S as sulphate is taken up by algae as part of their metabolism of amino acids and other S-containing biomolecules. These S moieties are then cycled through the chemical and biological intermediates. Algae also produce S-metabolites that are involved in a number of activities. Dimethyl sulfonium propionate (DMSP) is an osmo-protectant, antioxidant and buoyancy modulator. This algal-derived moiety has a central role in modulating the global climate. Some algae have the ability to deposit calcium carbonate. These calcifying algae contribute immensely to coral reefs and all the life within including the parrot fish. This fish grazes on the algae and ensures the health of the reefs and is also responsible for the formation of sand (**Figure 2.4**).

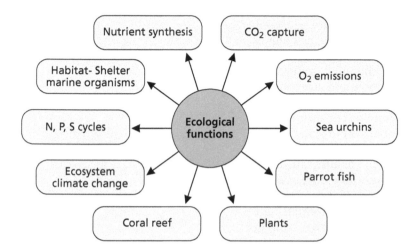

Figure 2.4. Ecological importance of algae. Algae are an important component of the ecosystem as they fulfil a range of functions including cycling of nutrients and are important parts of organisms like plants and corals. They serve as habitats for numerous organisms.

BOX 2.1 DIMETHYL SULPHIDE: A CLIMATE ACTIVE GAS OF ALGAL ORIGIN

Dimethyl sulphide (DMS) is the most abundant natural S gas in the atmosphere, the majority of which that originates in oceans is emitted by algae. Its oxidized products like SO_2 and SO_4 derivatives generate particulate materials responsible for condensation reactions resulting in cloud formation. These moieties also reflect sunlight a situation associated with a cooling trend that counters the warming features mediated by the greenhouse gases CO_2 and CH_4. Hence, the modulation of this sulphurous gas mediated by algae is an important contributor to conditions leading to the current climate change. The production of DMS is initiated by the decomposition of DMSP, a key metabolite in algae. This compound functions as an osmolyte, buoyancy regulator, cryoprotectant, predator deterrent, antioxidant, S storage and a senescence promoter. As methionine is a precursor of this molecule, its synthesis orchestrated by a series of enzymes including an aminotransferase, methyl transferase and a decarboxylase is dependent on the supply of N and P. Upon predatory grazing activity by marine herbivores, DMSP is released in the environment. This is a good source of S and C for bacteria that deploy their DMSP-lyases to generate DMS and acrylate. The volatile DMS serves as a key information chemical for birds and fish to locate their food. Furthermore, fish larvae utilize this signal to choose their habitat that can also help mitigate the feeding frenzy the algae are subjected to by the herbivores. Hence, it is clear that this physiologically critical molecule that algae have come to rely on also dictates the global weather pattern (**Box Figure 1**).

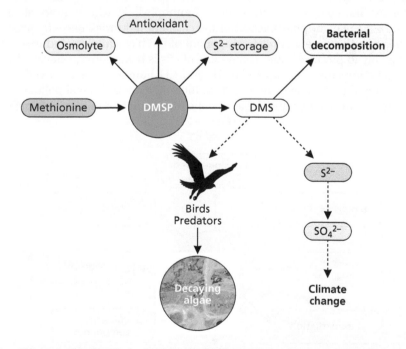

Box Figure 1. The dimethyl sulphide (DMS) derived from algae is a multi-faceted signalling metabolite.

2.5 The Economic Benefits Derived from Algae

Algae are not only essential for the survival of planet Earth, but they are also important contributors to our way of life. These living organisms and the products they generated are of major commercial interest as they are utilized for a variety of purposes. Algae, as a source of foods, have been consumed by humans since the dawn of civilization. The Chinese consumed *Nostoc* to survive famine more than 2,000 years ago while other societies utilized algae to savour the jelly-like 'pain des algues'. Today owing to its content in proteins, oils and vitamins, algae are an important component of modern cuisine and the food industry in general. Various algal species are utilized in soups and agar and as thickening agents in foods and beverages. In fact, algae are considered super foods due to their rich nutritional value. Proteins constitute 90% of the dry weight of *Chlorella* and are rich in essential amino acids such as lysine, leucine, tryptophan and methionine. The polyunsaturated fatty acids with omega-3 and -6 contents have numerous healthy attributes. The presence of vitamins A and B_{12}, iodine, manganese and other essential nutrients critical to human health has also increased algal consumption and has transformed these organisms and their products into nutraceuticals. *Laminaria*, a species of brown algae (kelp), contains numerous fold more calcium than milk. The complex carbohydrates that are an important constituent of algae are excellent source of dietary fibres responsible for nurturing the proper microbes in the digestive tract. In China, Japan and Korea, species such as *Porphyra*, *Ulva* and *Monostroma* are part of the regular food culture, and worldwide the algal industry has become a multi-billion-dollar enterprise.

As algae and the microbes they harbour are known to produce a plethora of bioactive metabolites, these organisms have also been known to be a source of medicinal products. Due to the presence of high levels of phycocyanin and iodine, algal cultures are used as a remedy against anaemia and goitre. The phycocyanin helps mobilize iron, while iodine aids the effective functioning of the thyroid gland. Laminarin and eckol synthesized by numerous algal species are a potent antibiotic and anti-hypertensive agents, respectively. Eckol also possesses anti-coagulative properties. Xanthophyll is widely used as a colouring agent in the pharmaceutical and cosmetic industries. Recently, algal cultures have become a critical component of the biotechnology industries. Their ability to produce copious amounts of oils is being commercialized by biodiesel companies. Biogases like (H_2 and CH_4) are also been derived by engineering their biochemical networks. Bioremediation of metal pollutants and organic wastes mediated by algal consortia is another industrial benefit that these organisms are providing (**Figure 2.5**).

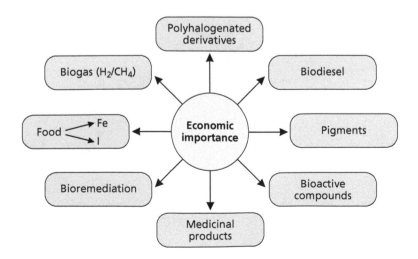

Figure 2.5. Algae, an important economic driver. Algae impart numerous benefits and are important source of food due to their high contents in essential nutrients like I and essential fatty acids. They are also involved in the production of biodiesel, biogas and medicinal products.

2.6 Algae and Their Microbiome

2.6.1 Intimate algae-cyanobacteria link

Algae are an essential part of the ecosystem, and the sustainability of human life depends on algal health. Hence, it is important that we understand what constitutes an alga and the molecular processes that drive algal physiology and biochemistry. Like all living organisms, they originate from bacteria and archaea as these prokaryotes are the building units of eukaryotic cells. The engulfing of a bacterium resulted in a mitochondrion, the energy generator of all eukaryotic cells. Such mitochondria-laden eukaryote then incorporated cyanobacteria via endosymbiosis and bestowed on algae the ability to capture light as they were eventually transformed into the light-harvesting organelle, the chloroplast. The intracellular conversion of solar energy into chemical energy provided a unique molecular machinery to power the synthesis of carbohydrate from CO_2 and H_2O. Thus, the engulfing of a cyanobacterium by a greedy eukaryote was the beginning of photoautotrophic life of all algae and plants. The establishment of a cyanobacterium into a foreign eukaryotic cell with the goal of seeking permanent residency may have necessitated a wide range of biochemical cross-talks and compromise to eliminate the antagonistic features associated with these two distinct organisms.

The protection of the nascent endosymbiont against the digestive machinery of the host, the elaboration of a metabolite exchange system, the tailoring of signalling mechanisms tending to the need of both partners and the exchange of genetic material to limit duplicative information are some of the governing rules to be adopted if this intracellular relationship is to survive. It has also been hypothesized that this intimate symbiotic relationship may be aided by the participation of an organism, a parasite that may have shepherded the cyanobacterium to establish without being degraded via phagocytosis. This ménage a trois hypothesis, although plausible awaits further elucidation. If this postulation is proven correct, it may become a template for other phenomena resulting in organelle formation. Thus, evolution can proceed by blending independent organisms into new life forms. The horizontal gene transfer (HGT) between the host and the endosymbiont results in functional protein products that are pivotal to the functioning of the whole entity, and the holobiont is a critical feature of this endosymbiotic relationship where one organism loses the majority of its characteristics in order to create a novel living system. For instance, numerous genes of cyanobacterial origin are housed in the nucleus and the proteins derived from these are targeted to the chloroplast (the original cyanobacterium). Algae and plants are such living organisms where the integration of a cyanobacterium and its transformation into an energy-making machine resulted in entirely new species with unique molecular processes. This intimate microbial connection is further strengthened by the presence of microbial life within algae as they contribute to a wide variety of functions that are essential for their proliferation and survival (**Figure 2.6**).

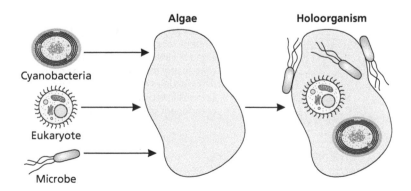

Figure 2.6. Eukaryote, cyanobacteria and microbes – the genesis of an algal holoorganism. Following the emergence of eukaryotes, cyanobacteria, bacteria and archaea assemble to result in the genesis of an algal holoorganism. The contribution of all the participants is essential for the survival of the host.

2.7 How Bacteria Benefit Algae? Bacteria within Algae

Although cyanobacteria evolved into a plastid (chloroplast) dedicated to perform photosynthesis within algae, these multicellular organisms have recruited numerous bacteria to execute a variety of functions that they cannot accomplish themselves. These tasks that would otherwise necessitate the development of specialized cells with their inherent shortcomings are farmed out to microbes residing with or within these organisms. The microbes may live in a symbiotic relationship modulated by an intricate communication network that requires decoding by both partners. In such situations referred to as mutualism both participants benefit. The supply of micro- and macro-nutrients like vitamin B_{12} and nitrogen (N) by the bacteria in exchange of fixed carbon illustrates this mutualistic connection. The presence of heterotrophic bacteria in *Chlamydomonas reinhardtii* helps either transport or synthesize vitamin B_{12}, while the *Rhizobium* species housed in *Chlorella vulgaris* provide easy access to N. Some bacteria residing on the surface of the algae have a commensal relationship whereby the microbes gain shelter and a source of nutrients without directly benefitting or harming the host. However, these ectobacteria are part of the algal phycosphere and are critical to the algae and the immediate ecological community. Hence, the multicellular life of algae consists of the endosymbiotic organelles (mitochondrion, plastid), the symbiotic bacteria, the commensal microbes and the algal cells (**Figure 2.7**).

2.8 The Microbial Contribution to the Algal Holoorganism

2.8.1 Biofilm formation and its multiple functions

Biofilm plays a pivotal role in the algal–microbial partnership as this polymer helps in the attachment of the microbes to the algal surface where the bacteria may find refuge, enter into a mutualistic relationship with the host and execute a variety of beneficial tasks. However, it is important to recognize that the extensive surface area associated with the algae, especially the macroalgae, presents an opportune environment for pathogenic microbes to utilize similar strategies to cling and inflict damage to algae. Hence, in marine environments, these unprotected submerged hosts become a fertile territory to be colonized by other organisms including bacteria. This phenomenon is referred to as biofouling and can lead to the demise of the host. Algae are known to invoke an intricate communication system to select the proper microbes and limit incursion by opportunistic ones. The epiphytic microbial communities are recruited by the algae specific and are tailored to reside on select surface of the host. Two main stratagems are employed for this purpose: nutrient release by the host that can only be utilized by the mutualistic microbes and signalling

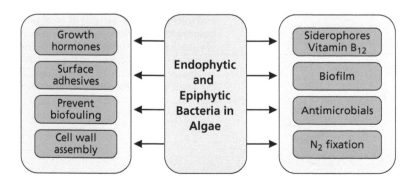

Figure 2.7. **Functions of algal microbiome.** Bacteria localized on the surface and within algae are involved in a variety of functions that enable the host to live. The microbes help in the acquisition of Fe, fix N_2 and dissuade colonization by pathogens and are an integral component of the multicellular host.

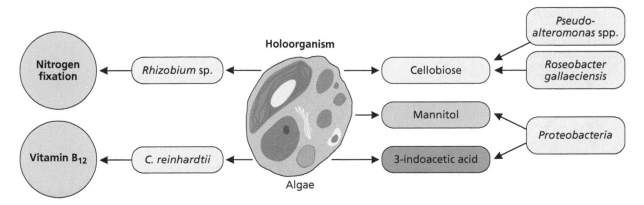

Figure 2.8. Algae are holoorganisms. Algae are known to secrete specific biomolecules to attract select microbes. Cellobiose helps attract *Pseudoalteromonas* spp., while 3-indole acetic acid recruits *Proteobacteria*. These microbes are assembled depending on the functions they perform for the host. They live within and/or on the surface of the algae.

molecules known as quorum sensors to control the population of microbes designed to reside on the host. For instance, the green alga *Ulva australis* lures the bacteria *Pseudalteromonas tunicate* and *Roseobacter gallaeciensis* with carbohydrate cellobiose. This saccharide is an excellent source of carbon for these microbes and quickly overwhelms other microbes in their midst. Brown algae like *Laminaria* have *Proteobacteria* as the most abundant group of microbes and mannitol secreted by the algae may influence this microbial selectivity. *Planctomycetes* appear to dominate the surfaces of other macroalgae. And once established as a community in the holobiont, these microbes have access to an abundant food supply that trickles from the cell wall of the host. These bacteria perform a variety of tasks including the synthesis of 3-indole acetic acid (IAA) from tryptophan supplied by the alga. The latter is a phyto-hormone that helps stimulate the growth of the host. This kind of cross-feeding as exemplified between the alga *Emiliana huxleyi* and the bacterium *Phaebacter inhibens* is widespread in nature and is critical in cementing the bond amongst the participants of the holoorganism (**Figure 2.8**).

2.8.2 Quorum sensing and microbial selection

Quorum sensing (QS) is a crucial aspect of algal–bacterial relationship resulting in the establishment of a proper density of microbial community beneficial to both partners. These quorum sensors that are information-laden chemicals dictate community organization and structure. Although acyl L-homoserine lactones (AHL) is one group of the signalling molecules that promote the appropriate bacterial density, peptides are also known to trigger biochemical processes conducive to promoting the desired microbial community. These signals control biofilm synthesis and induce phosphatase activity, an enzyme known to control phosphate levels and other enzymes responsible to liberate organic nutrients. The biofilms that are composed of polysaccharides not only help the establishment of episymbiosis but also are involved in colonizing the ecological niches that the holobiont can thrive in. Once this settlement is initiated, the biofilm may be modified chemically to aid in executing other tasks like enabling the uptake of nutrients, acting as a defence shield against pathogenic bacteria and acquiring genetic information to compete effectively in a given ecological niche. Although QS is an important channel via which bacteria and algae communicate in order to generate a functioning holoorganism, opportunistic microbes tend to mimic similar signalling pathways to gain access to their hosts. To thwart this stratagem, QS inhibitors are synthesized to blunt the signals of the invaders. These molecules can be analogues of the bona fide AHL like tumonoic acid or disparate metabolites like 2-*n*-pentyl-4-quinolinol designed to target the

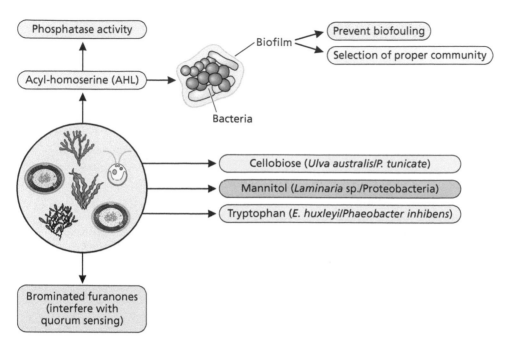

Figure 2.9. Microbial–algal cross-talk quorum sensing. Metabolites like mannitol and tryptophan attract select bacteria while brominated furanones interfere with quorum sensing. Acyl-homoserine regulates biofilm formation and controls microbial density by modulating phosphatase activity, a mediator of phosphate concentration.

QS receptors and block the reception of the QS cues. Enzymes with the mission to hydrolyse and modify the QS are also secreted. Haloperoxidases are known to introduce halogen moieties on the acyl group of the AHLs and render ineffective their signalling attributes. The deactivation of the AHL is a potent defence mechanism against the ability of pathogens to settle on the algal territory. Other algae tend to elaborate brominated derivative of furanone with the goal of interfering with the genetic responses elicited by the AHL. These strategies invoked by the microbial–algal holobiont help limit biofouling and promote the proliferation of these algae. Hence, the synthesis of biofilms and their regulation constitute a central aspect of the microbial–algal partnership. The important biological roles these polymers participate in and their ability to initiate microbial attachment on the algal surface make them an essential component of the holobiont. The modulation by the host of the synthesis of these carbohydrate-rich exopolymers guards against invasion by pathogens (**Figure 2.9**).

2.8.3 Nitrogen fixation: a key feature of the microbial–algal bond

Nitrogen is an essential nutrient as it is involved in a variety of biological functions including cellular growth. Biomass production is dependent on this ingredient as protein synthesis and cellular replication driven by nucleic acids are all nitrogenous biomolecules. Nitrogen gas is the most abundant component of air however, it is virtually inert and requires an exorbitant amount of energy and an intricate process to be reduced into ammonia (NH_3). The latter can then be integrated into biological systems with the aid of enzymes like glutamate dehydrogenase and glutamine synthetase. Only a few organisms, mostly microbes such as *Rhizobium*, *Azotobacter* and *Cyanobacterium*, have been endowed with the necessary biochemical machinery to accomplish this task. Most other organisms have to obtain this essential component from nutrients containing fixed nitrogen or live in symbiosis with microbes possessing nif attributes. Numerous algae appear to adopt the latter strategy and tend to harbour

BOX 2.2 BACTERIA AND THE KILLING OF ALGAE

QS is a potent communication strategy that organisms utilize to send signals aimed at selecting the proper community, strengthening antibiotic resistance, augmenting virulence, modifying feeding behaviour, promoting spore formation and eliciting senescence or episodic killing. For instance, the secretion of glycolate by algae helps privilege bacteria with the metabolic machinery to utilize this carbon source. In *C. vulgaris*, the resident bacteria suppress nitrogen fixation in response to high algal density. A variety of microbial constituents of algal communities is known to elicit killing tendency towards the host upon the detection of algicidal signals in the habitat. These information chemicals are produced by either host or the bacteria. The latter are sometimes even prompted by the host to generate deadly signals. *p*-Coumaric acid that is released by algae triggers pathogenic responses in bacteria directed at the host. The *Roseobacter* group of bacteria releases roseobactides, metabolites promoting algal lysis. *Pseudomonas protegens* associated with *C. reinhardtii* elaborates deadly lipopeptides that induce deflagellation and disrupt Ca^{2+} mobilization resulting in the immobilization and the eventual demise of the host. *E. huxleyi* is the most abundant coccolithophore in oceans and undergoes a periodic bloom dictated by high temperature, excess nutrients and sunlight. The alga plots its own death by sending chemical cues aimed at microbial executioners. While virus-induced glycosphingolipid synthesis by the host provokes a programmed

algal killing, the release of tryptophan by the host attracts *P. inhibens*, an algal killer. The bacterium produces IAA, a promoter of growth that kills when present in high amounts. Furthermore, the microbe generates an oxidative environment responsible for the dismantling of photosystem II and initiates an apoptotic cascade spearheaded by caspase culminating in the death of the alga. Hence, bacteria associated with algae are replete with attributes aimed at either producing or responding to death signals dedicated to their hosts. This arrangement allows the bacteria to benefit before and after the death of the algae (**Box Figure 2**).

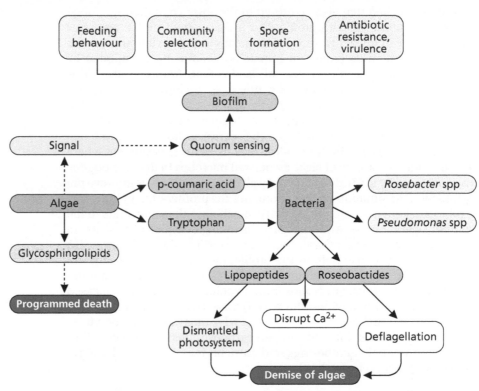

Box Figure 2. Molecular signals decoded by the bacteria that are responsible for the death/killing of the host.

nitrogen fixers on their surface in an episymbiotic partnership. Some, however, have evolved an endosymbiotic lifestyle. In exchange of organic carbon supplied by the host, the microbes provide fixed nitrogen in the form of NH_3 that is then channelled into numerous biomolecules. Free-living *Azotobacter* species associated with algae like the *Codium* becomes a source of nitrogen as they have the nif gene with the ability to transform N_2 into NH_3. *Mesorhizobium* and *Azospirillum* species that are known for their nif attribute are also prominent in numerous algae. The nitrogenase complex with molybdoferredoxin co-factor requires ATP and a complex network of electron transfer systems to reduce N_2. Some nitrogenases have also been reported to utilize vanadium or tungsten as cofactors. Recently, the occurrence of an endosymbiotic relationship involved in nitrogen fixation has been reported in an alga. In this instance, the endosymbiosis effecting the N_2 fixation is a cyanobacterium that has lost photosynthetic attribute and its ability to execute the tricarboxylic acid (TCA) cycle. The ability not to generate oxygen via photosynthesis may be contributing to the efficacy of nitrogen fixation. This genome reduction compels the cyanobacterium and the alga to live in an obligate relationship. The majority of the fixed nitrogen tends to be transferred to the host while only a small portion of the organic carbon synthesized by the host is allocated to the endosymbiont. The latter has a low growth rate and is basically recruited to provide bioavailable nitrogen. The nitrogen fixer seems to be trapped in an obligate lifestyle in the alga in exchange for a safe environment and a relatively limited supply of carbon. The latter may be pivotal to this partnership as the dependence on carbon-rich nutrients may have shaped the slow proliferative characteristic of the cyanobacterium and hence compelled it to live within the alga.

2.9 Microbes Promote Algal Growth

It is well established that algae cultured in the absence of their microbial partners have a poor growth profile. This is due to the fact that microbes supply a variety of nutrients, secondary metabolites and vitamins that are essential for numerous biological functions. Vitamin B_{12} is supplied by heterotrophic bacteria in exchange of fixed carbon from the algae. This vitamin is central in the metabolic processes including the effective utilization of fatty acids. These bacteria also secrete phosphatases that mediate the release of phosphate from organic compounds and from deteriorating biomass in the habitat. The increased level of bioavailable phosphate is a key contributor to algal growth. In an environment where CO_2 is limiting, the ability of algal-associated microbes to degrade organic carbon secreted by the host and its cell wall components helps improve the CO_2 budget. This situation further stimulates the proliferation of the algae. Microbes are known to secrete ancillary metabolites that contribute to the growth of algal biomass. Antimicrobials including enzymes like ascorbate oxidase and polyketide synthases that impede the invasion of the extensive surface of the algae also help in the proliferation of the host. A range of enzymes with the goal of enriching the ecological niche with nutrients and preventing opportunistic microbes from invading the territory is released by the bacteria residing within the host. Urease, laccasses, carrageenases and agarases can orchestrate the release of NH_3, bioavailable carbon from aromatic complexes, chitin and complex carbohydrates respectively. Thus, a wide assortment of microbe-triggered nutrients is at the disposal of algae to fuel their growth (**Figure 2.10**).

2.10 Siderophore – The Supplier of Iron – A Bacterial Contribution

Iron (Fe) is an essential commodity for virtually all organisms. Due to its ability to be involved in redox reactions mediated by the transfer of e– between Fe (II) and Fe (III), Fe participates in a wide range of life-supporting

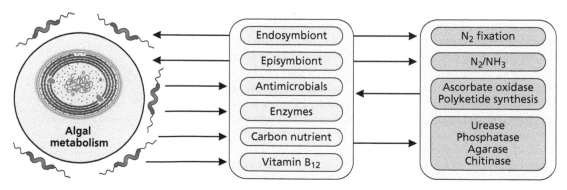

Figure 2.10. Algal microbiome: a source of nutrients and pathogen defence. Microbes residing with algae can be epicellular and endocellular. They supply nutrients by liberating enzymes such as chitinase, phosphatase and urease. They also produce antimicrobials in order to prevent the proliferation of pathogens and supply the host with fixed nitrogen.

biochemical processes. Production of ATP via oxidative phosphorylation and the replication of cells cannot occur without Fe. All the electron transfer complexes responsible for the synthesis of ATP are ineffective under Fe-deficient conditions. Ribonucleotide reductase, an enzyme that orchestrates the synthesis of deoxyribose nucleotide is central in the generation of DNA and subsequently cellular proliferation. However, acquiring Fe in the environment is a challenge as its solubility is relatively poor. Thus, organisms have evolved elaborate strategies to synthesize Fe-sequestering metabolites known as siderophores. These Fe chelators have a strong affinity for the trivalent metal that is taken up by the organisms via specific receptors and released following reduction to Fe (II). A complete set of proteins is needed to assimilate this micronutrient intracellularly. Algae are not commonly known to elaborate these small molecules aimed at immobilizing Fe. They tend to have ferroreductases that are enzymes involved in the reduction of Fe (Iii) into Fe (II). The divalent metal is then transported by the algae. The siderophore-bound Fe (III) is supplied by the microbes living in association of these marine algae.

Vibrioferrin, marinobactin and petrobactin are Fe chelators produced by bacteria belonging to *Marinobacter* species. These microbes reside in association with numerous algae and supply their hosts with iron in exchange of organic carbon. These siderophores sequester Fe with the aid of hydroxy/carboxylate moieties and release the micronutrient not via reduction but following photolysis. This unique property is bestowed on chelators due to their low iron binding capacity. Furthermore, the complexation of the metal with the siderophores makes the latter susceptible to photolytic degradation. Following the binding of the metal with for example vibrioferrin, the complex undergoes photochemical degradation releasing its cargo for the host. The micronutrient then becomes readily available to the algae as the photo-degraded product has relatively no affinity to the Fe. Hence, it appears that the synthesis of these carboxylate-type Fe scavengers has evolved to contribute to the supply of Fe that the algae need to proliferate. No elaborate enzyme system dedicated to the reduction of iron and its transport exists for these siderophores. They release their micronutrient upon photo-activation, a feature that enables the siderophore-producing microbe to fulfil the symbiotic duty within the holoorganism in exchange for living space and organic carbon moieties that trickle from the algal host (**Figure 2.11**).

2.11 Microbes Help Algae Survive Abiotic Stress

Algae are found in almost all ecological niches on this planet. From the frigid temperature of the Arctic to the saline environs of oceans, the metal-rich habitats of the polluted waterways and the driest place on Earth like the Atacama Desert, algae are known to proliferate in these diverse

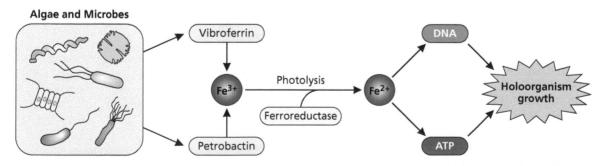

Figure 2.11. Siderophores of microbial origin: Fe suppliers to fuel growth. The microbial partners supply Fe with the aid of petrobactin and vibrobactin. The Fe is released following photolysis and incorporated in the algal host with the aid of ferroreductase. This mineral nutrient promotes cellular proliferation.

surroundings. To accomplish this extraordinary feat these organisms mainly adopt two strategies. First, algae enlist the support of microbes that reside within and tap on their wide array of biochemical networks to tame these seemingly hostile environments. The other stratagem is to acquire the genetic information predominantly from microbes living in these surroundings via HGT and enhance their survivability by adding to their genetic repertoire. Extremophilic red algae acquire their metabolic flexibility to thrive in acidic lakes by picking up genes from archaea found in these places. *Galdieria sulphuraria* isolated from acidic environments are known to harbour soluble ATPases and metal efflux pumps that are of archaeal origin. At elevated temperature, vitamin B_{12}-mediated methionine synthase is inactivated and algal growth can be seriously hampered. To circumvent this situation, algae are known to recruit vitamin B_{12}-producing bacteria to supply them with this essential ingredient for their growth. Hence, it is not surprising that the nature of bacteria associated with algae varies with temperature and weather.

The giant kelp, *Macrocystis pyriferia*, undergoes a decline in *Marinobacter* and *Pseudoaltermonas* spp. and an increase in such bacteria as *Colwellia* and *Saccharophagus* spp as the temperature rises. The production of exopolysaccharides (EPSs) by the bacteria in these harsh habitats allows these algal phototrophs to trap minerals that are essential in the fixation of CO_2. The organic nutrients secreted enable the community to thrive. A similar strategy is observed in glaciers and dry areas. In the former case, siderophores help sequester Fe, while a location-targeted decrease in pH aimed at solubilizing Ca and other micronutrients for the host are key for survival in these terrains. In deserts, the EPS provides protection against heat and evaporation. The production of carotenoids is aimed at limiting photo-inhibition and oxidative damage from UV radiation common in arid regions. In fact, obligate aerobes are promoted within the algal surface in an effort to maximize photosynthesis by restricting the access of O_2 to the photosynthetic machinery. Microbes like *Pseudomonas diminuta* requiring O_2 to grow have been observed in *Chlorella* sp. In the absence of this microbe, the algal photosynthetic power is markedly diminished. In extreme environments where nutrients are limiting, an efficient light-harvesting mechanism and CO_2 fixation are added adaptive benefits. Glycine and glutamate-derived osmolytes synthesized by the microbial partners are potent tools to survive saline conditions. For instance, the salt-tolerant species of *Picochlorum* possesses numerous functional genes dedicated to mitigate the negative biochemical influence of increased salt concentration in the environment. The cell wall constituents are modified in order to survive saline conditions. Survival is limited in the absence of microbial partners.

It is also important to note that during elevated temperatures, there is increased bacterial pathogenesis as the production of antimicrobials is hampered. Under such conditions, some algae are known to secrete *p*-coumaric acid, an infochemical that signals the *Roseobacter* spp. to kill the algae. Numerous troponoids, seven-membered aromatic ring compounds, are elaborated in order to eliminate the algae. Ice-residing algae such as *Raphidonema nivale* also live in association with heterotrophic bacteria as the latter provides a range of chemical nutrients including Fe that is required for the growth of the host. This type of cross-feeding is mutually beneficial as it allows the holobiont to colonize harsh habitats. The rise in atmospheric CO_2 also results in a marked change in the epibionts residing on the algal hosts. The introduction of these microbes in the holobiont is designed to combat the negative impact of this greenhouse gas. *Shewanella* spp. is more abundant in some macroalgae, and there is an increased expression of the numerous enzymes including carbon monoxide dehydrogenase that interacts directly with CO_2. Enzymes like alginate lyases, sulfatases and agarases involved in mediating the degradation of

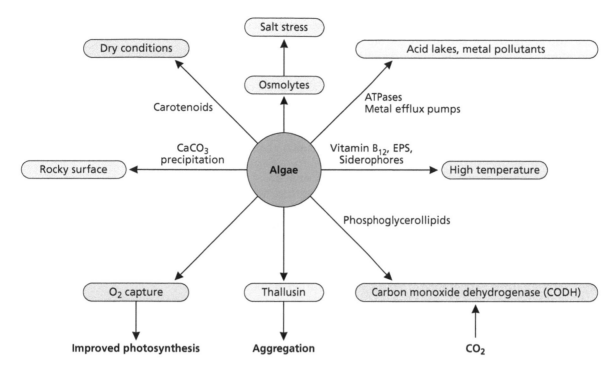

Figure 2.12. Microbiome and the ability of the host to adapt to abiotic stress. Microbes extend the adaptability of the host to a range of stressed environments. While metal efflux pumps allow algae to live in metal-polluted habitats, carotenoids help in the colonization of arid regions. Carbon monoxide dehydrogenase (CODH) fends against elevated CO_2 and C_aCO_3 promotes growth on rocky landscape.

cell wall are also augmented in an effort to increase the nutrient pool in the environment. Nitrate and nitrite reductases help maintain the supply of nitrogen. Intertidal seaweeds belonging to *Mastocarpus* spp. harbour distinct bacterial communities in order to stick to rocky surfaces or to live submerged in seawater. The calcified intertidal alga, *Corallina officianalis*, is home to at least nine microbial phyla that vary with intertidal and sub-tidal habitats. The specificity of these microbial residents is dependent on the functions to be performed and the surface area occupied. It appears that *Proteobacteria* are dominant, while *Chloroflexi* is least abundant. In these algae, the precipitation of carbonate by the microbes, a process mediated by carbonic anhydrase is critical for the survival of the holobiont. Similar recruitment of diverse microbial populations is invoked depending on the life phases of some algae. *Rhodobacter* spp. is prominent during the spore-forming stages, while an abundance of *Colwella* spp. is more evident in the gametophytes. Green algae like *Ulva* spp. and *Monostroma oxyspermum* are susceptible to the morphogenic ability of thallusin. This bacterial-derived cyclic molecule induces the formation of the thallus and orchestrates leafy morphology in these algae that otherwise grow as loose aggregates in its absence, a characteristic important depending on the habitat to be colonized. Hence, the algal microbiome is pivotal in the survival of the hosts exposed to abiotic stress (**Figure 2.12**).

2.12 Algae as Symbionts in Multicellular Organisms

2.12.1 Lichens and their algal partners

Bacteria constitute a crucial component of virtually all algae as they perform a variety of pivotal biological functions that are essential for the survival of the holoorganisms. However, it is important to note that algae themselves are also an integral part of numerous organisms where they are involved in a plethora of biochemical networks that make these organisms the way they are. In essence, if devoid of their algal partners, these organisms would not exist. The existence of the relatively sedentary lichens and corals to the highly motile salamander, would not be possible without the incorporation of algal metabolic networks within their midst. Even the slow-moving sloth harbours algae that act as camouflage and are used to feed the young ones. These algae may also provide protection against other pathogenic microbes.

Lichens are organisms that consist predominantly of filamentous fungi and photosynthetic algae that exist as discrete thalli. They are self-sustaining symbiotic systems occurring in most ecological niches. The fungal part termed as the mycobiont and the algal referred to as the photobiont are the main components of lichens. These organisms also harbour a variety of bacteria involved in a symbiotic partnership. Many of the fungi involved in this holobiont are obligate lichen forming. The host fungi from humid, temperate and boreal environments belong to the *Sordariomycetes* and *Lectiomycetes* class, while the *Eurotiomycetes* constitute the rock-inhabiting lichens. The latter contains the characteristic melanin in the cell wall, a moiety that is resistant to desiccation, radiation and elevated temperatures. The fungus provides an appropriate habitat for the alga where the latter fixes CO_2 via photosynthesis and in some cases also provides reduced nitrogen. Fixation of nitrogen into ammonia and release of proteases and peptidases by microbes are utilized to acquire this important nutrient in nitrogen-poor habitats. The specificity of these participants in the holobiont is tightly controlled as antibiotics like vancomycin and ristomycin are generated in an effort to fend off invasion by opportunistic invaders. These microbes also supply a range of important chemical tools that enable the holobiont to live and survive abiotic stresses. Quinone-derived cofactors like plastoquinone critical in photosynthesis and antioxidants such as tocopherol and glutathione enable the community within lichens to fulfil its requirement in organic carbon and fend against oxidative stress. Thiamine, an essential ingredient in numerous metabolic processes, is another moiety that is outsourced to the microbial partners.

The ability of lichen to colonize diverse ecological niches is also attributed to the algae and bacteria constituting these organisms. While the overexpression of arginine decarboxylase helps mitigate acidic stress, the presence of hydroxylases mediating the cleavage of aromatic compounds extends the resistance of lichens to areas contaminated with organic pollutants. Efflux pumps and enzymes involved in the methylation of arsenate located in the microbial dwellers are essential for the survivability of these organisms in metal-polluted habitats. The microbes within the lichens provide nutritional flexibility by extending the range of nutrients that can be accessed. Thus, soils, rocks and living or dead trees become a possibility as foods. However, when nutrients are scarce, microbes switch to ketone metabolism in an effort to supplement the energy need of the holobiont. Hence, it is clear that without the algal and microbial partners, lichens would not be able to conquer virtually all habitats on this planet. These suppliers of energy, protectors against microbial invaders and mitigators of inhospitable terrains enable the lichen to be a prolific organism (**Figure 2.13**).

2.12.2 *The bond between algae and corals*

Corals are organisms with a microbiome population consisting of algae, fungi, bacteria, archaea and viruses. These interconnected animal hosts, termed polyps, house algae with whom they share a unique endosymbiotic relationship. The algae are located in the gastrodermis of the animal hosts in specialized cells referred to as the symbiosome. It is here that they engage in a mutualistic relationship that makes each of the partners interdependent. While the corals provide easy access to CO_2, the algae capture light energy and transform CO_2 into organic carbon. A steady supply of algal nutrients propels corals optimal coralline growth. Corals

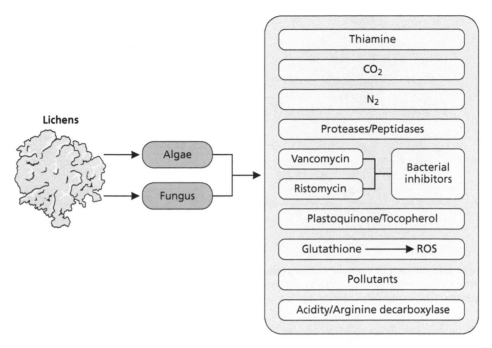

Figure 2.13. Algae within lichens: the tasks they accomplish. Algae play an important role in lichens where they live in association with fungi. They detoxify pollutants, regulate pH, thwart microbial invasion, supply nutrients, provide antioxidants and are a source of vitamins.

devoid of their photosynthesizing algae tend to experience abnormal reproductive tendency and die. Algae are so pivotal to the lifecycle of corals that they are transmitted via coral buds or fragments to the offspring. Corals involved in sexual reproduction transmit these essential partners either when eggs are deposited or the parents brood the larvae. There also appears to be a molecular communication between the larvae and the algae. Coralline algae tend to attract the larvae. Furthermore, the immune system of the host is modulated in order to accommodate the algae. This molecular adjustment allows the symbiont to reside within the host. The photosynthetic activity of the algae not only provides fixed organic nutrients but also promotes the deposition of $CaCO_3$, a key ingredient in the coral anatomy. Activating Ca^{2+} transport, creating the proper pH by consuming CO_2, generating precursors for the matrix essential in the mineralization of $CaCO_3$ and consumption of PO_4 are some of the molecular tasks algae help accomplish in this holobiont. Phosphate is an inhibitor of $CaCO_3$ formation and its removal is beneficial to this process. In a properly growing coral community, the density of the algal population remains relatively constant. An abundance of algae or a dearth of these photobionts or a dysbiosis of the microbiome referred to as coral bleaching can lead to the demise of the host animal. The population of the coral algae is usually kept in check by the host and marine grazers like the parrot fish (**Figure 2.14**).

2.12.3 Algae and sexual behaviour of sea anemones

The sexual preference of sea anemones is to some extent dictated by the algae they harbour, the intensity of sunlight and the environmental temperature. More gonad development is observed at moderate temperature in *Elliptochloris marina*, a sea anemone housing green algae and sexual reproduction is favoured. High temperature promotes asexual reproduction where fission (cloning) is the preferred proliferative mechanism. The energy output and the specificity of the resident algae are also important determinants in the sexual behaviour of this marine organism. Hence, the algal symbiotic state that modulates gonad development can tilt the reproductive behaviour towards a sexual or asexual route and this can have major implications for the survival of sea anemone as climatic conditions fluctuate. The algal–sea anemone

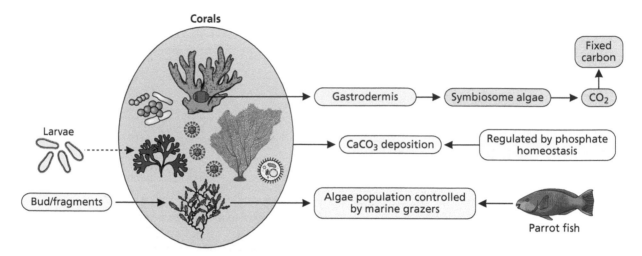

Figure 2.14. Algae are an integral part of the coral holoorganism. Corals cannot live without their algal partners that are part of the holoorganism during the developmental stages. Calcium carbonate ($CaCO_3$) deposition, fixing of CO_2 and exchange in NH4+ are central in this relationship. In this ecosystem, parrot fish helps control algal population in conjunction with the coral host.

symbiotic partnership occurs in the expandable tentacles and the oral discs where the photobionts are localized. The hosts adopt a lifestyle conducive to maximizing the ability of the algae to fix CO_2. They concentrate dissolved CO_2 as the level ($12\,\mu M$) is too low for ribulose 1,5 biphosphate carboxylase/oxygenase (Rubisco) to function effectively. This enzyme utilizes the ATP generated during photosynthesis to convert CO_2 into organic carbon like maltose, or glycerol. A proton efflux pump and carbonic anhydrase are utilized to accomplish this task. The former acidifies the impermeable bicarbonate (HCO_{3-}) to carbonic acid (H_2CO_3). This is then converted into diffusible CO_2 by carbonic anhydrase, a process that is activated by the operation of the photosynthetic machinery.

In order to enable the algal dwellers' access to an optimal amount of sunlight, sea anemones tend to live in shallow water. This habitat has its inherent risk as the hosts are exposed to the damaging radiation from the sun. To combat this situation, sea anemones produce protective pigments such as anthocyanins, β-carotene and pheophytin. These biomolecules tend to give these organisms their colourful characteristics. Furthermore, the mycosporine-like amino acids supplied by the algae to the sea anemone also act as a defence against sunlight as these moieties absorb ultraviolet radiation and dissipate energy. The O_2 released during photosynthesis creates a hyperoxic environment, a situation that exposes the host to oxidative stress. Superoxide dismutases, peroxidases and catalases aimed at neutralizing ROS are overexpressed in the host. The intimate molecular partnership exhibited by sea anemones and their resident algae allows this holobiont to maximize their lifestyle with minimal negative impact. This kind of mutual cooperation is also extended in order to manage the homeostasis of nitrogen. The NH_4+ secreted by the host is assimilated by the algae with the assistance of glutamate dehydrogenase and glutamine synthase. The former transforms NH_4+, α-ketoglutarate and NADPH into glutamate, while the latter fixes NH_4+ into glutamine with the participation of glutamate (**Figure 2.15**).

2.12.4 Algae in the life of salamanders

The spotted salamander (*Ambystoma maculatum*) is one of the known rare vertebrates to house algae (*Oophila amblystomatis*). The salamander lays eggs in open water where they are exposed to algae. It was initially postulated that the algae live on the surface and impart its green colouration; however, a recent study has revealed the location of the photobionts within

BOX 2.3 CORAL BLEACHING: A CASE OF ALGAL EVICTION TRIGGERED BY CLIMATE CHANGE

In the coral holoorganism, the host forges a very intimate relationship with the algal symbiont. The latter does not only supply fixed organic nutrients, but it also contributes to a range of chromophores that gives coral reef its colourful hue. This arrangement is severely jeopardized by the rising temperature of oceans culminating in the expulsion of the photosynthetic partners, the bleaching and eventually the starvation of the coral. The perturbation of the metabolic networks governing this relationship is an important contributing factor in the demise of the holobiont. The increased metabolic demand provoked by the heat stress compels the host to degrade amino acids in effort to fulfil its energy need. This situation is punctuated by the release of NH_4+, a nutrient readily utilized by the resident algae. During the normal symbiotic process, nitrogen nutrients as well as the organic products from photosynthesis are supplied by the algae to the coral. This switch in nutrient exchange triggered by heat stress prompts algal proliferation and retention of nutrients usually earmarked for the host. To counter this situation, the host resorts to the breakdown of amino acids leading to an increase in NH_4+ that fuels the growth of the algae. There is a marked increase in the gene system mediating amino acid degradation in the coral coupled with a decrease in the genetic fingerprint responsible for the TCA cycle. Hence, the cessation of nutrient translocation from the algae to the host creates a carbon-limited condition within the latter and hence prompts a metabolic response favouring the proliferation of the algae. This metabolic disconnect can also generate excessive reactive oxygen species (ROS) leading to the activation of the immune system orchestrating the eviction of the algae from the holobiont that results in coral bleaching. Climate change evokes metabolic dysfunction in the host and its microbial partners endangering the holoorganism and eventually the marine ecosystem (**Box Figure 3**).

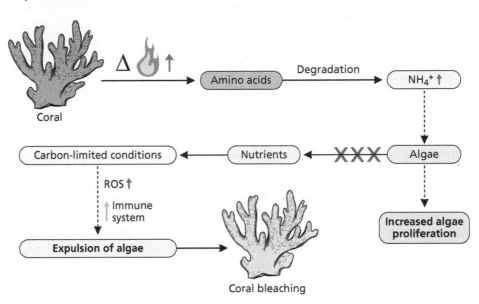

Box Figure 3. A decrease in nutrient exchange prompted by a rise in temperature results in the expulsion of the algae by the hosts.

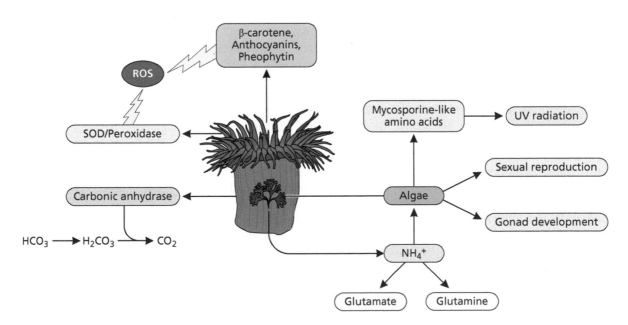

Figure 2.15. Algae and sea anemones living in partnership – an intimate molecular bond. In order for the algae to nourish the sea anemone with fixed organic carbon, the host helps concentrate CO_2 and provides a range of antioxidants to combat oxidative stress. The algae also assimilate NH_4+ secreted by the sea anemone into glutamate and glutamine.

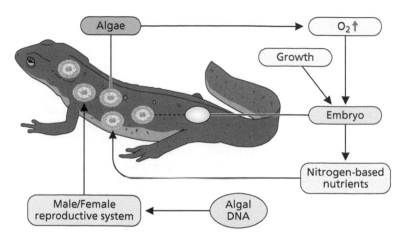

Figure 2.16. Algae in the development of salamanders. Algae located in the embryo promote the growth and development of salamanders. These algal species are vertically transmitted.

the embryos where they are responsible for the characteristic green colour. It is hypothesized that the nitrogen-rich wastes generated by the embryo become a source of algal nutrients, while the algae may be increasing the O_2 content of the embryo and supplying organic carbon. Embryos devoid of algae develop at a slower rate compared to those with their algal symbionts. The freshly laid eggs tend to have these algae. These CO_2-fixing organisms are transmitted to the embryos during reproduction as the oviducts and the male reproductive system are associated with algal DNA. And this transmission strategy enables the establishment of the algae within this vertebrate as the embryonic stage has a poorly developed immune system and hence can tolerate the inclusion of a plant-like organism. The possibility that the immune response may be fine-tuned to accommodate this alga may also be operative and may assist in nurturing this partnership (**Figure 2.16**).

2.12.5 Algae: an integral component of plant microbiome

Some plants are known to harbour algae either in close proximity or as residents within. These associations impart numerous benefits to both the hosts and the algae since this arrangement helps fend off a variety of biotic and abiotic stresses. As the algae, a constituent of the plant microbiome are prone to changes in a timely fashion, they are essential for the fitness and adaptability of the holoorganism. The microalgae usually can gain access to the plants through the stomata and colonize the intracellular space where they exist as endosymbionts in well-defined organs like in the case of *Gunnera* spp. (rhubarb). The *Nostoc* spp. housed herein is host specific and a genetic system homologous to the rhizobial Nod genes seems to be operative. Roots of wheat and cotton plants do also have an appreciable population of algae. These organisms are known to combat pathogens by producing antifungal compounds and by priming the immune system of the host, a process promoted by the D-lactate. They elaborate a range of range molecules like diterpenoids, chitinase and glucanases aimed at bacteria and fungi, respectively. Macrocystins and nodularins originating from algae are potent insecticides. This pathogen-fighting attribute is coupled to the ability of the algae to contribute to the development of the plant. They are a source of numerous phytohormones such as auxins and cytokinins that promote growth of the host and impede senescence. Ascorbate peroxidase, superoxide dismutase and catalase help combat oxidants, a feature associated with the anti-ageing process. The elaboration of micronutrients like vitamin B improves the metabolic fitness of the host. Salt stress and drought are mitigated by polysaccharides derived from algae. Thus, algae residing within plants are involved in a plethora of functions essential for the health and survivability of the host (**Figure 2.17**).

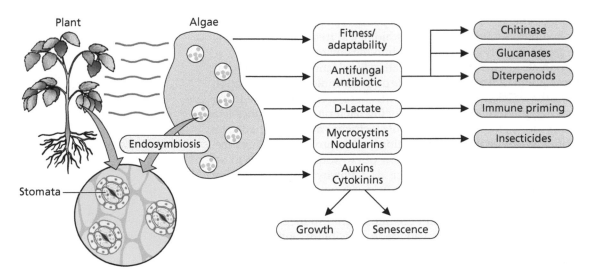

Figure 2.17. Algal functions in plant holoorganisms. Algae live in association with different plant tissues and fulfil numerous functions including immune priming and combatting fungal infections.

2.13 Algae and Their Microbiomes: A Biotechnological Boon

As microbes are an integral constituent of algae where they perform a variety of critical functions, the molecular understanding of this relationship will enable the effective utilization of algae in commercial activities that will be beneficial to society. Algae are currently utilized as foods, and its products are used as medications and cosmetics. They are also involved in environmental clean-up, the bioenergy industry and synthesizing bioplastics. For all these applications, it is pivotal to obtain optimal amounts of algae, inexpensive technologies for harvesting them and effective means to lyse the cells. Understanding how algae and their microbiome communicate will enable the efficient harnessing of algal biochemistry for societal benefits. The microbes not only optimize the growth of algal biomass but also synthesize valuable chemicals during their association with the holobiont. For instance, the supplementation of *Chlorella* spp. cultures with *Rhizobium* spp. is known to increase algal yield by up to 3-fold. They tend to eliminate algicidal organisms. The downstream processing activities like harvesting and disrupting cell wall can also be optimized. Numerous microbial species produce bioflocculants like exopolysaccharides that are ideal to harvest microalgae. For example, *Bacillus licheniformis* can be a cost-effective harvester. The ability of microbes to secrete hydrolytic enzymes that interact with algal cell walls can be very economical as a cell disruptor and will enable the release of value-added chemicals stored within the cells.

The algal–microbial bond can be harnessed in the removal of phosphorus, nitrogen and organic carbon from agricultural and municipal wastewaters. Algae are avid consumers of these nutrients that are either shared with the microbes or supplied to each other. Unravelling this symbiotic relationship will be an asset in environmental clean-up. Furthermore, the ability of the constituents of the algal holobiont to degrade pesticides and aromatics and to neutralize toxic metals can be utilized in bioremediation technologies. These features are usually attributed to the diversity of the genetic information associated with the algal community. The cross-feeding nature of the holoorganism is also exhibited in the joint detoxification of chemical pollutants. In this instance, each species contributes partially to the elimination of the contaminant whereby the participation of the holobiont is required for complete detoxification. While metals can be absorbed, precipitated or biotransformed, the recalcitrant organic pollutants can be mineralized with the concomitant release of O_2. This biological treatment will enable the discharge of relatively clean effluents into the environment.

Algal nanofactories involved in generating bioplastic, biodiesel, biogas (H_2, CH_4) and bioethanol will be profitable once the molecular information orchestrating the efficiency associated with these communities of living organisms is deciphered. Methane-producing bacteria and microbes like *Geobacter* spp. known for the generation of electricity can be coached to be part of the holobiont with the aim of creating a select product. For instance, activation of hydrogenase and elimination of O_2 would beneficial to an algal community designed to produce H_2. Polyhydroxybutyrate, an ingredient for bioplastic production and starch granules for ethanol fermentation, will be within reach. Sustainable aquaculture where a proper amount of nutrients is supplied and chemicals are produced to prevent opportunistic microbial invasion will be a possibility. Environmental monitoring and control of algal blooms will be more effective. *Roseobacter* spp. and *Rhodococcus* spp. that are known to produce algicides can be recruited to limit the uncontrolled proliferation of algae. The dynamics between predatory microbes and the holobiont will inform the efficacy of this anti-algal bloom technology.

The interaction of algae with other organisms can provide clues to mitigate the negative impact of climate change on coral reef. An increase in temperature is associated with bleaching of corals. This is a phenomenon where the coralline host expels the algae, an event that has a drastic influence on marine life. Hence, repopulating the reef with algae and the microbial partners known to be part of this ecosystem can provide a remedy to this situation. Recently, probiotics were introduced in marine habitats in the Pacific Ocean in an effort to combat the deterioration of the coral reef. Similar strategies can be initiated to optimize the density of sea anemones, organisms that are dependent on their algal microbiome for an array of biological functions including reproduction. Lichens the symbiotic life forms comprising of host fungi, algae and microbial constituents are known to occupy virtually all ecological habitats on Earth. The molecular networks that enable these disparate biological entities to live in harmony can unleash technologies to colonize barren lands, deserts, even glaciers and hot springs. The spotted salamander, the only known vertebrate so far to interact directly with algae is an intriguing model system to unravel the relationship between an organism engaged in the conversion of light energy into fixed carbon and an animal. This will pave the way for endless possibilities on the need for chemical energy to sustain life in higher organisms. Furthermore, it will establish a direct intimate connection between solar energy and animals. Currently, this occurrence appears to be extremely rare in nature (**Figure 2.18**).

2.14 Conclusions

Despite the multicellular nature of algae, these organisms evolved with bacteria where the latter occupy a central role. They are parts of the machinery that enables the hosts to improve their fitness and extend their survivability. Algae can thus colonize almost all ecological niches. This biological partnership among seemingly disparate species is essential for adaptability and the physiological well-being of the holoorganism. Although the hosts have acquired specialized cells and organs dedicated to specific functions, they have forged a unique relationship with microbial partners endowed with a range of molecular flexibility in an effort to augment their capacity to live. In fact, the presence of microbes in algae and algae in numerous other hosts are designed to ensure maximal biological efficacy to survive and reproduce. When dysbiosis occurs due to environmental stress, this arrangement collapses and the demise of the holoorganisms ensues

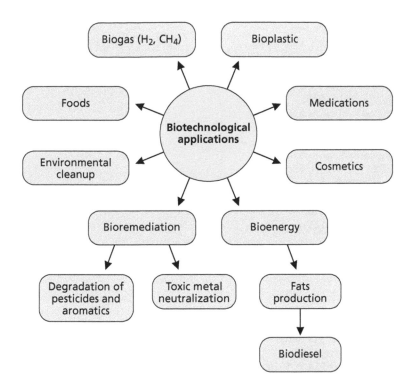

Figure 2.18. Algae and biotechnological importance. Algae are utilized in the production of value-added products like biogas, biodiesel and foods. They are also involved in pollution management and in the cosmetic industry.

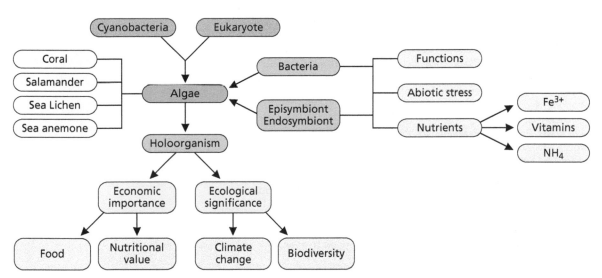

Figure 2.19. Microbes within algae and algae within other organisms. This Figure encapsulates the roles of microbes in algae and also depicts how algae are important components of other holoorganisms. Multicellular organisms need their microbial partners.

like in the case of coral bleaching where the survivability of all the constituents is compromised. In this instance, the negative influence can be devastating to the whole global community. The biological flexibility imparted by the microbiome does not only help the host but also benefits the whole metagenomic community and beyond (**Figure 2.19**).

SUGGESTED READINGS

Amin, S et al. (2009). Photolysis of ion – siderophore chelates promotes bacteria – algal mutualism. *Proceedings of the National Academy of Sciences, 106*(40), 17071–17076. doi:10.1073/pnas.0905512106.

Anesio, A. M., Lutz, S., Chrismas, N. A., & Benning, L. G. (2017). The microbiome of glaciers and ice sheets. *Npj Biofilms and Microbiomes, 3*(1). eCollection 2017. doi:10.1038/s41522-017-0019-0.

Appanna, V. D., & Viswanatha, T. (1986). Effects of some substrate analogs on aerobactin synthetase from Aerobacter aerogenes 62-1. *FEBS Letters, 202*(1), 107–110. doi:10.1016/0014-5793(86)80658-5.

Bingham, B. L., Dimond, J. L., & Muller-Parker, G. (2014). Symbiotic state influences life-history strategy of a clonal cnidarian. *Proceedings of the Royal Society B: Biological Sciences, 281*(1789), 20140548. doi:10.1098/rspb.2014.0548.

Bramucci, A. et al. (2018). The Bacterial symbiont *Phaeobacter inhibens* Shapes the Life History of Its Algal Host *Emiliania huxleyi. Frontiers in Marine Science, 5*, 1–12. doi:10.3389/fmars.2018.00188.

Brodie, J., et al., (2017). Biotic interactions as drivers of algal origin and evolution. *New Phytologist, 216*(3), 670–681. doi:10.1111/nph.14760.

Burns J.A et al. (2020) Heterotrophic carbon fixation in a salamander-alga symbiosis. *Frontiers in Microbiology,* https://doi.org/10.3389/fmicb.2020.01815.

Cirri, E., & Pohnert, G. (2019). Algae–bacteria interactions that balance the planktonic microbiome. *New Phytologist, 223*(1), 100–106. doi:10.1111/nph.15765.

Deng, X., Chen, J., Hansson, L., Zhao, X., & Xie, P. (2020). Eco-chemical mechanisms govern phytoplankton emissions of dimethylsulfide in global surface waters. *National Science Review, 8*(2), 1–8. doi:10.1093/nsr/nwaa140.

Egan, S., Harder, T., Burke, C., Steinberg, P., Kjelleberg, S., & Thomas, T. (2013). The seaweed holobiont: Understanding seaweed–bacteria interactions. *FEMS Microbiology Reviews, 37*(3), 462–476. doi:10.1111/1574-6976.12011.

Gabay, Y., Weis, V. M., & Davy, S. K. (2018). Symbiont identity influences patterns of symbiosis establishment, host growth, and asexual reproduction in a model cnidarian-dinoflagellate symbiosis. *The Biological Bulletin, 234*(1), 1–10. doi:10.1086/696365.

Gregor, R., David, S., & Meijler, M. M. (2018). Chemical strategies to unravel bacterial-eukaryotic signaling. *Chemical Society Reviews, 47*(5), 1761–1772. doi:10.1039/c7cs00606c.

Kaup, M., Trull, S., & Hom, E. F. (2021). On the move: Sloths and their epibionts as model mobile ecosystems. *Biological Reviews.* 1–23. doi:10.1111/brv.12773.

Kerney, R., Kim, E., Hangarter, R. P., Heiss, A. A., Bishop, C. D., & Hall, B. K. (2011). Intracellular invasion of green algae in a salamander host. *Proceedings of the National Academy of Sciences, 108*(16), 6497–6502. doi:10.1073/pnas.1018259108.

Kneip, C., Lockhart, P., Voß, C., & Maier, U. (2007). Nitrogen fixation in eukaryotes – New models for symbiosis. *BMC Evolutionary Biology, 7*(1), 55–67. doi:10.1186/1471-2148-7-55.

Lemay, M. A., Martone, P. T., Hind, K. R., Lindstrom, S. C., & Parfrey, L. W. (2018). Alternate life history phases of a common seaweed have distinct microbial surface communities. *Molecular Ecology, 27*(17), 3555–3568. doi:10.1111/mec.14815.

López, Y., & Soto, S. M. (2019). The usefulness of microalgae compounds for preventing biofilm infections. *Antibiotics, 9*(1), 9–25. doi:10.3390/antibiotics9010009.

Matsuo, Y., Imagawa, H., Nishizawa, M., & Shizuri, Y. (2005). Isolation of an algal morphogenesis inducer from a marine bacterium. *Science, 307*(5715), 1598–1598. doi:10.1126/science.1105486.

Meyer, N., Bigalke, A., Kaulfuß, A., & Pohnert, G. (2017). Strategies and ecological roles of algicidal bacteria. *FEMS Microbiology Reviews, 41*(6), 880–899. doi:10.1093/femsre/fux029.

Orsi, W. D. (2018). Ecology and evolution of seafloor and sub-seafloor microbial communities. *Nature Reviews Microbiology, 16*(11), 671–683. doi:10.1038/s41579-018-0046-8.

Ramanan, R., Kim, B., Cho, D., Oh, H., & Kim, H. (2016). Algae–bacteria interactions: Evolution, ecology and emerging applications. *Biotechnology Advances, 34*(1), 14–29. doi:10.1016/j.biotechadv.2015.12.003.

Segev, E., et al. (2016). Dynamic metabolic exchange governs a marine algal-bacterial interaction. *Elife, 18* (5), e17473. doi:10.7554/elife.17473.025.

Singh, R. P., & Reddy, C. R. (2016). Unraveling the functions of the macroalgal microbiome. *Frontiers in Microbiology, 6.* 1488–1494. doi:10.3389/fmicb.2015.01488.

Thompson, A., et al. (2012). Unicellular cyanobacterium symbiotic with a single-celled eukaryotic alga. *Science, 337*(6101), 1546–1550. doi:10.1126/science.1222700.

Wells, M. L. et al., (2016). Algae as nutritional and functional food sources: Revisiting our understanding. *Journal of Applied Phycology, 29*(2), 949–982. doi:10.1007/s10811-016-0974-5.

Yellowlees, D., Rees, T. A., & Leggat, W. (2008). Metabolic interactions between algal symbionts and invertebrate hosts. *Plant, Cell & Environment, 31*(5), 679–694. doi:10.1111/j.1365-3040.2008.01802.x.

Yoshioka, Y., et al., (2021). Whole-genome transcriptome analyses of native symbionts reveal host coral genomic novelties for establishing coral–algae symbioses. *Genome Biology and Evolution, 13*(1), evaa240. doi:10.1093/gbe/evaa240.

Zhou, J., Lyu, Y., Richlen, M. L., Anderson, D. M., & Cai, Z. (2016). Quorum sensing is a language of chemical signals and plays an ecological role in algal-bacterial interactions. *Critical Reviews in Plant Sciences, 35*(2), 81–105. doi:10.1080/07352689.2016.1172461.

FUNGAL MICROBIOME AND MYCOBIOME WITHIN OTHER ORGANISMS

3

Contents

Keywords

- Fungal Microbiome
- Mycobiome
- Corals
- Plants
- Humans
- Immunity
- Digestion

3.1 Fungus: An Omnipresent Organism

The fungus is an organism that is widespread in nature and occupies almost all ecological niches. It is found on the soil, underground, in freshwater, in marine environment, in the air, in the house, in arid habitats, in regions with disparate temperatures, within the human body and in other organisms like corals and plants. Although fungi exhibit limited mobility and tend to resemble plants, they are more animal-like. For instance, their cell wall is made up of chitin, a polymer of N-acetyl glucosamine common in insects and lobsters. Unlike plants, they cannot fix CO_2. Hence, fungi have to depend on their partners they live with in symbiosis or have to secrete enzymes to hydrolyse and acquire nutrients from their surroundings. They are an integral part of human culture as they have been utilized as foods, beverages and medicines since 6,000 years ago. In fact, yeasts were among the first fungus to be domesticated by humans. The present-day fermented alcoholic drinks, cheese and delicacy mushrooms are just some of the examples of the influence of fungi in our daily life.

DOI: 10.1201/9781003166481-3

3.1.1 Classification and common attributes

The fungal family comprises eight phyla with the *Ascomycota* and *Basidiomycota* being the most diversified. *Ascomycota* possesses approximately 90,000 species while around 50,000 species are assigned to the *Basidiomycota* phylum. Yeasts and moulds are part of the former group while mushrooms belong to the latter. Fungi display an immense diversity of shape, size and colour. They may have a stalk with a bulb on the top, can be filament-like, can be microscopic, may possess gigantic structures and can have interconnected hyphae. In fact, the honey fungus (*Armillaria* spp.) a member of the *Basidiomycota* group measures up to 4 km and can grow as long as they have nutrients. These organisms that may have evolved 1 billion years ago depend on their ecosystem for foods where they live as symbionts, parasites or saprophytes. If nutrients are not available in a mutualistic fashion, fungi secrete digestive enzymes like cellulases and ligninases to hydrolyse cellulose and lignin that are then absorbed. This task of digestion and absorption is usually confined to the tubular hyphae. In fact, these protoplasm-filled structures form an underground network with plant roots in an effort to engage in a symbiotic relationship beneficial to all the participants. Fungi tend to proliferate via sexual and asexual reproduction. Oxylipins generated via the peroxidation of fatty acids help modulate the various reproductive and development stages. Spore formation and fragmentation of hyphae are important reproductive strategies that enable fungi to proliferate rapidly. Hyphae, the hallmark of fungi do not only allow these organisms to acquire nutrients by interacting with the soil or wood, but they also form fruiting bodies within which spores are produced. Spores are generated in a prodigious manner and are released in response to heat, moisture and interaction with targeted hosts (**Figure 3.1**).

3.1.2 Fungi: an essential component of the global ecology

Fungi not only shape the community dynamics and ecology with the organisms they interact, but they also play a pivotal role in the cycling of nutrients. Fungi are amongst the few organisms that produce cellulases and ligninases, enzymes mediating the hydrolysis of wood debris. Without the efficacy of this hydrolytic process, nutrients like carbon and nitrogen will be locked up in this biomass and unavailable for consumption by other organisms. Fungi as a human nutrient are part of a $42 billion industry. Edible mushrooms constitute a significant food source for humans as they

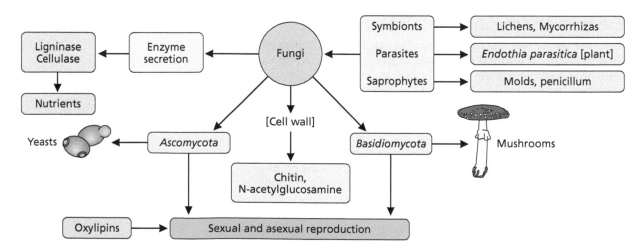

Figure 3.1. Fungi and their general characteristics. Fungi are classified in numerous phyla including *Ascomycota* and *Basidiomycota*. Fungal cell wall consists of chitin, a polymer of *N*-acetyl glucosamine. Fungi secrete enzymes to acquire nutrients and can undergo both sexual and asexual reproduction. They live as symbionts, parasites or saprophytes.

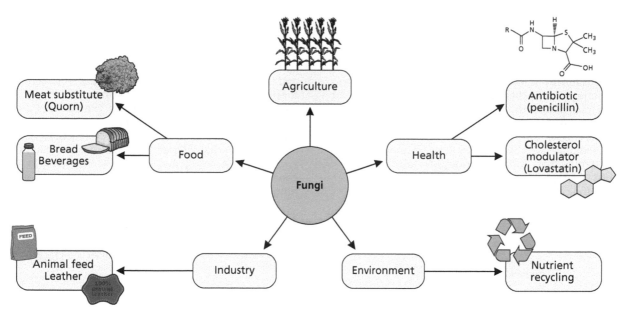

Figure 3.2. Economic importance of fungi. Fungi are an important economic contributor as they are a source of food and are involved in the production of medicines. They play a pivotal role in agriculture and in nutrient recycling.

are an excellent source of proteins and minerals. For instance, the Portobello mushroom contains more potassium (K) than a banana and is laden with fibres and proteins. *Saccharomyces cerevisiae* that is a must in the processing of a variety of foods including bread and beverages has become a staple worldwide. Meat substitute (Quorn) derived from the fungus *Fusarium venenatum* known to generate 300 kg of product in a controlled environment can indeed play a major role in decreasing the carbon footprint. Fungal species like *Penicillium camemberti* and *Penicillium roqueforti* due to their ability to produce flavours and create various physical transformations upon interacting with the constituents of milk are central in ripening cheese. Numerous medications including penicillin, an antibiotic and lovastatin, a cholesterol modulator have been initially uncovered in fungal systems. As fungi live in a competitive environment and are restricted in their mobility, chemical defence is a strategy of choice that is deployed in order to survive. Fungal-derived enzymes such as lipases, cellulases and catalases are widely utilized in the leather, paper and cotton industries (**Figure 3.2**).

3.2 Fungal Microbiome: Their Bacterial Partners

3.2.1 Endosymbiosis: the microbes living within

Fungi like all multicellular organisms live and thrive in an ecological community where they are associated intimately with a variety of microorganisms including bacteria. This type of partnership is usually symbiotic in nature whereby all participants benefit. The bacterial residents can be either intracellular or extracellular. The former occurrence referred to as endosymbiosis is characterized by the residence of the microbes within the fungi while during ectosymbiosis the microbes live on the surface of the fungal structures. In such holoorganisms, the constituents tend to avoid duplication and work together to benefit all the constituent members. They do poorly or die if they live separately. The species belonging to the fungal phylum of *Mycoromycoca* are known to have evolved an endosymbiotic lifestyle with the microbes from the *Betaproteobacteria* and *Mollicutes* group. *Mycoromycoca* is a plant decomposer and is part of the arbuscular mycorrhizal fungi where its endosymbionts are *Candidatus*

BOX 3.1 FOOD FLAVOURS AND THE FUNGAL CONNECTION

Fungi and their resident microbes are known to produce a variety of volatile secondary metabolites dedicated to numerous functions including signalling and elimination of pathogens. These chemicals have become an important driving force behind the aromas we are accustomed to in the food industry. Flavours associated with daily staples like wine, cheese, dairy products and edible mushrooms are indeed generated by the organisms responsible for the elaboration of these foods. Alcohols, phenols, carbonyls, aldehydes, fatty acids and sulphur derivatives that are some of the constituents imparting these appetizing aromas are concocted by a range of enzymes harboured by the fungal holobionts. Although humidity, temperature and the nature of the substrates are important in creating these savoury products, the organisms are the main architects of these succulent edibles. The oxidases, reductases and decarboxylases these microbes possess are the prime mediators of the biochemical reactions resulting in these flavourful delights. For instance, acetoin in butter, tyrosol in wine, tryptophol in beer and methionol in cheese are some of the flavourful chemicals generated by fungi (**Box Figure 1**).

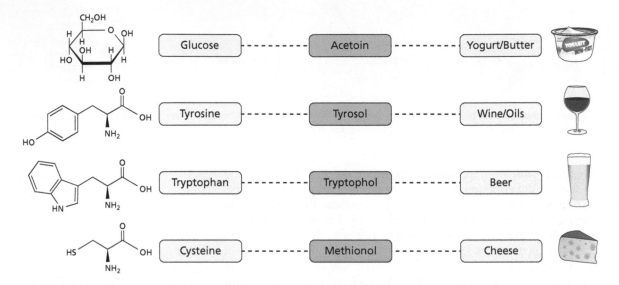

Box Figure 1. The generation of flavours by the fungal holoorganism. Foods such as cheese, wine, beer and yoghurt are dependent on microbial activity for the flavours they possess. Here, fungi and bacteria work in tandem to unleash a diverse array of aromas.

Glomeribacter gigasporarum (CaGg; *Betaproteobacteria*), *Candidatus Moeniiplasma glomeromycotorum* (CaMg; *Mollicutes*), *Burkholderia* (*Betaproteobacteria*) or *Mycoavidus cysteinexigens* (*Betaproteobacteria*). CaGg is a rod-shaped stable component of the arbuscular fungus *Gigaspura margarita*, while *Burkholderia* is an integral part of the soil fungus *Rhizopus microspores.* CaGg is Gram-negative bacteria and occurs individually or in groups in vacuoles localized in the spores or the hyphae. Their cytoplasm is rich in ribosomes and these bacteria are transmitted vertically through fungal generations. In fact, fungi devoid of their bacteria are unable to reproduce, thus compromising the stability and longevity of this holobiont (**Figure 3.3**).

Endobacteria are part of the arbuscular mycorrhizal fungi that colonize plant roots and facilitate the uptake of mineral nutrients in exchange of photosynthetic-derived metabolites like monosaccharides and fatty acids. Although fungi devoid of CaGg can grow, they tend to have reduced hyphae and a decline in oil droplets. This is due to the absence of the fatty acid synthase complex that is located in the bacterium. In this tripartite arrangement, both the plant host and the bacterium may be contributing to the fatty acid budget of this oleaginous fungus. The biomass of the latter can be up to 20% oil droplets. Strigolactone, a plant-derived hormone known to stimulate the colonization by arbuscular mycorrhiza (AM) fungi has a positive influence on this endosymbiont, Spores treated with this hormone tend to harbour more of the bacteria. It appears that the

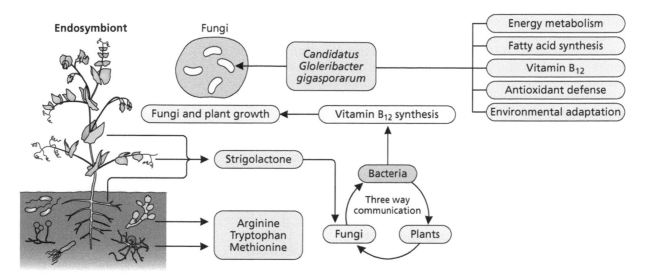

Figure 3.3. Endosymbiosis: bacteria within fungi and interaction in the rhizosphere. Microbes live within fungi that are part of rhizosphere where they accomplish numerous functions for the host. Strigolactones have a positive influence on this three-way cross-talk and enable the storage of oil droplets within the fungi. The microbes depend on the fungi for amino acids such as tryptophan, arginine and methionine.

endosymbiont is involved in energy metabolism within the fungus, an attribute that promotes its association with the plant host. This tripartite communication is central to the growth, development and proliferation of the holobiont. Although the microbe is dependent on the fungus for such amino acids as arginine, tryptophan and methionine, it does possess the genetic machinery dedicated to the synthesis of vitamin B_{12} and a variety of transporters aimed at recruiting amino acids. The vitamin B_{12} may potentially be dedicated to the other residents of the mycorrhizosphere. The fungal mitochondrion seems to be a target of this endosymbiotic lifestyle as the endosymbiont may be performing the task of an energy-generating machinery. The microbe is known to increase the bioenergetics of its host by augmenting respiration and ATP production. It is also well equipped with molecular tools to combat oxidative stress triggered during energy production and is a stable component of the fungus *G. margarita*. A fungal spore can harbour as many as 20,000 of these microbes. CaMg is an obligate coccoid symbiont residing in numerous fungi belonging to the family of *Glomerycotina* and *Mortierellomycotina*. It has a markedly reduced genome and, however, it can also acquire genes from its hosts. The remarkable information plasticity inherent in the mollicute may be a stratagem it exploits to acquire genetic information from its environment via active recombination machinery and horizontal gene uptake, a feature that provides a window on inter-domain communication. The intracellular bacteria, the fungus and the other occupants of mycorrhizosphere live in an intricately modulated ecosystem where each partner contributes to the proper functioning of the holoorganism in a manner analogous to all the specialized organs or cells that operate in harmony in a multicellular organism.

3.3 Living Together: The Partnership between *Rhizopus microsporus* and *Burkholderia*

R. microsporus is essentially a saprophytic fungus and can become an opportunistic pathogen in plants and humans. It lives in an endosymbiotic relationship with the bacterium *Burkholderia* spp. However, unlike other such mutualistic arrangements where the partners cannot manage well independently, both the host and the bacterium can grow independently. The bacterium can live outside the fungus and can recolonize the host if it wants to opt for an endosymbiotic lifestyle. It possesses sufficient genomic material to support its functional capabilities and to communicate effectively with the host to seek a residence within and to contribute to the holobiont. This microbe is a facultative endosymbiont in contrast to the CaGg and CaMg that are obligate endosymbionts.

Burkholderia spp. has an intricate secretory apparatus that enables it to access the fungus. The type II secretory pathway is a vehicle aimed at transporting chitinase and chitosanase, the mediators of fungal cell degradation. This activity is responsible for the subsequent entry and establishment of the bacterium in the host. The secretion of exopolysaccharide may also aid in this process. Effectors delivered intracellularly by the type III secretory help manipulate the host metabolism in order to accommodate the bacterium. Diacyl glycerol kinase contributes to the lipid homeostasis, an event pivotal to the energy supply required to maintain this endosymbiosis. Perturbation of this delicate lipid-balancing act is known to shift this symbiotic interaction into an antagonistic one. *Burkholderia* spp. has a repertoire of genetic information mediating the synthesis of a variety of secondary metabolites that host utilizes to fend off any microbial invasion and to infect plants. The anti-mitotic polyketide rhizoxin is generated cooperatively by the two partners via epoxidation to render the phytotoxin potent. The microbe also has an influence on its transmission within the fungal host as it utilizes sporangiospores during asexual reproduction and the zygospores during sexual reproduction for the vertical transmission to the progenies. In fact, when the fungus is devoid of the bacterium, asexual reproduction is abolished. Hence, this co-dispersal strategy that is tightly controlled by the symbiont ensures that the genetic information of the holobiont is maintained (**Figure 3.4**).

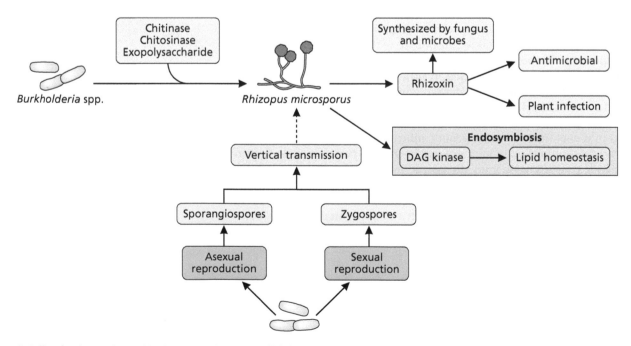

Figure 3.4. Facultative endosymbiosis: establishment and biological functions. The degrading enzymes are involved in the symbiotic process and allow the bacterium (*Burkholderia* spp.) to access the fungus (*Rhizopus microsporus*), even though they can live independently. They subsequently elaborate rhizoxin jointly. This toxin is an antibiotic and can also infect plants. Diacyl glycerol kinase (DAK) plays a key role in lipid homeostasis and in the success of this endosymbiosis.

3.4 Ectobacteriome: Exocellular Bacterial Residents

Fungi do live in association of bacteria that also adhere on the surface of the hyphae. In this living arrangement, the microbes contribute to a variety of functions ranging from generating volatile organic chemicals (VOCs), long-distance communicators, to synthesizing numerous secondary metabolites, participants in a variety of essential functions. These moieties help in signalling processes, act as antibiotics or modulate cellular morphology. For instance, *Laccaria bicolor* found in mycorrhizosphere accumulates trehalose, a disaccharide that attracts and promotes the proliferation of the microbe *Pseudomonas fluorescens*. In exchange, *P. fluorescens* and other similar ectobacteria secrete such organic acids as malic acid and citric acid, stimulants of fungal growth. These carbon sources may be acting both as infochemicals and nutrients. *Aspergillus nidulans* and *Aspergillus niger* harbour numerous actinomycetes, the filamentous bacteria. They are known to participate in primary and secondary metabolism beneficial to both fungi and bacteria. The lignocellulose decomposer *Clitocybe* tends to live in association with nitrogen-fixing bacteria. This arrangement supplies the latter with carbon nutrients while the microbes provide the fungi with ammonia and organic nitrogen. Metabolic and nutritional dependency coupled with the elaboration of protective barriers provides proper incentives for this fungal–bacterial co-habitation to emerge. Numerous mushrooms are associated with growth-promoting bacteria that not only stimulate fungal growth but also play an important role in dictating their morphological features. The edible fungus *Agaricus* spp. found in arid regions grows underground in a saline environment with elevated pH and produces large fruiting bodies. It usually proliferates in the presence of such microbial partners as *Pseudomonas alcaliphila*, *Pseudomonas putida*, *Halomonas* spp. and *Thiobacillus* spp. While *P. alcaliphila* shorten harvest time, *Halomonas* spp. play a key role in nitrogen metabolism. Sulphur homeostasis is maintained by *Thiobacillus* spp. The ability of *P. putida* to degrade 1-octen-3-ol, an inhibitor of the germination process, contributes to the increased fungal yield (**Figure 3.5**).

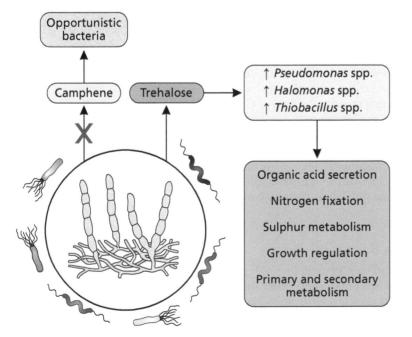

Figure 3.5. Fungi and their ectomicrobiomes. Chemical signals such as trehalose attract the appropriate bacteria on the surface of the host while camphene dissuades the establishment of opportunistic microbes. These bacteria participate in numerous functions like sulphur metabolism essential for the growth of the fungus.

3.5 Fungal–Bacterial Communication

3.5.1 Short- and long-distance dialogues

Bacteria and fungi utilize a variety of diverse communication channels to signal their presence and to interact with each other. This dialogue may proceed via metabolite exchange, secretion of enzymes, physical adhesion and the release of VOCs. The former three strategies are intended for close proximity interactions while the latter is designed for long-distance exchanges. Small signalling molecules like farnesol and acyl homoserine lactones (AHL) involved in quorum sensing help maintain population density, coordinate activities and promote development. Although farnesol is a known quorum sensor in fungi, it can also be utilized by bacteria to control their population density. Similarly, bacterial AHL can be decoded by some fungi like *Candida albicans*. Oxalic and citric acids exuded by fungi in their immediate vicinity are discerned by targeted bacteria like *Collimonas* spp. that subsequently are attracted to the hyphae that also serve as a microbial dispersal vehicle. Polyols do also harbour important informational cues. For instance, glycerol is both a chemotactic agent and a nutrient for numerous bacteria. Chemical signals can be laden with information aimed at dissuading other organisms not to join the fungal-microbial partnership. Anti-microbial peptides, biosurfactants, siderophores and quorum sensor inhibitors are released into the environment to ward off microbial intruders seeking to disrupt the holobiont. Long-distance communication is usually effected by VOCs. Terpene derivatives produced by the fungus *Fusarium* spp. are known to modulate the mobility of bacterium *Collimonas pratensis*. Similar motility-modifying behaviour has been observed and is utilized by the fungal host to attract the coveted microbe. Growth can also be impacted by VOCs as observed in *Bacillus* spp. In this instance, fungal-derived octanol arrests the proliferation of these bacteria, an event aimed at dissuading these microbes to be a constituent of the holobiont.

3.5.2 Habitat changing signals

Bacterial–fungal interaction can also be facilitated by the modification of the habitats where these organisms reside. The genesis of the specific ecosystems favours select groups of bacteria. Fungi sense and modify soil pII to promote their symbiotic partners at the expense of other microbial intruders. The fungal mycelia not only supply nutrients to the desired microbes but can also modulate the water content tailored to a particular consortium of microbial symbiont. Regulation of oxygen tension is another strategy utilized to goad the microbes in a mutualistic relationship. The preference of anaerobes by *Candida* spp. compels these fungi to generate an anaerobic environment. This habitat is achieved by the formation of biofilms and the release of compounds like phenazines to curb aerobic respiration. As biofilm may sometimes retard fungal growth, the fungus *C. albicans* recruits *P. aeruginosa* to both produce the exobiopolymer and phenazine by releasing ethanol. The presence of this alcohol in the habitat signals an anaerobic environment and favours the synthesis of these two moieties, a situation that dissuades aerobic microbes. Hence, both chemical and habitat-modifying strategies are invoked to promote the proper mutualism to emerge in an effort to generate the right holoorganism suitable for the ecosystem (**Figure 3.6**).

3.6 Truffles: Aromatic Fungi with a Bacterial Touch

Truffles belong to the *Ascomycota* phylum and are known to lead a subterranean lifestyle in association with plant roots. They form an extensive ectomycorrhiza where they live in harmony with bacteria, viruses, yeasts and other filamentous fungi. These aromatic fungi are

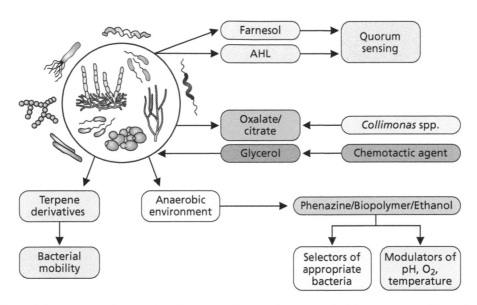

Figure 3.6. Fungal–bacterial communication. Fungi and bacteria utilize a variety of chemical signals short and long-distance communication. While farnesol and AHL (acyl homoserine lactone) mediate proximal interactions, terpenes modulate contact with distant microbes. Citrate, oxalate and glycerol attract select bacteria; modulation of pH and O_2 gradient is also involved in the assembly of fungal microbiome.

classified as *Tuber* spp. that includes *Tuber (T.) melanosporum, T. borchii* and *T. magnetum*. They are characterized by signature odorous attributes elaborated with the participation of the microbial community residing in the spores. These aromatic and flavourful fungi are luxury foods that are well-sought after. For instance, the price of the black Perigold truffle (*T. melanosporum*) and the white Piedmont truffle (*T. borchii*) can range from $2,000 to 5,000 per kg.

Bacteria are an important contributor to the characteristics that make truffles so attractive. *Proteobacteria, Bacteroidetes, Firmicutes* and *Actinobacteria* are essentially the main groups that comprise the complex communities of bacteria associated with these fungi. However, the fungal lifecycle, temperature, CO_2 and properties of the soil tend to dictate the abundance and the specific nature of bacteria, these fungi possess. For instance, *Actinobacteria* are dominant in the ectomycorrhizae but relatively rare in the fruiting bodies of *T. melanosporum* that are associated with an elevated amount of *Proteobacteria*. The microbial population can vary between 10^6 and 10^9 cells per gram. *Bradyrhizobium* spp. appears to be the dominant bacterium in all species due to its pivotal role in supplying fixed nitrogen. The fruiting bodies that develop underground are formed during the mating of a maternal and paternal counterparts residing in symbiosis with plant roots in the ectomycorrhizae. These fruiting bodies replete with spores of both mating types do house a variety of bacteria as they mature. The inner part of the organ known as the gleba and the outer part referred to as the peridium are heavily colonized by a disparate group of microbes. The melanization of the peridium in *T. melanosporum* imparts its black hue. The captivating and enticing aromas emanating from these fruiting organs are due to numerous volatile alcohols, aldehydes, ketones and aromatic and sulphur derivatives. These odorants that are a product of numerous factors including the fungi and the resident bacteria are aimed at small mammals involved in the dispersal of the spores. Only a small fraction of these odorants is perceived by human smell receptors. Volatile compounds like 2-methyl butanal, 3-methyl butanal and 3-methyl butanol are common in most truffles. Flagrance-rich compounds such as 2-phenylethanol and 2-methoxy-4-methyl phenol are generated during the sexual phase of the fungi, The aroma released by 2,4-dithiapentane and 3-methyl 4,5-(2*H*) thiophene is found only in *T. borchii* and *T. melanosporum*, respectively. The scent emanating from these truffles is a

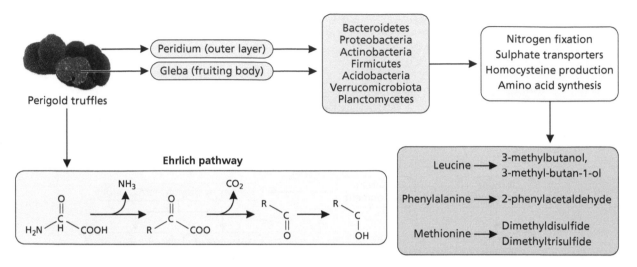

Figure 3.7. Truffles and their microbes – a flavourful partnership. Truffles are expensive edibles due to the flavours produced in collaboration with their microbial partners. The bacteria assist in a variety of functions and also produce amino acids that liberate a range of captivating odours, mediated by the Ehrlich pathway. Homocysteine and methionine are processed into volatile sulphur products.

product of numerous effectors including the microbiome. The unique time-sensitive smell transmitted by these fungal organs is indeed part of the range of attributes that makes them prized delicacies.

Amino acids metabolized via the Ehrlich pathway are central in the synthesis of these volatile derivatives characteristic of truffles. This catabolic process proceeds by the deamination of the amino acid into a keto acid. The latter is then decarboxylated into an aldehyde followed by its subsequent reduction or oxidation into an alcohol or acid respectively. Methionine and homocysteine that are precursors to the thiophene derivatives can be supplied by the bacteria residing in these fruiting bodies. Furthermore, the synthesis of these odorants may be jointly undertaken by the truffles and their microbial associates. In fact, the microbiome is known to actively participate in the production of sulphur metabolites during the sexual stage of the fungal life cycle. Upregulation of homocysteine synthase, sulphate transporters and enzymes mediating the synthesis of various amino acids is a common biochemical strategy adopted during this reproductive phase. Dimethylsulphide (DMS), one of the most common sulphur volatiles found in truffles, can readily be derived from methionine. DMS is readily recognized by various mammals including pigs known to consume truffles. Dogs are trained to locate this prized underground commodity. It is critical for the survival of the fungi as the intense smell attracts small mammals responsible for the dissemination of spores. The microbiome is recruited to elaborate chemical cues designed to lure these mammalian vectors. Insects like beetle can also locate the fruiting bodies. The faecal pellets laden with spores of the hypogeous fungi subsequently help in the establishment of plants and vegetation. On the other hand, similar volatile signals that are perceived by the human nose have catapulted these fruiting bodies quite high in the hierarchy of luxury edibles. Thus, a truffle orchard is indeed a keeper of tightly guarded smelly secret concocted by the fungal–bacterial partnership (**Figure 3.7**).

3.7 Mycobiome: The Fungal Partners Residing in Other Organisms

In the preceding section, the microbial residents (primarily bacteria) living within and on fungi have been explored. There is, however, a range of other multicellular organisms that harbours fungi in order to live and proliferate effectively. This population of fungi that constitutes an integral component of these organisms is referred to as the mycobiome and contributes to a variety of physiological functions ranging from acquiring essential nutrients like phosphate to priming the host immune system. These fungal entities also play a critical role in attracting other microbes thus contributing to the establishment of holoorganism, visible to the naked eyes.

3.7.1 Plant mycobiome

Mycorrhiza is literally the network of fungi and roots that permeates underground below most plants. It is derived from myco (fungus) and rhizae (roots). The plant and fungus live in a symbiotic relationship where the former provides carbon nutrients while the latter supplies essential ingredients like phosphate and nitrate that are made easily accessible to the roots. The fungal hyphae are interconnected into a web-like extensive structure called mycelium that interacts with the roots and can extend as long as hundreds of kilometres especially in forests where this arrangement can provide a communication channel to all the organisms living in the ecosystem. The fungus can connect to the roots either through the ectomycorrhiza whereby the fungus surrounds the roots or the endomycorrhiza, a situation where the fungal hyphae grow within the cell wall and cell membrane of the root. The AM is of the endo type and is the association adopted by approximately the majority of land plant species including most of the agriculturally important crops such as soybean, corn, wheat and rice. The AM fungus is an obligate biotroph and depends on its autotrophic host to survive and proliferate. The fungal spores respond to root exudates such as the strigolactone mediators of hyphal development and branching. The hyphae emerging from the hyphopodium penetrate into the root through the pre-penetration apparatus of the root and are guided through the root cells towards the cortex and the apoplast where further branching occurs. This arrangement involving the intra-radical and extra-radical mycelium not only allows for an intimate physical contact between the soil, the plant and the fungus but enables an efficient exchange of nutrients and information among the partners. In this instance, the fungus is supplied directly the organic carbon that the plant produces and in exchange has access to a variety of nutritional elements like H_2O, Pi and nitrogen derivatives the fungus help trap. The bidirectional exchange of nutrients across the mycorrhizal interface is central to this symbiotic relationship.

The AM fungus produces signalling molecules termed (mycorrhizal) Myc factors aimed at informing the plant to prepare the roots for accommodating the fungal symbiont. Lipochitooligosaccharide (LCO) secreted by the fungus prompts the receptor-like kinase to interact with the cytoplasmic components within the root and permits the intracellular establishment of the hyphae. Recently, the role of extracellular ATP in promoting the root and fungal association has been proposed. Following the establishment, this physical association, high-affinity Pi, SO_4, NO_3 and NH_4^+ transporters are expressed in the mycorrhizal roots while the fungal membrane is punctuated with uptake proteins programmed to select such monosaccharides as glucose and xylose. Acid phosphatases and organic acids are secreted to solubilize the Pi. Once the Pi is transported in the cells, it is incorporated into phospholipids, DNA, RNA and polyphosphates. The latter may act as a Pi reserve. NO_3 is assimilated by nitrate and nitrite reductases and subsequently converted into glutamate, glutamine and arginine with the aid of enzymes involved in the urea cycle.

The ectomycorrhizal (EC) symbiosis is less common than the AFM and occurs only in 3% of plant species. Wooden perennials and shrubs from boreal and montane forests are the prominent plants that are involved in this mutualistic relationship with fungi belonging predominantly to the *Basidiomycota* family. These fungi tend to have a dual lifestyle as they are facultative saprotrophs. They secrete a variety of hydrolytic enzymes aimed

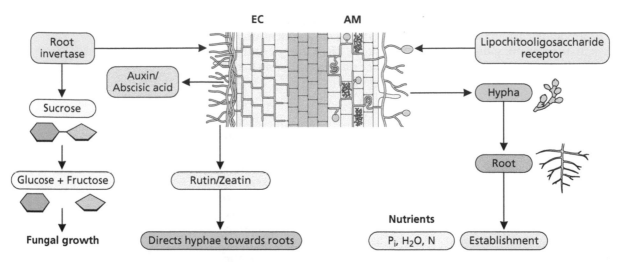

Figure 3.8. Mycorrhizas and its biochemical features. Arbuscular mycorrhiza (AM) is the intracellular association between plants and fungi, while ectomycorrhiza (EC) is the exocellular interaction between the hosts and symbionts. Invertase secreted by roots converts sucrose into hexoses responsible for the attraction and growth of the fungi. The binding of fungal lipochitooligosaccharides to the corresponding receptors promotes the intracellular attachment of the fungi within the host. Root exudates such as abscisic acid, zeatin and rutin also contribute to these symbiotic relationships.

at degrading litter polymer and other organic nutrients. As these fungi cannot metabolize plant cell walls, they are limited to live outside the roots. The phytohormones such as auxins and abscisic acid secreted by the EC fungi help communicate with the root. This conversation results in short root development and other morphological changes conducive to the promotion of this exocellular partnership. The fungus also responds to the root exudate consisting of rutin and zeatin that directs the growth of the hyphae towards the root. The exchange of nutrients and minerals occurs in a manner analogous to the AFM. However, in this instance, EC fungi rely on the host invertase to convert sucrose into glucose and fructose. The upregulated expression of monosaccharide transporters in the fungus facilitates the uptake of these hexoses that subsequently fuel fungal proliferation. Another important benefit derived from this fungal interaction with the host relates to the superior immune response elicited by the plant. The fine-tuned communication that enables the association of these AM and EC fungi with the plants primes the immune system to be extremely vigilant to any foreign invader. Consequently, the molecular surveillance mechanisms of the host are sharply boosted (**Figure 3.8**).

3.8 Endophytes: A Unique Fungal–Plant Relationship

Some plants harbour a different variety of fungi referred to as endophytes. These are non-mycorrhiza fungi that associate with plants for at least part of their life cycle. They live within the plant and may be distributed in the leaves, stems, roots, flowers and even seeds. Although this is not a symbiotic interaction, none of the partners involved is harmed. The benefits derived from this arrangement can be mutual or tilted more towards one of the partners. For instance, the fungus may have a safe place to live and the plant can count on the numerous bioactive chemicals produced by the fungus to defend against intruders. In some cases, the fungus may lead a dormant life and become active only when the senescence of the plant is triggered. This relationship involving the non-obstructive and symptomless existence of the fungi inside plants was first observed in the 19th century and occurs in plants ranging from the tropic to the arctic regions. The significance of this arrangement was revealed when the death of numerous cattle due to 'fescue toxicosis', a condition attributed to the fungus

Neotyphoium coanophialum present in the grass (*Festuca arundinacea*) the animals were fed. This fungus produces toxic alkaloids not aimed at the host but at the herbivores. Thus, it became clear this association is designed to extend the genetic pool of the host and elaborate chemical tools to facilitate their growth.

Although this relationship is dictated by a variety of factors including biotic features, abiotic stress, geographic parameters, genetic composition of the host, plant tissue composition and plant age, both the host and the fungus have to modulate their defensive mechanisms in order to establish the chemical entente enabling the latter to live within the former. The fungus has to avoid the host immune response and 'pretend' to be a component of the host. Furthermore, the fungus has to grow within the plant without causing any visible manifestation of injury that is the fungal colonization has to be asymptomatic. Hence, a fine-balancing act between virulence and defence has to be 'de rigueur'. Following the physical contact triggered by plant exudates like nutrients and VOCs, the chemical barriers set up by host are carefully circumvented by the fungus, an event that may be mutually programmed. The expression of the mitogen-activated protein kinase is pivotal to the establishment of the endophytic fungus (*Epichloe festucae*) within rye grass (*Lolium perenne*). The endophyte *Fusarium solani* escapes the topoisomerase inhibitor camptothecin produced by the happy tree (*Camptotheca acuminate*) by modifying the amino acid residues in the catalytic domains in its own topoisomerase. This strategy that is reminiscent of the acquired immune system of some invasive viruses allows the fungus to proliferate within. The host may also signal the fungus to perform a specific task by releasing such amino acids like homoserine and asparagine and also may be involved in tripartite communication with their bacterial partners. The response to this relayed information appears to be operative only in planta as the axenic environment does not trigger any action. Once, the signal is decoded, the fungus (e.g. *Balansia* spp.) can establish within the intracellular spaces and in the shoot meristem of the host (grass). The fungus may immediately participate in a variety of physiological functions or assume a quiescent state until it is called into action like decomposing senescent tissues. In fact, virtually all plants house fungal partners. The *Glomeromycota* and *Ascomycota* groups are the most prominent fungi that are engaged in this lifestyle.

3.8.1 How endophytes perform for their hosts?

It is clear that the fungal endophytes and all the other microorganisms together with the host constitute a holobiont that works for the growth and proliferation of the holoorganism. The contribution of fungi to this living system is multifaceted. They produce an array of bioactive chemicals aimed at fortifying the defence arsenal of the host and augmenting their fitness. They elaborate terpenoids, peptides, alkaloids and a plethora of toxins designed to fend against such encroachers as microbes, insects and herbivores. They synthesize antioxidants to tolerate any oxidative stress triggered by abiotic stress like drought, salt and heat. The ability of barley to resist saline conditions has been attributed to the root endophyte *Piriformospora indica*. The fungus *Curvularia protuberta* tends to colonize plants growing in geothermal soils and helps their host survive elevated temperatures. The fungal endophytes also increase CO_2 uptake, promote nitrogen homeostasis, catalyse the acquisition of Pi and synthesize siderophores to solubilize Fe. The development of the plants is also aided by their endophytes as they are known to produce growth promoters like auxins, adenine derivatives, acetoin and 2,3-butane diol. Some *Piriformospora* spp. are known to stimulate enzymes nitrate reductase and glucan-water dikinase mediating cellular growth. Desert plants such as the *Agave* spp. are able to survive in arid and semi-arid conditions due to the ability of their endophytic partners to modulate stomatal activity. In *Agave*

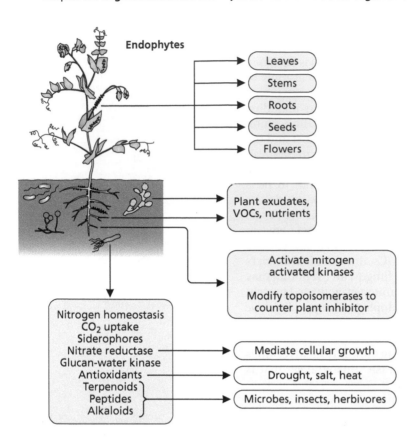

Endophytes

- Leaves
- Stems
- Roots
- Seeds
- Flowers

Plant exudates, VOCs, nutrients

Activate mitogen activated kinases

Modify topoisomerases to counter plant inhibitor

Nitrogen homeostasis
CO_2 uptake
Siderophores
Nitrate reductase
Glucan-water kinase
Antioxidants
Terpenoids
Peptides
Alkaloids

- Mediate cellular growth
- Drought, salt, heat
- Microbes, insects, herbivores

Figure 3.9. Fungal endophytes and the tasks they perform. Endophytes are fungi living in various plant tissues. The exudates secreted by the hosts and defensive strategies elaborated by the fungi enable this relationship to survive. The fungal symbionts allow the hosts to extend their adaptive attributes.

deserti, the fungi *Entrophosphora* and *Glomus* spp. are located underground and the aerial part of the plant is home to *Capnodiales* spp. Despite the diverse nature of the fungal endophytes and the varied plant hosts they associate with, the protection they get, the nutrients they are provided, and the spatial structure they enjoy makes this partnership an attractive one. Furthermore, the presence of the fungi within the seeds enables a safe and guaranteed pathway to ensure the survival of their lineages. This vertical genetic transmission strategy reveals the meta-genomic nature of this living organism and the need for disparate genetic entities to live as a holobiont (**Figure 3.9**).

3.9 The Human Fungal Partners

Fungi are also an integral component of the human body. Although they are less abundant than other microbes like bacteria and viruses, fungi do fulfil important functions in the human holobiont. One can safely say that no human is free of fungi and humans devoid of fungi will have difficulty in surviving. Human stool samples contain 0.01% to 0.1% of genes that are of fungal origin. Humans are constantly exposed to fungi in the environment and from the foods they consume. Bread, cheese, beer and other fermented foods do have an appreciable amount of fungi. They are part of the commensal microbial flora that reside within different sites in the body including the mouth, skin, lungs, gut and vagina. Like their other microbial counterparts, fungi occupy the human body during its development and after birth. For example, the gut comprises *Cryptococcus* spp. and *Saccharomyces* spp., while the mouth lodges *Rhodotorula* spp. and *Candida* spp. The latter is the most abundant fungus in the oral cavity. The fungal–bacterial and fungal–fungal interactions are critical for oral health. *Pichia* spp., *Cladosporium* spp. and *Fusarium* spp. are found in healthy individuals, those infected with HIV possess *Alternia* spp., *Epicoccum* spp. and *Trichosporon* spp. The presence of *Candida* in oral biofilms is known to promote the proliferation of *Staphylococcus mutans*. The adherence of

BOX 3.2 THE MOLECULAR LANGUAGE OF FUNGI AND ITS PARTNERS

Fungi and their partners communicate via a range of chemicals designed to promote interaction leading to close intimacy. Arbuscular and ectomycorrhizal fungi secrete LCOs that allow them to associate and seek residence within their hosts. These chitin-derived molecules decorated with select fatty acids and oligosaccharides initiate interaction and promote spore germination, growth and hyphae development. They are information-laden moieties referred to as mycorrhizal factors usually comprising 3–5 N-acetyl glucosamine residues with lipids and oligosaccharide substituents programmed to recognize receptor kinases in plants. Upon binding to the extracellular LysM domain, the host innate immune system is suppressed with the concomitant activation of the genetic apparatus responsible for the partnership of the fungi and the hosts. It is important to note that these short-chain LCOs trigger the Ca^{2+} oscillatory signal in order to initiate symbiosis while longer chains with 8 N-acetyl-glucosamine elicit the release of defensive strategies designed to eliminate fungal intruders. In this instance, chitinase, phytoalexins and reactive oxygen species are released. This molecular language is precisely calibrated to distinguish the partners working for the holobiont from the foes bent on its demise.

This molecular communication channel is effective when the organisms are in close proximity and may not be effective if the partners have to be recruited from afar. To rectify this situation, long-distance conversation is promoted by the elaboration of volatile organic compounds like terpenes. These molecules not only evaporate at normal temperature and pressure, but they can also readily diffuse through water and gas-filled pores and reach their far-flung targets. Depending on the information they exude, molecules like terpenes and camphenes can stop invaders on their tract while enticing symbionts to proceed towards their hosts. Bacteria belonging to *Collimonas* spp. are attracted to fungi where they colonize the hyphae. Isopentenyl diphosphate is a key precursor for the synthesis of terpenes, a biochemical feature that has been inherited from bacteria (**Box Figure 2**).

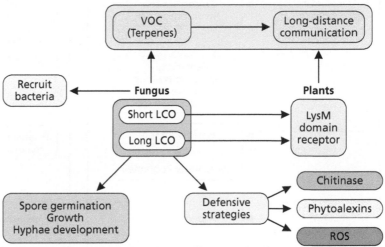

Box Figure 2. Communication between fungi and plants. Lipochitooligosaccharides (LCOs) are key in this conversation and allow the entry of the symbionts via the LysM receptor. For long-distance signalling, terpenes and other volatile organic compounds (VOCs) are utilized.

bacteria to fungal hyphae helps promote the distribution of bacteria within this organ. The *Candida* spp. and *Phialemonium* spp. located in the stomach can survive low pH.

The gut of healthy individuals tends to have an abundance of *Candida*, *Saccharomyces*, *Trichosporon* and *Cladosporium*. Obese individuals are characterized by a higher amount of low-density lipoprotein, cholesterol and triglycerides. They tend to have a low abundance of *Mucor* spp. while the elevated quantity of *Penicillium* spp. correlates well with increased levels of high-density lipoprotein, a marker for healthy individuals. Irritable bowel syndrome is characterized by a reduction in the amounts of *Saccharomyces* spp., *Penicillium* spp. and *Kluyveromyces* spp. with the concomitant increase in *Candida* spp. and *Malasseziales* spp. Hepatitis B infection is characterized by an increase in *Candida* spp. and *Saccharomyces* spp. *Aspergillus* spp. dominate in healthy lungs. The presence of fungi in humans allows the priming of the host's immune system. The exposure of macrophages to fungal cell-wall component (β-glycan) triggers a stronger response to any fungal intrusion and modulates the secretion of pro-inflammatory cytokines. *C. albicans* can evoke innate immunological memory in myeloid cells and block the synthesis of nitric oxide, (NO), a pivotal intracellular signalling molecule. The ability of *Saccharomyces boulardii*, a common probiotic to mitigate *Clostridium difficile*-evoked colitis by stimulating the production of immunoglobulin A (IgA) and promoting the synthesis of anti-inflammatory cytokine clearly validates

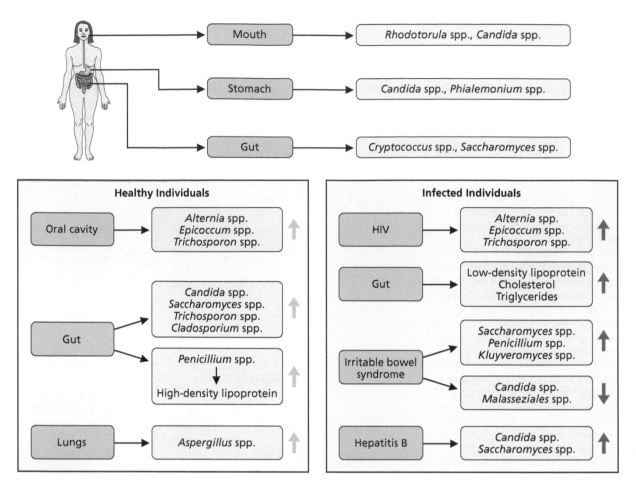

Figure 3.10. Distribution of fungi within humans and their functions. Fungi are housed in various organs in the human body and contribute to the well-being of the host. They perform numerous functions and fungal dysbiosis is associated with different diseases. High amount of *Penicillium* spp. is linked to the increased level of high-density lipoproteins, a characteristic of healthy individuals, while irritable bowel syndrome is associated with a high concentration of *Malasseziales* spp.

the notion that the mycobiome contributes to the immunological fitness of the host and highlights the dialogue between the host and the fungal partners. Furthermore, this fungus secretes phosphatases and proteases that contribute to neutralize toxins elaborated by infectious bacteria (**Figure 3.10**).

3.10 Mycobiome: Gut–Brain Axis

The human mycobiome not only calibrates the immune system, but it also plays an important role in modulating the activity of the central nervous system as numerous fungal residents are known to synthesize neurotransmitters. While *S. cerevisiae* can produce norepinephrine, a mediator of brain activity, *C. albicans* synthesizes histamine, a modulator of appetite and cognitive response. Hence it is conceivable that a bidirectional communication network exists between the gut mycobiome and central nervous system via the vagus nerve that connects these two organs. The ability of these fungi to also synthesize short-chain fatty acids like butyrate may also contribute to this signalling network. An abundance of *Candida* spp. has been reported in patients with autism spectrum disorders. This increase in fungal population can also affect 'normobiosis' and may result in the perturbation of the microbial harmony critical for a healthy condition. In fact, supplementation with probiotics like *Lactobacillus rhamnosus* and *Bifidobacterium animalis* has been shown to rectify this imbalance.

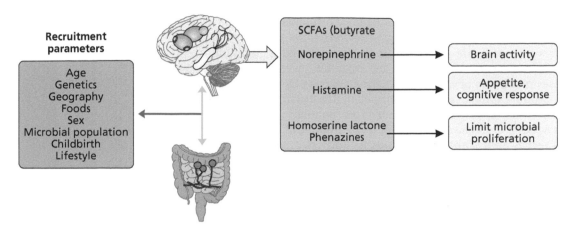

Figure 3.11. Gut–brain axis cross-talk promoted by the mycobiome. The gut fungi produce a range of signalling molecules and/or their precursors that are known to communicate with the nervous system. Histamine plays a role in cognitive response and norepinephrine modulates brain activity. The fungal and bacterial interactions help control the microbial population. The assembly of the mycobiome within humans is dependent on a range of factors including age, genetics, food and lifestyle.

These information-rich chemicals released by the mycobiome can also participate in fungal–fungal interaction and in bacterial–fungal cross-talk. For instance, *C. albicans* promotes the growth of *Staphylococcus mutans*, a microbe known to utilize the fungal hyphae to spread at a site. *Streptococcus gordonii* secretes auto-inducer peptides to stimulate hyphal development and to repress farnesol, a moiety known to elicit fungal morphogenesis. *Pseudomonas aeroginosa* elaborates compounds like 3-oxo C_{12} homoserine lactone and phenazines that play a pivotal role in modulating microbial proliferation, an event that can result in diseases. The establishment of the mycobiome depends on a variety of factors including age, genetics, geography, foods, sex, other microbial population, mode and place of childbirth and lifestyle. The gut mycobiome of infants has an abundance of *Malassezia* and *Saccharomyces* spp. that changes with age. Individuals deficient in mannose-binding lectin are colonized by *C. albicans*, while those lacking STAT 3 (signal transducer and activator of transcription 3) tend to experience an increase in *Aspergillus* spp. on the skin. This transcription factor modulates genes involved in proliferation, cell differentiation and immune responses. Urban dwellers have a reduced skin and gut fungal population compared to rural residents. Wayampi Amerindians tend to possess lower amounts of *C. albicans* than their Western counterparts. The presence of bacteria within the same organ also has an impact on the fungal inhabitants. As vaginal *Lactobacillus* produces reactive oxygen species, lactic acid and H_2O_2, the level of *C. albicans* within this organ is drastically reduced. These moieties have fungicidal attributes (**Figure 3.11**).

3.11 Fungal Community: A Key Component of Wood-Eating Organisms

3.11.1 Xylivory in fish

The ability of fungus to synthesize a variety of enzymes involved in the degradation of wood is an important characteristic that some fishes have exploited to feed themselves. Wood is rich in cellulose, hemicellulose and lignin. Only a few organisms like fungi have evolved the capability to produce cellulases and ligninases. These enzymes liberate monomers that can be readily utilized as nutrients. The monosaccharides such as glucose and fructose are subsequently integrated into the regular metabolic pathways to generate anabolic and catabolic substrates. The Amazonian catfish (*Panaque nigrolineatus*) is a xylivore that is a wood-eating organism, a nutritional habit made possible due to the fungal residents it harbours in its gastrointestinal (GI) tract. The fungi *Aureobasidium pullulans* and *Debaryomyces prosopidis* are prominent in the foregut, mid-gut and hind gut of the fish. They secrete endocellulases, exocellulases, ligninases and β-glucosidases mediators of wood degradation ingested by the organism. Xylivory, the wood-eating lifestyle has compelled this fish to have a sucker mouth armed with spoon-shaped teeth. This anatomical adaptation enables the aquatic organism to gobble a large amount of wood. The GI tract of this fish is approximately 10 times its body size and provides a surface area rich in micro-environments conducive to colonization by both bacteria and fungi. In fact, such bacteria as *Clostridium* spp. and *Aeromonas* spp. known for their cellulose-degrading activity cohabit with the mycobiome. The nitrogen-fixing microbe belonging to *Bradyrhizobium* spp. is also an inhabitant of this holoorganism. Hence, these microbes are an important component of the fish and participate in the day-to-day activity of the holobiont that has evolved in tandem with the resident microbial community.

3.11.2 *Wood-consuming insects and birds: their fungal partners*

In a similar fashion, birds and insects have recruited fungi to manipulate wood in their environment. In these cases, the fungi are also intimately associated with these organisms and are involved in the metabolism of the lignocellulolytic biomass. The red-cockaded wood-pecker (*Picoides borealis*) enlists the help of a variety of fungi including *Perodaedalea pini* to bore cavities in trees. The fungus is involved in softening the wood with the participation of its cellulose and lignin-degrading enzymes. This process facilitates the excavating task undertaken by the bird in an effort to construct its nest. In exchange, the wood-pecker contributes to the dispersal of the fungal spores far and wide. The ambrosia beetles are known to colonize decayed wood with the assistance of wood-degrading fungi known for their enzymes involved in the catabolism of lignin and cellulose. These mycosymbionts are mostly located on the exoskeleton and lodged within a specialized organ referred to as the mycangium, located in the head region of the insect. The fungi are shuttled from tree to tree to generate nutrients for the hosts and in return, they not only enjoy a safe environment to live but also find an efficient means of being propagated. This organism plays a pivotal role in the dispersal of the fungus. In the beetle *Ambrosiodmus minor*, the fungus *Flavodon* spp. can be isolated from the mycangia where it grows its hyphae and orchestrates the decomposition of rotten wood into carbon sources palatable to the insect. The fungus is like a head-gear the beetle wears to spray the wood-decomposing enzymes to bore trees and extract nutrients.

These three disparate multicellular living systems have all decorated their anatomical organization with members of the fungal family in order to perform a select task critical for their development and survival. It is important to note that the wood-metabolizing enzymes are restricted to very few organisms and are primarily limited to the fungal kingdom. Hence, whenever the necessity to interact or live with a wood-containing ecosystem arises during evolution, this situation compels organisms to acquire wood-degrading enzymes. They enlist the assistance of fungi. Thus, the incorporation of fungi with wooding-eating attributes became an effective evolutionary trajectory to follow. The hosts have developed strategies to acquire these fungi and make them part of their anatomical features. This eliminates the need for further specialization, a permanent development that may be rendered redundant fast. The cross-talk between the molecular architecture of the fungi, other microbes and the hosts creates holobionts that are suited to the environment where they proliferate. The holobiont is indeed the hallmark of all multicellular organizations where seemingly disparate organisms with their unique genetic make-up converge to give rise to highly adaptable living entities.

3.12 Fungi in Marine Organisms

Sessile organisms residing in a marine environment are confronted with a variety of biotic and abiotic stress on an ongoing basis. To help thwart such dangers that can threaten their existence, they have incorporated fungi into their daily activities and the fungal partners are a constitutive part of how they live. Fungi can tolerate marine ecosystems and are known to proliferate within numerous organisms where they assist in a plethora of biochemical activities. Fungi and other microbial residents are essential components of such multicellular organisms as corals, sponges and ascidians. The anatomical features of corals that include tentacles, the mouth, digestive filaments and the mineral-rich skeleton provide diverse micro-environments where fungi can be housed. The organic matrix of the carbonate skeleton is an ideal entry point that fungi utilize to grow inside corals as endolithic saprotrophs. *Aspergillus* spp. and *Penicillium* spp. are some of the fungal members of this diverse microbial community including algae and bacteria that live within this holobiont.

The ability of fungi to secrete lignin and cellulose-hydrolysing enzymes helps generate nutrients that supplement this ecosystem and play a pivotal role in aiding skeletal biomineralization, a crucial component of the coral life cycle. Some fungi are known to participate in the nitrogen budget of the holobiont by converting nitrite and nitrate into ammonia. The latter can subsequently be trapped into glutamine with the aid of glutamine synthetase. They utilize their vast array of genetic information to produce secondary metabolites aimed at fending off any pathogenic intruders and contribute to the defence arsenal of the host. Damage by ultraviolet (UV) radiation, a situation that is exacerbated due to the inability of the host to seek shelter, is mitigated by the mycosporin-like amino acids elaborated by the fungi. These moieties not only act as a screen against UV radiation but also protect against oxidative and salt stress. Furthermore, the energy associated with the UV light may contribute to enhance photosynthetic activity of the resident algae that supply the host with an array of carbon-rich nutrients. Other marine organisms like sponges, ascidians and sea cucumbers are also known to incorporate fungi as one of the partners in their lifecycles. The fungal population associated with these marine organisms varies with environmental conditions such as temperature, CO_2, salt concentration and presence of other microbes (**Figure 3.12**).

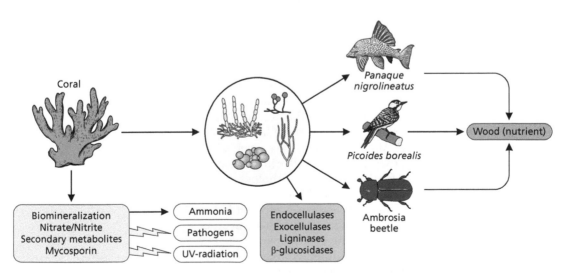

Figure 3.12. Fungal input in the lives of other holoorganisms. Fungi are an integral part of wood-consuming organisms. Fishes, termites and birds that require wood to survive recruit fungi to hydrolysis cellulose, hemicellulose and lignin. The monomers released are utilized as carbon sources. Corals too possess their mycobiome as fungi help these sedentary organisms in multiple tasks including production of secondary metabolites, biomineralization, prevention of infection by pathogens, and nitrogen metabolism.

BOX3.3 ORCHIDS DEPEND ON THEIR MYCOBIOME TO LIVE

Orchids, the largest of all plant families with an excess of 30,000 species, need their fungi usually belonging to the *Basidiomycota* phylum to live and proliferate. Orchids are flowering plants that require their microbes to complete their lifecycle. They cannot germinate in the absence of specific microbial communities. These plants grow in diverse habitats ranging from the tropics, temperate regions to cold ecosystems. They can be terrestrial (growing on land, *Goodyera pubescens*), epiphytic (growing on the surface of other plants, *Dendrobium* spp.) or lithophytic (growing on rocks, *Phalaenopsis* spp.). This relationship starts from the beginning when the germination process is initiated. Germination cannot occur in the absence of a fungal partner. These seeds are very tiny ranging in size from 0.3 to 14 μg with a coat and an embryo, essentially containing protein bodies as the main nutrient. Following the release of these dust-like seeds from the orchid fruits, they are dispersed by the wind. The formation of seeds is usually associated with the recruitment of fungi such as *Tulasmella* spp. and *Ceratobasidium* spp. This interaction results in the formation of a specialized tissue referred to as protocorm whereby the fungal and the seed components are in an intimate symbiotic relationship. As the seeds are tiny, they depend on these fungi for their nutrition to propel their development.

To germinate, these seeds rely on an obligatory relationship with fungi. Ammonium (NH_4+) is the attractant that lures the fungi to enter this relationship as revealed by the increased expression of NH_4+ transporters, arginine-succinate lyase and glutamine synthetase in the symbiont. This enables the fungus to grow within the seed as coiled complexes known as peloton and starts to supply C, H_2O and Pi to the latter. The unidirectional transport of nutrients from the fungus to seed during the early stages of life is later transformed into a bidirectional transaction. Orchids have evolved in close association with fungi and are unable to survive without fungal input.

Before the appearance of the plantlet above ground, the seeds engage in mycoheterotrophy and the fungi are utilized as a source of food. This nutritional habit may persist even when the fully developed plants are highly photosynthetic. These orchid mycorrhizal fungi participate in the initiation of germination and promote the differentiation of the embryo and the growth of the emerging plant. This symbiotic relationship continues in the fully grown plant and involves the participation of other microbes including bacteria. Hence, the lifecycle of this family of flowering plants is dependent on the microbial symbionts that are an integral component of the orchid hologenome and seeds devoid of these fungi are unable to germinate. This association between two disparate organisms where the development of one is dependent on the other reveals that the protocorm, the fusion of the fungus and the seed is part of the same holobiont (**Box Figure 3**).

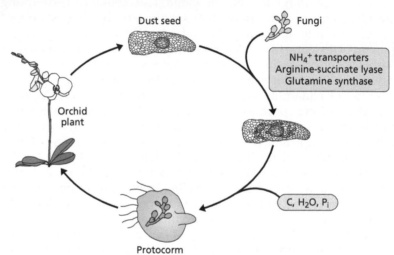

Box Figure 3. **Orchids need their fungi to grow.** Orchid seeds utilize nutrients from the fungi to develop and germinate. The fungi supply nitrogen-rich metabolites, Pi, C and H_2O. This mycophagy supports the emergence of the photosynthetic plant. The nutrient exchange may become bidirectional as the orchid matures.

3.13 Conclusions

Bacteria are an important aspect of the multicellular lifestyle of fungi where they execute numerous functions including providing vitamins and nitrogen-derived products to support the holobiont. This relationship is guided by a precise molecular conversation that allows the host and the microbial residents to live together. Although fungi have evolved to accommodate their bacterial partners, they are known to live in association with other organisms where their input for the survival of the holoorganism is crucial. For instance, plants will be unable to lead a life usually rooted in their habitats without the assistance of their mycorrhiza. Fishes, birds and insects have also incorporated fungi into their anatomical features in order to satisfy their wood-degrading habit. Rather than evolve novel biological networks to deal with this carbon source in their ecosystems, these organisms have recruited fungi with the unique ability to release nutrients locked in the wood in a manner analogous to fungi housing N_2-fixing bacteria. Developing specialized cells or organs to tackle a specific challenge is relegated in favour of acquiring microbes with the biochemical machineries primed for such task. Thus, the fungal microbiome and the mycobiome evolved in tandem with their hosts to generate holoorganisms with a wider range of adaptive characteristics.

SUGGESTED READINGS

Ainsworth, T. D., Thurber, R. V., & Gates, R. D. (2009). The future of coral reefs: A microbial perspective. *Trends in Ecology & Evolution, 25*(4), 233–240. doi:10.1016/j.tree.2009.11.001.

Anke, T., & Schüffler, A. (2018). *The Mycota: A Comprehensive Treatise on Fungi as Experimental Systems for Basic and Applied Research*. Cham: Springer.

Bonfante, P., & Anca, I. (2009). Plants, mycorrhizal fungi, and bacteria: A network of interactions. *Annual Review of Microbiology, 63*(1), 363–383. doi:10.1146/annurev. micro.091208.073504

Chialva, M. et al., (2020). Gigaspora margarita and its endobacterium modulate symbiotic marker genes in tomato roots under combined water and nutrient stress. *Plants, 9*(7), 886. doi:10.3390/plants9070886

Crosino, A. et al., (2021). Extraction of short chain chitooligosaccharides from fungal biomass and their use as promoters of arbuscular mycorrhizal symbiosis. *Scientific Reports, 11*(1), 1–12. doi:10.1038/s41598-021-83299-6

Deka, D., Sonowal, S., Chikkaputtaiah, C., & Velmurugan, N. (2020). Symbiotic associations: Key factors that determine physiology and lipid accumulation in oleaginous microorganisms. *Frontiers in Microbiology, 11*, 1–8. doi:10.3389/ fmicb.2020.555312

Deveau, A. et al., (2018). Bacterial–fungal interactions: Ecology, mechanisms and challenges. *FEMS Microbiology Reviews, 42*(3), 335–352. doi:10.1093/femsre/fuy008

Dzialo, M. C., Park, R., Steensels, J., Lievens, B., & Verstrepen, K. J. (2017). Physiology, ecology and industrial applications of aroma formation in yeast. *FEMS Microbiology Reviews, 41*(1), 95–128. doi:10.1093/femsre/fux031

Hardoim, P. R. et al., (2015). The hidden world within plants: Ecological and evolutionary considerations for defining functioning of microbial endophytes. *Microbiology and Molecular Biology Reviews, 79*(3), 293–320. doi:10.1128/ mmbr.00050-14

Huseyin, C. E., O'Toole, P. W., Cotter, P. D., & Scanlan, P. D. (2017). Forgotten fungi— the gut mycobiome in human health and disease. *FEMS Microbiology Reviews, 41*(4), 479–511. doi:10.1093/femsre/fuw047

Jambon, I., Thijs, S., Weyens, N., & Vangronsveld, J. (2018). Harnessing plant-bacteria-fungi interactions to improve plant growth and degradation of organic pollutants. *Journal of Plant Interactions, 13*(1), 119–130. doi:10.1080 /17429145.2018.1441450

Jia, M et al., (2016). A friendly relationship between endophytic fungi and medicinal plants: A systematic review. *Frontiers in Microbiology, 7*(906), eCollection 2016. doi:10.3389/fmicb.2016.00906

Jia, Q. et al., (2019). Terpene synthase genes originated from bacteria through horizontal gene transfer contribute to terpenoid diversity in fungi. *Scientific Reports, 9*(1), 1–11. doi:10.1038/s41598-019-45532-1

Jusino, M. et al., (2016). Experimental evidence of a symbiosis between red-cockaded woodpeckers and fungi. *Proceedings of the Royal Society B: Biological Sciences, 283*(1827), 20160106. doi:10.1098/rspb.2016.0106

Kusari, S., Hertweck, C., & Spiteller, M. (2012). Chemical ecology of endophytic fungi: Origins of secondary metabolites. *Chemistry & Biology, 19*(7), 792–798. doi:10.1016/j.chembiol.2012.06.004

Marden, C. L., Mcdonald, R., Schreier, H. J., & Watts, J. E. (2017). Investigation into the fungal diversity within different regions of the gastrointestinal tract of Panaque nigrolineatus, a wood-eating fish. *AIMS Microbiology, 3*(4), 749–761. doi:10.3934/ microbiol.2017.4.749

Pawlowska, T. E. et al., (2018). Biology of fungi and their bacterial endosymbionts. *Annual Review of Phytopathology, 56*(1), 289–309. doi:10.1146/annurev-phyto-080417-045914

Rush, T. A. et al., (2020). Lipo-chitooligosaccharides as regulatory signals of fungal growth and development. *Nature Communications, 11*(1), 1–10. doi:10.1038/s41467-020-17615-5

Schmidt, R. et al., (2016). Microbial small talk: Volatiles in fungal–bacterial interactions. *Frontiers in Microbiology, 6*(1495), eCollection 2015. doi:10.3389/fmicb.2015.01495

Van Der Heijden, M.G.A., Dombrowski, N., & Schlaeppi, K. (2017). Continuum of root–fungal symbioses for plant nutrition. *Proceedings of the National Academy of Sciences, 114*(44), 11574–11576. doi:10.1073/pnas.1716329114

Ward, T. L. et al., (2018). Development of the human mycobiome over the first month of life and across body sites. *MSystems, 3*(3), e00140-17. doi:10.1128/msystems.00140-17

Witherden, E. A., Shoaie, S., Hall, R. A., & Moyes, D. L. (2017). The human mucosal mycobiome and fungal community interactions. *Journal of Fungi, 3*(4), 56. doi:10.3390/jof3040056

Yarden, O. (2014). Fungal association with sessile marine invertebrates. *Frontiers in Microbiology, 5*(3895), 1–6. doi:10.3389/fmicb.2014.00228

Yeh, C., Chung, K., Liang, C., & Tsai, W. (2019). New insights into the symbiotic relationship between orchids and fungi. *Applied Sciences, 9*(3), 585. doi:10.3390/app9030585

You, L. et al., (2015). New fungus-insect symbiosis: Culturing, molecular, and histological methods determine saprophytic polyporales mutualists of ambrosiodmus ambrosia beetles. *Plos One, 10*(9), e0137689. doi:10.1371/journal.pone.0137689

Zambonelli, A., Iotti, M., & Murat, C. (2018). *True Truffle (Tuber spp.) in the World Soil Ecology, Systematics and Biochemistry*. Cham: Springer International Publishing.

Zhang, Y., Kastman, E. K., Guasto, J. S., & Wolfe, B. E. (2018). Fungal networks shape dynamics of bacterial dispersal and community assembly in cheese rind microbiomes. *Nature Communications, 9*(1), 1–12. doi:10.1038/s41467-017-02522-z

Zivanovic, A., & Rodgers, L. (2018). The role of fungal endophytes in plant pathogen resistance. *Bios, 89*(4), 192–197. doi:10.1893/0005-3155-89.4.192

MICROBIOMES OF SEDENTARY AQUATIC ORGANISMS

4

Contents

Keywords

- Sessile Organisms
- Photobionts
- Methanogens
- Biomineralization
- Ageless Hydra

4.1 Corals and Their Microbial Communities

4.1.1 What are corals?

Corals are aquatic invertebrates that belong to the phylum Cnidaria, a classification shared by jelly fish and sea anemones. They are very colourful animals and come in a variety of shapes and sizes. However, they are sessile, i.e. they are attached to a surface in the water body and acquire their nutrients through different strategies including hosting numerous microbial organisms with whom they are engaged in symbiotic partnerships. Corals are known to grow in shallow or deep water and can occupy a diverse ecosystem ranging from tropical to arctic waters living either in groups or a singular lifestyle. Even though they are spread out in less than 0.2% of the sea floor, they provide a living environment for than 30% of marine life. These marine habitats are created in the coral reef where

DOI: 10.1201/9781003166481-4

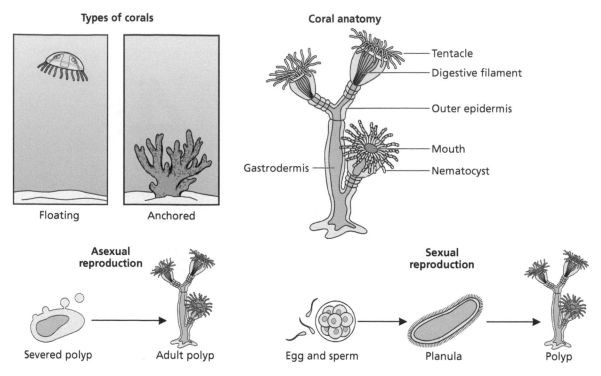

Figure 4.1. Habitats, anatomy and physiology of corals. Corals are located in diverse habitats where they are either anchored or float. They have a mouth surrounded by tentacles and a digestive tract. The nematocysts are sting cells utilized to immobilize prey. These animals can reproduce sexually and asexually.

tiny corals known as polyps live in colonies and construct large carbonaceous structures rich in calcium carbonate ($CaCO_3$). This is home to a large biodiversity including sea horses, turtles and fishes.

The coral-engineered reef is also a rich source of organic and inorganic nutrients released by the tissue and calcareous skeleton that help support planktivores and other organisms, pivotal components of the marine food chain. The biological activity within the coral reef is intimately linked to climatic conditions around the globe. This calcareous skeleton can be either hard or soft. Sea fans and sea plumes contain soft version and tend to adapt to the action of sea waves whereas the hard skeleton prominent in reef is more or less well rooted. Each coral polyp is enclosed in a $CaCO_3$ cup-like structure and cemented with other encased polyps. This arrangement may amount to millions of polyps that grow on an ongoing basis.

4.1.2 Coral anatomy and physiology

Coral polyps have a simple anatomy constituted of a single mouth opening surrounded by tentacles and a closed basal plate attached to a surface ranging in diameter from 1 to 3 mm. The stomach occupies most of the central space and is lined with the gastrodermis coupled with the digestive filaments. The ring of tentacles aids in the intake of food that is subsequently digested and the waste is expelled again through the same opening. The epidermis lines the outer tissues and possesses the nematocysts (cnidocytes) that sting the prey before it is ingested. Upon being properly stimulated, the cnidocyte ejects a reactive and single-use mixture of biochemicals aimed at capturing a prey ready to be utilized by the polyp. The basal plate is formed by the calyx resulting from the deposition of $CaCO_3$ by the polyp, a process aided by the resident microbes.

The generation of new polyps can proceed via either asexual or sexual reproduction. During the asexual process sometimes referred to as budding, a part of the polyp is severed off and grows as an adult polyp. In this instance, clones with the exact genetic replica of the adult are formed.

BOX 4.1 THE CHEMICAL ARSENAL AT THE DISPOSAL OF CORALS TO ATTACK AND DISSUADE PREDATORS

Corals like most members of the *Cnidaria* phylum have to resort to a predominantly sedentary lifestyle. They are usually rooted to ocean floors. They need to hunt to feed themselves and defend against predators in search of food. To overcome this dilemma, they have evolved specialized cells termed nematocysts or cnidocytes often located in the tentacles charged with the task of elaborating a range of programmed molecular strategies to immobilize prey and dissuade predators. Upon sensing of mechanical or chemical signals, these cells discharge toxins that render prey incapacitated and ready to be swept by the tentacles into the mouth. These venoms that are enzymes, pore-forming moieties, channel blockers, metalloproteases and phospholipases are dedicated to distress the cellular systems of the subjects. The porin-forming proteins like actinoporins self-assemble within membranes and create pores thus inducing osmotic shock and death. While metalloproteases degrade proteins and cause tissue damage resulting in cellular dysfunctions, phospholipase A$_2$ is associated with cytotoxic properties due to its ability to hydrolyse phospholipids. Inhibitors of Na$^+$- and K$^+$-gated channels and serine protease have a paralytic activity to subdue the victim. Bioactive amines such as serotonin and histamine are vasodilators. Although corals accommodate numerous organisms in their midst, they ferociously hunt for food and defend against aggression with the assistance of these chemical weapons.

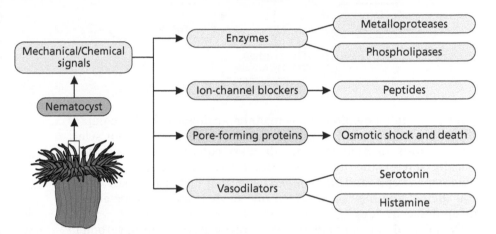

Box Figure 1. Nematocyst producing a diverse array of biochemicals aimed to immobilize the prey and predators.

This type of body fragmentation occurs when the polyp reaches a certain size. Sexual reproduction starts with the release of eggs and sperms in the water where they fertilize to form a larva known as the planula. The latter is attracted to a hard surface where it attaches and grows into a polyp. In the brooder species of corals, the male gamete is picked by a female coral containing egg cells. Following fertilization inside the organism, a planula develops and is released from the mouth. Corals have a tendency to spawn en masse at the same time each year, an event that appears to coincide with the lunar cycle.

4.1.3 Corals and their microbial partners

As all multicellular organisms irrespective of their degree of specialization or the amount of organs they possess, they all depend on their microbial partners to survive. This holds true for corals. Their inability to be involved in any meaningful locomotion has compelled them to recruit an array of microorganisms that enable corals to live effectively. These microbes that include algae, bacteria, archaea, fungi and viruses reside within and on the coral and perform a variety of biological functions crucial for the survival of the host. The establishment epi- and endo-symbionts are intricately regulated and their spatial distribution is precisely monitored. For instance, the photosynthetic dinoflagellates responsible for supplying fixed carbon nutrients to the host are lodged in the gastroderm in a specialized tissue referred to as symbiosome. In fact, up to 90% of the organic carbon of the holobiont is supplied by these photosynthetic partners. A significant portion of this carbon source is earmarked for the synthesis of the mucus that forms a protective layer on the surface of corals. This carbohydrate-rich coating does not only provide a shelter against UV radiation or pathogens, but it also serves as a residence to numerous episymbionts. The nature of the nutrients it has in store has a discerning function in selecting specific

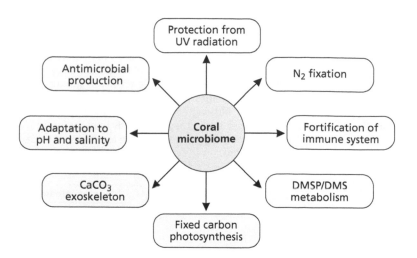

Figure 4.2. **Functions of the coral micro-biome.** The microbes constituting the coral holobiont are involved in a variety of functions essential for the survival of the host. They supply nutrients, contribute to the formation of the calcium-rich exoskeleton, combat pathogens, help select the microbial community and are involved in N and S metabolism. They enable the host to adapt to a range of abiotic stress. (DMSP: dimethylsulphoniopropionate; DMS: dimethyl sulphide.)

bacteria for the holobiont in a manner analogous to the complex carbohydrates produced by human colon, moieties tailored for select microbes with the ability to degrade these compounds. This strategy basically starves the undesirable microbes. In fact, microbes like *Pseudomonas* spp. and *Alteromonas* spp. that called this part of the coral their home may be attracted due to their ability to metabolize mucopolysaccharides.

The microbiome not only provides a ready access to fixed carbon in nutrient-poor ecosystem in particular, but its members also produce numerous antimicrobials that help fend off pathogens. This is critical for survival of the coral as it is constantly exposed to opportunistic microbial settlers in an open aquatic environment. Removal of toxic NH_3, a metabolic by-product, fixation of N_2, metabolism of dimethyl sulphide, enhancing the formation of the $CaCO_3$ structure and fortifying the immune system are some of the benefits the microbiome imparts to the host. In exchange, the microbes are provided a relatively safe home and occasional nutrients to live on. However, it is important to note that this mutually favourable relationship is tightly guarded and is established only following the input from all the partners (**Figure 4.2**).

4.2 The Algal Coral Symbiosis

This is a critical relationship that is the very essence of the life of this holobiont. The coralline way of living is dependent on the presence of symbiotic algae, the supplier of a major portion of the energy the host requires. In fact, the success of coral reefs is reliant on the photosynthetic capacity of these dinoflagellate algae known as *Symbiodinium*. The disappearance of these photosymbionts, a phenomenon referred to as coral bleaching results in serious damage to the marine ecosystem. The *Symbiodinium* converts sunlight and CO_2 into organic carbon, O_2 and energy that fuel the growth of the host, a process aided by the enhanced calcification induced by the photobionts. This intimate partnership akin to a specialized organ occurs in a dedicated space within the gastroderm in membrane vesicles that border the gastric cavity, an anatomical arrangement that promotes the effective exchange of metabolites between these two entities. The symbiotic interaction starts with the initial interaction of the *Symbiodinium* with the host, followed by engulfment and intracellular translocation. Subsequently, the establishment and proliferation of the symbiosome proceed.

The communication between the alga and the host is intricately guarded in order to prevent the fortuitous entry by unwanted microbial invaders. The specific signal secreted by the alga is recognized by the host that in turn modulates its immune system in an effort not to eliminate the symbiont. The phagocytic machinery is arrested, lysosomal fusion is impeded, and proteins like acid phosphatases and ferritin that participate in phagocytosis are downregulated. The host surveillance mechanism is usually fine-tuned upon the reception of the algal α-mannose or α-glucose glycans that are recognized by lectin receptors. This signal activates the symbiotic genes resulting in the dynamic remodelling of the host cytoskeleton, the entry and establishment of the symbiont. The association between the alga and the coral is very specific and symbiont is vertically transmitted. The larva, however, acquires the *Symbiodinium* from the environment and the successful interaction of these two entities triggers molecular events responsible for metamorphosis to occur. This starts following the formation of an open mouth and the development of the gastric cavity where proper symbiosis is initiated. The population of the endosymbiotic algae is kept in check by their active degradation mediated by the host in an effort to maintain the desired amount of photosynthetic activity.

4.3 Photosynthesis: A Mutually Beneficial Affair

Although the photosynthetic activity resides primarily with the dinoflagellates equipped with the light-harvesting complexes and all the enzymes mediating the production of ATP and NADPH and the fixing of CO_2, the host ensures an ample supply of CO_2 and contributes to maximize light input to the photosynthetic apparatus by expanding the range of absorbed light by synthesizing chromophoric biomolecules. The corals also help guard against oxidative stress, a known photosynthetic hazard by overexpressing enzymes like superoxide dismutase and catalase responsible for the neutralization of superoxide (O_2-) and hydrogen peroxide (H_2O_2). In seawater, where the pH fluctuates around 8.0–8.2, the bicarbonate concentration is ~2.2 mM, while the CO_2 level is at ~30 μ M. This can make photosynthesis relatively ineffective, a situation that is further compounded by the intracellular location of symbiont not directly exposed to the water containing CO_2. To rectify this problem, the host has evolved an effective CO_2-concentrating mechanism (CCM), aimed at keeping the CO_2 supply line proficient. Carbonic anhydrase involved in the catalysis of carbonic acid to H_2O and CO_2 and H+/ATPase dedicated to the acidification of the micro-environment work in tandem to fuel the photosynthetic process. The fixation of CO_2 into organic carbon nutrients is further amplified by the expression of ribulose 1,5 biphosphate carboxylase/oxygenase (Rubisco) with a high affinity for this gaseous substrate. The glycerol, succinate, malate and glycolate generated by this metabolic engine are readily shuttled into the host's gastric cavity to be readily utilized. Ammonia, a metabolic waste excreted by the host is converted into amino acids. These symbiont-derived amino acids are supplied to the host to be utilized. The enzymes glutamate dehydrogenase (GDH) and glutamate synthase (GS) are two important participants in this metabolic undertaking. If this endosymbiosis that provides the host with most of its nutrients is perturbed, the life of the coral is at risk (**Figure 4.3**).

4.4 The Bacterial Residents and Their Contribution to the Coral Holobiont

Corals are populated by a wide assortment of microorganisms including algae and bacteria. While algae comprise approximately 2–3% of the microbial inhabitants, the bulk of microbes is constituted by residents of bacterial origin. These bacteria are localized within the surface mucus

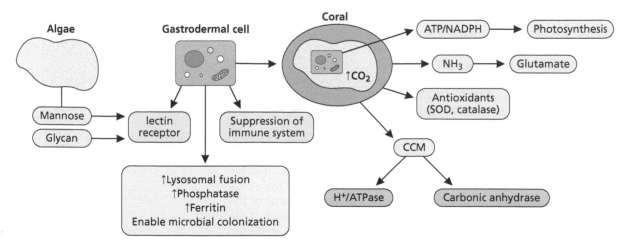

Figure 4.3. Symbiotic association between algae and corals. The mannose and glucose-rich algal glycan are recognized by the lectin receptors of the host. This signal triggers the suppression of the immune system and the downregulation of hydrolytic enzymes, events that enables the alga to establish within the gastroderm of the coral. To facilitate the photosynthetic process, the host activates its CCM and the anti-oxidant defence system. The alga does not only produce fixed carbon but also utilizes the excreted NH_3 to generate amino acids for the host; CCM: CO_2-concentrating mechanism.

layer that interfaces with the external environment and the epidermis, the outer coral layer. They play a critical role in pathogenic defence and are susceptible to physico-chemical changes in the ocean. The pH, nutrient levels, O_2 content, temperature fluctuations and salinity of the water all appear to dictate the microbial colonization of this coral anatomy. All these factors contribute to render these microbial communities to be highly dynamic, even though they fulfil some critical physiological functions. The host invests through its immune effectors and nutrient secretion to ensure the proper mix of bacteria can call this part of the coral body their home. The dynamic nature of its residents enables the host to adapt to changing environmental conditions and recruit microbes more apt to tolerate a change in pH or salinity. The coral tissues that include the mesoglea, gastroderm, epithelium and calcoblast also provide residential space to the microbes. Microbes like *Propionibacterium* spp. *and Ralstonia* spp. have been spotted in the gastroderm, a tissue where the dinoflagellate endosymbionts reside. On the other hand, *Endozoicomonas* spp. has been reported in both the epidermal and the gastrodermal tissues. The coral skeleton the component attached to the surface is an attractive habitat for non-photosynthetic algae, fungi and bacteria. The bacterial constituents of this part of coral play an important role in the biomineralization of $CaCO_3$. Hence, it is clear that the bacteria are specific to the tissue they colonize and the functions they perform. Some of them appear to be susceptible to elimination especially those residing on the mucus surface while those associated with the other coral tissues tend to be more permanent.

The duration and the nature of association of the microbes have led to the postulation that there may be three categories of microbial residents in the coral holobiont – the core microbiome, the resident microbiome and the microbiome responsive to environmental changes. The core microbiome is highly structured, i.e. tissue specific, does not vary significantly amongst different species and is relatively stable over a diverse geographical area. For instance, the coral *Porites astreoides* from disparate ecosystems is associated with *Endoziocomonas* spp. and stable symbionts like *Actinobacter* spp. and *Ralstonia* spp. They do tend to assemble, disassemble and re-assemble depending on the ecological dynamics as do the other transient microbial dwellers. The core microbiome is usually constituted by 2 or 3 phyla such as *Proteobacteria, Actinobacteria* and *Bacteroidetes. Proteobacteria* is the most dominant amounting to 60–70% of the microbes. *Alpha-Proteobacteria* and the *Gamma-Proteobacteria* are the dominant classes irrespective of

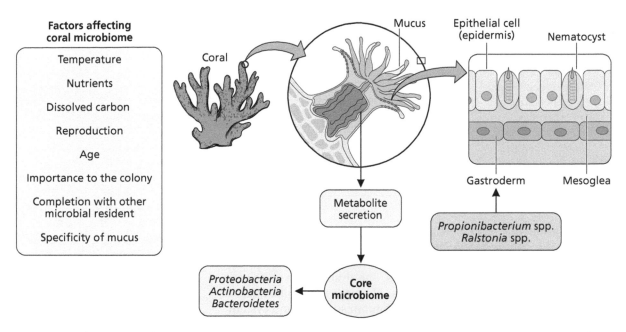

Figure 4.4. Factors affecting microbiome colonization of the host. The assembly of the microbiome is governed by numerous factors including nutrients, reproductive route and temperature. The host also secretes different metabolites aimed at attracting select microbes that are harboured epicellularly and endocellularly. The epidermis, gastrodermis and nematocyst are some of the specialized cells with microbes.

environmental drivers. The resident microbiome varies among different species while the microbiome modulated by biotic or abiotic changes is very dynamic in its constituents it harbours (**Figure 4.4**).

4.5 Functions Executed by the Coral Microbiome

Microbes residing within corals are cells with specialized attributes that enable the coral to execute various physiological activities. Nitrogen fixing is a very important biological activity assigned to only a few organisms. In nutrient-poor habitats, some corals develop a symbiotic relationship with diazotrophic bacteria in order to satisfy their need for fixed N_2. *Gluconacetobacter*, *Cyanobacteria* and *Rhizobia* spp. are known to be involved in such arrangements. *Pseudomonas* spp., *Vibrio* spp., *Spongiobacter* spp. and *Roseobacter* spp., with their propensity to metabolize dimethylsulphoniopropionate (DMSP) and dimethyl sulphide (DMS), contribute to the biogeochemical cycle of S. Bacterial residents assist corals in acquiring phosphate, an essential nutrient that tends to be scarce in seawater. The biosynthesis of the coral skeleton is aided by the energy generated by the dinoflagellates and by the ability of the resident bacteria and fungi to deposit $CaCO_3$. The general metabolic performance and the fitness of the immune system are strengthened by the presence of the microbes within the holobiont. The microbial residents also contribute to the ability of corals to proliferate in diverse ecosystems ranging from tropical to cold oceans. They produce antioxidants to thwart oxidative challenges in tropical waters where photosynthesis is very active. The antimicrobials like isatin derived from bacteria such as *Alteromonas* sp. keep pathogenic microbes in check (**Figure 4.5**).

4.6 Dysbiosis and Coral Health

This holoorganism consists of the coral as the host and the diverse microbial constituents as the partners work in a highly orderly fashion in order to maintain the vast coral reef system that is home to 30% of the oceanic biodiversity. The genesis of this holobiont is tightly controlled by all the participants in order to protect against undesired intruders. Even the

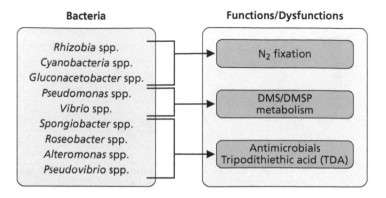

Figure 4.5 **Specific bacteria in corals and their roles.**

population of bona fide members is tightly guarded and as is their contribution to the complex cellular organization. An imbalance in this harmonious relationship can lead to disastrous consequences to the whole community and the biodiversity it harbours. The dysbiosis in the coral holobiont is punctuated by various diseases. Although numerous factors including environmental changes and a compromised immune system may contribute to ill health in the host, the disruption of microbial homeostasis is an important contributing cause. The black band disease whereby corals acquire characteristic dark spots is caused by a polymicrobial lesion. In this instance, the infection by the cyanobacterium *Roseofilum reptotaenium* leads to microbial disharmony resulting in a drastic increase in sulphate-reducing *Desulfovibrio* spp. and *Acinetobacter* spp.

The high photosynthetic activity mediated by the cyanobacteria creates an organic carbon and oxygen-rich environment that fuels the proliferation of various aerobic bacteria. The rapid consumption of O_2 triggers micro-anaerobic conditions where *Desulfovibrio* spp. thrives. This provides a niche for sulphate reducers to generate sulphide, a moiety responsible for the black colour of the coral during this disease. The bacterium *Vibrio shiloi* is responsible for the bleaching of corals, a condition where the algae are expelled from the host. The microbe produces a toxin that targets the intracellular *Symbiodinium*. Increase in temperature enables the microbe to adhere to the coral and penetrate the epidermis making its way to the inner tissue. The toxic secretion disrupts the membrane and perturbs the pH gradient, thus drastically affecting photosynthesis. This results in the demise of the photobiont. Owing to the shared metabolic networks among the participants of the holobiont, the whole organism succumbs. It is important to note that *Symbiodinium* produces copious amounts of DMSP, a moiety that is an important source of S for the marine community. Furthermore, it is also utilized in the synthesis of tropodithietic acid (TDA) by the communal *Pseudovibrio* spp. This is an important antibiotic involved in shaping the microbial residence of the coral community. Hence, it is clear that the perturbation of a core entity in the meta-organism can have a dramatic impact on the whole community. Understanding how this multi-partite entity comes together and how this delicate balance can be deranged is critical in identifying solutions to preserve ocean biodiversity (**Figure 4.6**).

4.7 Ascidians and Their Microbial Constituents

4.7.1 Biological properties, anatomy and physiology

Ascidians are a class of tunicates that belong to the phylum chordate and are found in marine ecosystems worldwide. Although they are part of the chordate group, they are soft and boneless. The only period in their lifecycle they exhibit a backbone known as the notochord is when they exist as the short-lived tadpole-like larvae. In fact, they are primitive

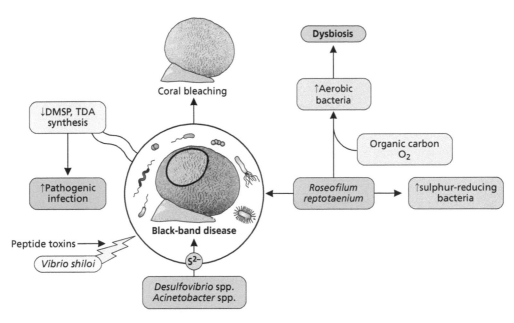

Figure 4.6. Dysbiosis and coral health. Coral health is dependent on the harmonious relationship between the host and the microbial partners. Infection by *Roseofilum reptotaenium* results in an increase in sulphur-reducing bacteria leading to the black band disease characterized by the deposition of S^{2-}. In the presence of *Vibrio shiloi*, there is a decrease in antibiotic production and the disruption of the algal–host relationship. The eviction of the algae is responsible for bleached corals, a situation detrimental to ocean biodiversity. (DMSP: dimethylsulphonionpropionate; TDA: tropodithietic acid.) Nematocyst producing diverse biochemicals and their activities.

BOX 4.2 TROPODITHIETIC ACID – A DRIVER OF MICROBIAL RESTRUCTURING

It is critical to have an assemblage of proper microbial partners if a holoorganism is to make good use of its constituents and be able to live effectively. The possibility of opportunists, pathogens and freeloaders seeking residence has to be curtailed and the population of the regular residents has to be tightly regulated. To achieve this fine balance, communication between the host and all the desired residents has to be intricately calibrated and precisely acted upon. TDA, a seven-membered ring structure with an adjacent disulphur bridge is secreted by numerous microbes that are integral constituents of a variety of marine holobionts including corals. This compound belonging to the tropolone family not only shapes the density of microbial partners that comprise corals, but it also fends off pathogens due to its antibiotic attribute reflected in its ability to perturb ATP production. TDA is also a quorum sensor involved in gauging the concentration of prokaryotes in the environment and modulating their growth. This cross-species/genus signal elaborated by such genera as *Roseobacter*, *Ruegeria* and *Pseudovibrio* is pivotal in the initiation and development of the numerous marine holoorganisms. The sulphur required for the biosynthesis of this information-laden compound is usually acquired from DMSP and DMS. These S-rich molecules released by the hosts also serve as attractants that guide the TDA-producing microbes to their holoorganisms where they participate in various tasks. Phenylalanine and histidine are also known to stimulate

the production of TDA, a molecule that orchestrates potent changes in metabolism, mobility, biofilm formation and attachment to a host. It is an autoinducer and modulates amounts of *N*-acyl homoserine lactone (AHL) and cyclic di GMP, moieties responsible for behavioural changes in microbes. The decrease in biofilm formation triggered by TDA promotes balanced colonization by impeding aggregation on crowded surfaces and favours attachment onto moving hosts. This pluripotent moiety rich in diverse messages permits only microbes that are tolerant to its toxicity to colonize the host and supports microbial proliferation in a manner amenable to the holoorganism.

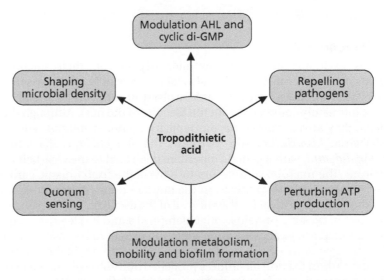

Box Figure 2. The multi-faceted functions of TDA.

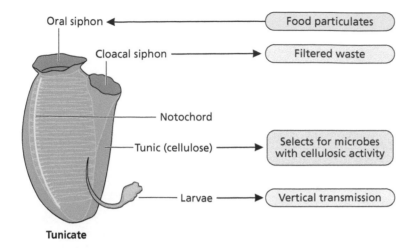

Figure 4.7. Ascidians: anatomy and physiology. Ascidians are tunicates with two openings referred to as the oral siphon and the cloacal siphon involved in uptake of foods and elimination of wastes. The outer layer rich in cellulose known as a tunic helps in the selection of microbial residents that are also vertically transmitted.

chordates. The larvae attach to rocks or any solid support and grow into adults. They are sessile organisms with very limited mobility. Ascidians can lead a solitary life or can choose to live in colonies. They are surrounded by an outer skin referred to as a tunic that is essentially composed of cellulose, a component relatively stable and resistant to degradation. Their anatomy is simple with a tubular or globular shape containing two openings dedicated to bringing in food and eliminating waste respectively. As they are filter feeders, water laden with food particulates is siphoned through the oral siphon into the digestive chamber. A water current is generated through the body by fine hairy structures referred to as cilia. This morphological arrangement allows food particles to be filtered with the aid of a ring of tentacles containing mucoid nets before entering the mouth. The filtered wastes are expelled via the cloacal siphon. Ascidians possess male and female sexual organs that release sperms and eggs in the oceans that are subsequently fertilized and morphed into free-swimming larvae. The colonial tunicates usually retain and brood their eggs. Larvae and juveniles are a source of foods for numerous marine organisms while adults are known to be an important modulator of coral reefs. The latter is also consumed in some countries. They are valued as producers of chemicals with medicinal properties, an attribute usually assigned to their microbial inhabitants (**Figure 4.7**).

4.8 The Microbes within Ascidians

4.8.1 Photobionts

Numerous ascidians like the didemnids depend on their microbial symbionts not only for nutrients but also for a variety of functions essential for their survival. These partners are involved in an array of metabolic networks intimately connected with the lifecycle of the host. Although they filter feed, they also acquire their nutrients from photosymbionts such as cyanobacteria. *Lissoclinum* spp. harbours the obligate cyanobacterium *Procholon didemni*, with a genetic repertoire dedicated to the well-being of the tunicate. The microbe supplies up to 100% of the fixed organic carbon the host needs via photosynthesis. In the nutrient-poor environment, this relationship may be critical for the survival of the ascidian. The presence of unique light-harvesting proteins with chlorophyll a and b, a feature unique to this cyanobacterium allows a wide spectrum of light to be absorbed and enables the organism to thrive in diverse habitats including shallow tropical waters. The carbon fixation machinery propelled by Rubisco is lodged in the β-carboxysomes where it is protected from the toxic influence of O_2. In some cyanobacteria residing in symbiosis with ascidians, CCMs mediated by high-affinity CO_2/HCO_3- transporters help manage the supply of this ingredient essential during the dark reaction. Glycolate and other carbon-containing metabolites are shared with the host.

4.8.2 Nitrogen fixers, sterol producers and metal concentrators

The nitrogen budget is supplemented by the utilization of NH_3 derived from the tunicates. The NH_3 is converted into glutamate and glutamine with the participation of the transaminase and GS pathways. Some of these photosynthetic microbes can also reduce NO_3 into NH_3 and fix atmospheric N_2 into NH_3. To protect against UV radiation, the bacterium elaborates mycosporine-like amino acid, a feature important to combat photo damage. *P. didemni* contributes to the homeostasis of fatty acids in the host as it produces sterols like lanosterol, an amphipathic molecule which is part of the membrane systems of the organism. Alkanes and other hydrocarbons that are associated with tunicates are also synthesized by the communal microbes. Polyketide synthesis mediated by polyketide synthase is part of the genetic information these symbionts harbour. An obligate endosymbiont, *Candidatus Endolissoclinum faulkneri*, is the main architect in the synthesis of this moiety and is dependent on the host to survive. This antibiotic is one of the several bioactive chemicals the resident microbial community in the ascidian elaborates to help guard the host from microbial invaders. Cyanobactins, a group of cyclic peptides consisting of 6–8 amino acids, are microbially derived compounds. They can bind metals such as copper and zinc and may have a role in CO_2 hydration and O_2 activation.

Proteobacteria is one of the largest groups of microbes associated with ascidians. *Endozoicomonas* spp. that belongs to the *α-Proteobacteria* group possesses enzymes with the ability to degrade complex polysaccharides including cellulose, a component of the outer layer in tunicates. This polymer of β-glucose may be a molecular attractant that allows only a select class of bacteria to seek residence within this host. *Firmicutes* like *Halobaccilus* spp. are also part of the holobiont and their molecular machinery designed to tolerate elevated salt concentration enables the host to adapt to changing salinity in oceans. Enzymes such as lipases and ligninases that are part of the genetic repository of the microbiome inhabiting ascidians allow the degradation of compounds that are relatively recalcitrant to metabolic manipulation. Metal biotransformation is another important characteristic that these microbes participate in. This attribute enables the hosts to fend metal pollution, a situation they are constantly confronted with as a consequence of changing water currents in oceans. The metal-processing trait of the microbes also helps fulfil the need for mineral nutrients like Fe, Cu and Va in these organisms. In fact, ascidians are known to accumulate an enormous amount of Va in their blood stream, a feat accomplished by the microbial symbionts with the aid of their ATP-driven Va pumps. In tunicate *Ascidia sydneiensis samea*, microbes from the genera *Vibrio* and *Shewanella* are the main agents involved in the concentration of this mineral that is known to be a component of numerous biological processes in the host and various microbial life.

The high degree of specificity, a hallmark of the microbial ascidian interaction, may be a result of the unique carbohydrate-rich tunic and physiological features associated with these sessile organisms. The distinct microenvironments nurtured by these inherent attributes allow only a select group of bacteria to populate the holobiont. It is quite likely that some members of this microbial population that call tunicate their home are vertically transmitted as bacteria are found in larvae. The colonial life of ascidians may help in this transmission as the organisms are in close contact. Only microbes with the ability to utilize the cellulose containing outer layer will be privileged in this part of the host while the oxygen gradient maintained due to the physiological activity will only attract anaerobic organisms in the oxygen-deprived parts of the body. Although colonization of the host depends on a variety of biotic and abiotic modulators, the emerging view appears to favour the presence of a core group of microbes that is present in ascidians. For instance, *Reichenbachiella* spp., a constituent of this core family produces carotenoids, metabolites central in guarding against phototoxicity, while *Pseudomonas* spp. provides reprieve against metal and hydrocarbon pollution.

Microbes like *Rhodobium* spp. and *Novispirillum* spp. are all part of the holobiont as they are excellent manipulators of nitrogen-rich compounds. They are programmed to transform NH_3, NO_2 and NO_3 both under oxic and normoxic conditions. The presence of such archaea like *Nitrosopumilus* spp. with their ammonia and denitrification capacities is also recruited to manage the nitrogenous wastes that the tunicates are exposed to an ongoing basis. The diversity of microbes housed within the ascidians may also be harbinger of the level and type of pollution in the oceans as the abundance of different genera of bacteria may be indicative of the specific toxicant the organism is exposed to. Hence, these sessile organisms can be an excellent biomarker for environmental changes. The variety of microbial participants in these meta-organisms that include such genera as *Acinetobacter*, *Bacillus*, *Endozoicomonas* and *Staphylococcus* enables the elaboration of a wide assortment of biochemical molecules that are utilized by the host to maintain its integrity in the open oceans. These bioactive compounds range from peptides to alkaloids and have numerous commercial potentials. Thus, these holoorganisms are not only excellent indicators of climate change but they are an important source of valuable chemicals (**Figure 4.8**).

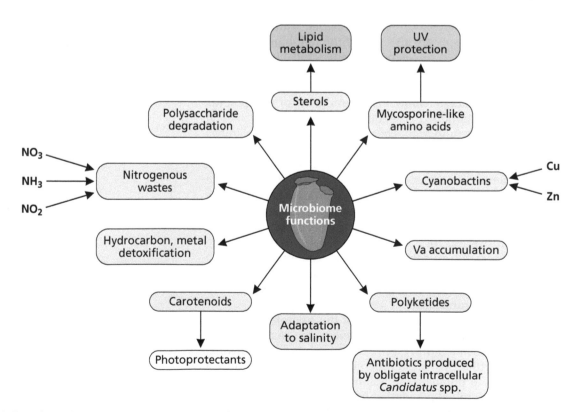

Figure 4.8. Functions of the microbiome in ascidians. The microbial residents are involved in a range of activities aimed at helping the host survive in open water. They help produce biomolecules contributing to photo-protection in the acquisition of nutrients, in fending pathogens and in the synthesis of membranes. Elimination of pollutants and extending survivability in a saline environment are also functions attributed to the microbial partners. The microbiome provides both organic carbon and nitrogen-rich compound.

4.9 The Lifestyle of Sponges – A Microbial Perspective

4.9.1 Sponges and their anatomical components

Sponges are grouped in the phylum *Porifera* literally meaning pore bearers and are known to occupy a diverse ecosystem. They are present in intertidal zone, shallow coastal water and deep sea. They are virtually immobile and are attached to the sea floor. Although the majority of sponges are marine organisms, they are also known to proliferate in fresh water. They are among the oldest animal that can lead a solitary lifestyle or live in a colony. They possess no tissue or organ but have specialized cells independent of each performing a variety of functions in a concerted manner. They come in different shapes and sizes with tubular, globular, tree-like and amorphous arrangements. The structural design is aimed at maximizing the efficiency of water flow, a feature critical in food intake and waste extrusion. Structures like cavities, canals and chambers punctuating the body allow water to move through all cells and thus obviate the need for a circulatory system or organ.

The central cavity of the sponge is referred to as the spongocoel through which water enters via the ostia, the body pores. The filtered water devoid of sponge nutrients and the waste products is eliminated with the aid of the osculum. While the pinacocytes line the outer layer, the choanocytes border the inner side of the sponge body where the mesophyll is located. The choanocytes are decorated with flagella that play an important role in generating water currents and trapping food particles. The mesophyll contains the amoebocytes, cells that move freely and are involved in the immune system, transport nutrients and generate reproductive cells. They produce spicules, sponging, spongin and collagen-like biomolecules to maintain the structural integrity of the mesophyll. Sponges reproduce asexually, a process mediated by budding and fragmentation. Sexual reproduction is fuelled by the release of gametes, an event that is dependent on numerous factors including temperature. Fertilization and development of the larvae occur within or outside the sponge's body. The viviparous and oviparous reproductive strategy is a common trait associated with these organisms. Sponges are basically a federation of specialized cells loosely held together in order to engage in a lifestyle important for the ecosystem. However, the input of its microbial residents is pivotal to make this holobiont a proper biological working machine.

Sponges play a very important role in the ecosystem and are involved in the biogeochemical fluxes of Pi, N, C and Si. They facilitate the consumption of nutrients and their ability to release NH_3 provides a nitrogen source to the habitat they are associated with, an activity pivotal in maintaining biodiversity. Their ability to populate the deep sea (~8,000 m) is a biological asset that helps benthic life. Spongin (modified collagen) harvesting is an ancient tradition that humans have engaged in for centuries and sponges are currently harvested in many countries. They are used for bathing and in the cosmetic industry. Sponges are also an important source of bioactive compounds; an attribute drug companies are exploiting (**Figure 4.9**).

4.9.2 Microbes are a pivotal component of sponges

Sponges are composed of specialized cells that work as close-knit multicellular machines. However, the diversity of specialized cells they possess is unable to accomplish all the house-keeping tasks these organisms need to do to bring their lives to fruition, i.e. to reproduce and fend all the challenges they encounter on an ongoing basis. To help achieve these objectives, sponges have resorted to the functional versatility of microbes. They have recruited a diverse population of microbial residents with unique biochemical capabilities to enable them to live and propagate effectively. Furthermore, these microbes allow sponges to adapt relatively

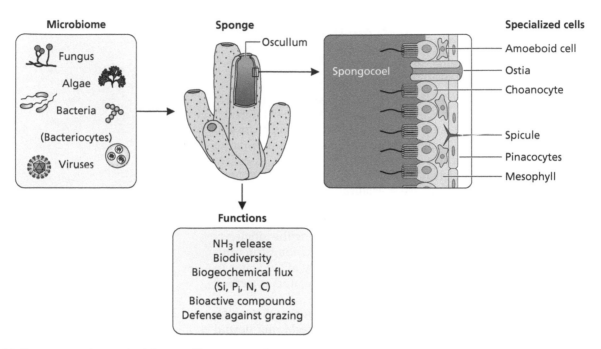

Figure 4.9. Sponges: anatomy, physiology and importance. Sponges contain numerous specialized cells like the spicule and choanocyte and a microbiome with its bacteria, viruses, fungi and algae. The microbes participate in the biogeochemical cycle of Pi, N, Si and C. Their structural architecture is designed to maximize water flow as the intake of foods and extrusion of wastes occurs in an aqueous environment.

BOX 4.3 DEPOSITION OF SILICA: SPONGE'S BONE STRUCTURE

Sponges utilize silica as their skeletal backbone, a structural arrangement critical in maintaining the morphology of the organisms. This mineral is deposited mainly in the extracellular space but is also localized in the cytoplasm and the nucleus. In these internal compartments, silica is associated with the maintenance of osmotic pressure and ionic balance. The deposition of this skeletal body is promoted by the presence of Si in the oceans, a condition that prompts the expression of proteins like silcatein and aquaporin. While the former mediates the polycondensation of Si, the latter helps manage the concentration of H_2O, a by-product of this reaction. Following the uptake of

dissolved Si by aquaglyceroporins found in the specialized spicule-forming cell known as sclerocyte or scleroblast, an axial canal is initiated. This event is preceded by the release of silicasomes replete with silicic acid and silcatein that attach to the gel-like galectin and is transported to the extracellular space around the elongating spicule. This high-energy demanding polymerization is mediated by a constant supply of ATP, an ingredient replenished with the aid of arginine kinase. Fe^{3+} is known to promote this process that results in the cellular growth and deposition of silica. The catalytic triad of His, Asn and Ser components of the silcatein mediates the polymerization of and the mineralization of silica.

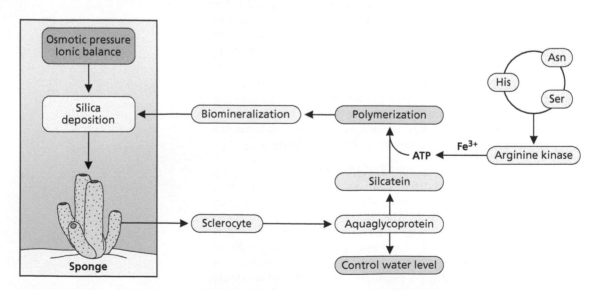

Box Figure 3. Biomineralization of silica assisted by arginine kinase, an ATP provider.

quickly to the fluxes in the environment and have become an integral constituent of the sponge. The sponge will not exist only with its own cells; it has entered into a partnership with microbes and enhanced its survivability. The associations with the microbial constituents can be obligatory, facultative or transient in nature. In fact, the microbial residents comprise up to 35% of the host's biomass. Fungi, unicellular algae, archaea, bacteria and viruses are among the most prominent residents. High microbial abundance sponges harbour 10^9 cells/cm^3 while those having a low abundance of microbes are associated with 10^{5-6} cells/cm^3. They are located in the extracellular matrix and in the mesophyll around the choanocyte. The endosymbionts are housed in specialized cells known as bacteriocytes and in the pinacoderm layer where they have maximum access to sunlight. The colonization of sponges by bacteria depends on the response of the immune system to the microbes and the cross-talk they engage in. The spatial differentiation as modulated by the pH, O_2 and the nutrient gradient also impacts on the selectivity of the symbionts.

Sponges are associated with microbes with a high phylum diversity and specificity. Microbes belonging to 52 phyla have been uncovered with the following phyla being most prominent *Proteobacteria* (gamma and alpha), *Actinobacteria, Chloroflexi, Cyanobacteria, Nitrosospirae, Poribacteria* (candidatus phylum) and the archaea *Thaumarchaea*. For instance, the cyanobacterium *Synechococcus spongiarum* has been detected in a wide variety of sponges. As in corals, the microbial population can be categorized as core, generalist and transient. The latter is more or less responsive to environmental cues. The acquisition of the microbial constituents takes place both vertically and horizontally. The gametes tend to carry microbes. For example, the sperms of the sponge *Chondrilla australiensis* tend to carry cyanobacteria. The availability of the desired microbes from the environment can be a risky proposition as it depends on the environmental conditions and is also prone to be hijacked by opportunistic bacteria 'cheaters'. However, it is important to note that symbionts secrete a variety of proteins and carbohydrate-rich compounds to interact with their eukaryotic host in order to establish initial contact and modulate the immune response evoked via the amoebocytes. Pattern recognition receptors may play a role in deciphering these molecular signals and promote the selection of the bacteria targeted to live within the sponge. The nucleotide oligomerization domain-like receptors coupled with the scavenger receptor cysteine-rich proteins may be providing this sensor mechanism designed to sort the microbial settlers and the invaders. Quorum-sensing mechanisms spearheaded by a plethora of biomolecules help check the stability of the microbiome. The host can also expel the excess amount of microbes, activate its phagocytic machinery and deprive the symbionts of nutrients (**Figure 4.10**).

4.10 Tasks Executed by the Microbes

The microbes that have a protected space to live are also supplied with a healthy dose of nutrients like NH_3 in exchange for numerous tasks they execute for the well-being of the holobiont. The photosynthetic microbes contribute up to 50% of the organic carbon budget of the host. In this instance, glycerol is the main organic nutrient the sponge derives from the microbes and a metabolic coupling between the partners helps in rendering this process efficient. Metabolic networks like glycolysis, tricarboxylic acid cycle, oxidative phosphorylation, pentose phosphate pathway and 3-hydroxypropionate cycle are activated to promote the metabolite exchange. The cyanobacterium *Candidatus Synechococcus spongiarum* possesses a photosystem II that allows to capture light and promotes photosynthetic activity. Its dependence for methionine on the host further attests to the molecular adaptation operative in this holoorganism. This

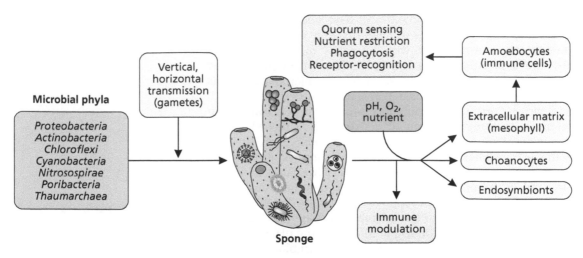

Figure 4.10. Sponge microbiome, acquisition and population control. Sponges are densely populated by microbes and the host utilizes a range of strategies to maintain its microbial populations. Quorum sensing, activation of the immune response, nutrient limitation, phagocytosis and receptor recognition are some of strategies at the disposal of the host to modulate its microbial partners. Up to 52 phyla have been identified in sponges. Both vertical and horizontal microbial transmissions are at work.

symbiont that is important for the energy need of the holobiont does not have lipopolysaccharide (LPS), an attribute that allows the host to discriminate this microbe from pathogens. This molecular strategy provides for an effective entry and establishment of the symbiont within the host. Nitrogen cycling is an important activity of sponges where the microbial constituents play a pivotal role. NH_3 oxidation in both aerobic and anaerobic environments in tandem with the reduction of NO_2 and NO_3 mediated by reductases participates in maintaining the global nitrogen level. This denitrification process is further aided by the involvement of urease, an enzyme that degrades urea, a metabolite abundant in oceans into CO_2 and NH_3. While the latter can be utilized as a source of nitrogen nutrient or denitrified, the former can be fixed by photosynthetic microbes.

Sulphur-oxidizing bacteria are a common occupant of this holobiont as they contribute to the oxidation of reduced sulphur moieties generated by the sulphur-reducing microbes. The occurrence of polyphosphate granules within sponges is a tell-tale sign of microbial activity orchestrating the synthesis of this energy-rich compound that can also be part of the metal-processing arsenal these organisms have at their disposal. In fact, polyphosphate accounts for 25–40% of the phosphate concentration in the holobiont and may enable the survival of the organism during phosphate-deprived conditions in a manner analogous to the utilization of glycogen, a polymer of glucose in humans. When glucose is low, humans are known to activate glycogen phosphorylase via a glucagon-mediated route. The microbes associated with this holobiont are also engaged in synthesizing vitamins and amino acids. Vitamin B_{12} is a prominent cofactor elaborated by the microorganisms as it is involved in metabolic activities including the Wood–Ljungdahl pathway, a metabolic network dedicated to the conversion of CO_2 into acetyl-CoA. In fact, some sponges have only catabolic enzymes mediating the degradation of vitamins and amino acids while their biosynthesis is essentially taken over by the microbiome. This arrangement vividly illustrates why sponges are a meta-organism and cannot survive without their partners as some of the critical tasks dedicated to upholding the sponge lifestyle are implemented by other discrete entities working in harmony with the rest of the cellular community.

Sponges are known to be a treasure trove of a plethora of bioactive chemicals, a fact that has been actively embraced by pharmaceutical companies. Sponge mining in an effort to uncover unique chemicals with antibiotic, anti-cancer or anti-obesity properties is an ongoing medical

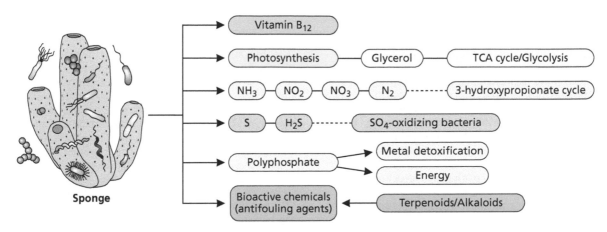

Figure 4.11. Interdependence of sponges and their microbes. The microbes within sponges produce vitamins, supply nutrients and contribute to the nitrogen cycle. The polyphosphate that acts as an energy reserve and metal-detoxifying agent is elaborated with the assistance of the resident microbes. The antifouling compounds, antibiotics, silica and calcium deposits aimed at dissuading predation are produced in collaboration with microbial partners. The photosynthetic activity mediated by the photosynthetic bacteria are well integrated into the holobiont and enables the host to acquire organic carbon. (TCA: tricarboxylic acid.)

pursuit. However, it is important to note that a significant portion of the biological compounds is elaborated by the microbial community made up of fungi, algae and bacteria that the host houses. Predation is a major problem that sessile organisms like sponges have to face. Their inability to move compels them to arm themselves with a rich chemical arsenal to defend against fishes and other marine predators keen on sponge food. Toxins like terpenoids and alkaloids are known to be a potent deterrent against grazing by marine organisms. The presence of calcareous spherules and siliceous spicules elaborated by sponge-dwelling bacteria add to the defence and protect sponges from being utilized as food by predators. Space occupied by sponges can be subject to invasion by other organisms. To guard from such territorial poachers, the microbial community within sponges secrete antifouling agents. These chemicals comprising various halogenated moieties and cyclic peptides are uniquely conceived by the microbes to fulfil this duty and maintain the structural integrity of the host. The inability of human body to inhibit the 'biofouling' of its cellular space results in the growth of cancers (**Figure 4.11**).

4.11 Microbes at the Rescue of Sponges Living in Disparate Habitats

In an effort to adapt to diverse geographical regions where the nature and availability of foods vary, sponges have resorted to hosting disparate microbial associates aimed at fulfilling their nutritional needs. In shallow water where light is plentiful and in the ocean deep where CO_2 is abundant, sponges are associated with higher amounts of photosymbionts like *Synechococcus*. Here, the biotransformation of light energy into fixed organic carbon is maximized. However, in deep seas usually more than 8,000 m, where sunlight is very scant, sponges have to adopt an entirely different lifestyle tailored to the extreme environment. In this habitat they become carnivorous. The sponges are armed with a less aquiferous system and are equipped with anatomical and morphological features dedicated to trapping and engulfing prey in the surroundings. They tend to have an adhesive surface where the prey can be immobilized before being ingested. The symbiotic microbes that aid in trapping and digestive duties are compartmentalized in the root and the digestive section of the host's distinctive body parts. These carnivorous organisms are rich in *Flavobacteria* followed by *Proteobacteria* (γ, α). They possess genera like *Nitrospira*, *Colvellia* and *Reichenbachiella*; many of these species are

BOX 4.4 NUTRIENT CYCLING, MICROBES, SPONGES AND HEALTH OF THE ECOSYSTEM

Sponges and their resident microbes are essential for the well-being of the ecosystem as they contribute to cycling of numerous nutrients including C, N, P and Si. These filter feeders process large volumes of water daily usually in excess of 10,000 times their size to acquire their nutrients. This activity concentrates some of the sparsely found nutrients in the oceans that are then readily utilized by organisms. The cellular materials are consumed by detritivores while Si becomes an important source of nourishment for diatoms where this mineral serves as an important structural support. These diatoms are known to be a major CO_2 sink and modulators of climatic change. In fact, sponges are pivotal in the accumulation and cycling of Si, an element that aids the proliferation of numerous marine organisms dedicated to the capture of CO_2. Following the death of the diatoms, they sink on the ocean floor with their trapped CO_2, thus diminishing the atmospheric content of this greenhouse gas. Nitrogen is also an essential ingredient required for cellular growth. Sponges not only produce NH_3 that can be incorporated by other organisms, but they also harbour microbes with the ability to fix N_2 and transform it into

NH_3, NO_3 and NO_2 moieties mediating the global N cycle. *Crenarchaeota* like *Nitrosopumilus maritimus* is endowed with ammonia oxidase mediating the aerobic conversion of NH_3 into N derivatives. The ether-linked lipids of these microbes impermeable to ions provide an effective environment for this oxidative process. The endosymbionts localized in sponges contribute to the accumulation of polyphosphate that are stored as granules. The concentration of this anion that occurs in very small amounts in sea water helps trigger the proliferation of a diverse ecosystem as all organisms are reliant on this compound for energy and DNA production. The polyphosphate kinase genes found in *Synechococcus* spp. and *Rhodopirellula* spp. are the main architects dedicated to the synthesis of these phosphoanhydride bonds that form the backbone of this phosphate reserve. Hence, organisms vital in maintaining the global climate are nourished due to their ability to concentrate nutrients and they are aided in this task by the microbial partners.

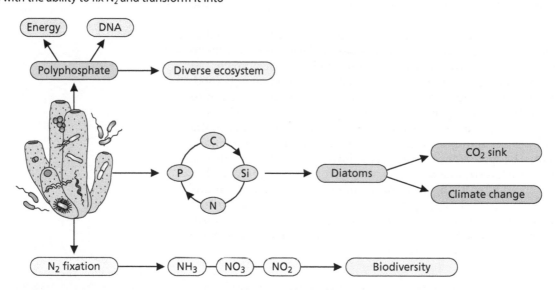

Box Figure 4. Climate change, sponge microbiome and essential nutrients for diatoms.

involved in the polysaccharide-degrading activity as they possess chitinases and other glycosidases. These microbial communities are transmitted by vertical transmission as the symbionts are found even at the larval stage.

In marine areas such as in the Gulf of Mexico that is characterized by asphalt flow, oil seepage and gas venting, sponges are known to utilize these reduced carbon sources to fulfil their energy requirement. They harbour methanogenic bacteria that facilitate the oxidation of this gaseous hydrocarbon. These microbes are host specific and are found in the embryos. The bulk of the methane-oxidizing bacteria is housed in the gelatinous matrix between the epidermis and gastrodermis. They have a coccoid-like shape and are usually located in close proximity to the flagellated choanocytes that provide a constant supply of methane and oxygen via their water-translocation activity. These methanogens are obligate as they lack the genes that are typical in their free-living counterparts. Hence, they have undergone a gene reduction process in an effort to integrate their metabolism with that of the host, a phenomenon that is evident in their highly expressed methane-metabolizing genes. The

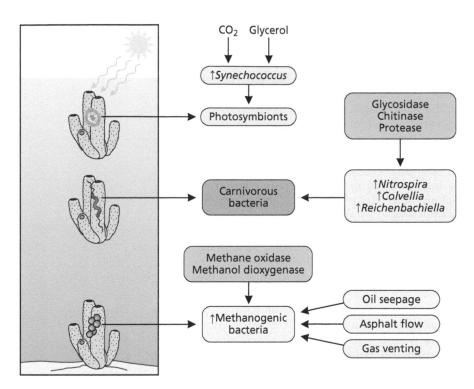

Figure 4.12. Microbes aid sponges colonize diverse habitats. The ability of sponges to live in disparate habitats is dependent on the microbes they house. In regions where sunlight and CO_2 are plentiful, they harbour photosynthetic bacteria that supply them with fixed carbon. In deep oceans where light is minimal, the sponges tend to be carnivorous. In this instance, microbes possessing elevated amounts of proteases, lipases and glycosidases are abundant. At the bottom of oceans where hydrocarbons abound, obligate methanogens are part of the specialized partners sponges accommodate in order to fuel their growth. Undoubtedly, the microbiome extends the range of ecosystems where sponges can survive.

acquisition of these bacteria by the sponges has led to the fine-tuning of their metabolic pathways. While the former is streamlined to possess genes dedicated to engage in methanotrophy in an effort to supply organic nutrients to the host, the latter has evolved its biochemical systems to use these metabolites for anabolic and catabolic needs. The methanogenic activity is propelled by a minimal suite of enzymes made up predominantly of methane oxidase and methanol dehydrogenase that are expressed in abundance. Biochemical effectors involved in cell motility, inorganic ion transport and membrane biogenesis, required for independent living are absent in the methanogenic symbionts. The host, on the other hand, is endowed with a set of genes responsible for the incorporation of the metabolites generated from CH_4 into its lipids and other components. Hence, the methane-derived carbon that the host utilized is elaborated by the bacteria. In these holobionts, methanogenic bacteria are the main supplier of nutrients that fuel the lifestyle of the holoorganisms and allow them to colonize a geographic terrain rich in gaseous hydrocarbon. This again reveals that the association of the host organism with microbes enables the former to adapt to diverse habitats effectively without major anatomical adjustment (**Figure 4.12**).

4.12 The Lifestyle of Hydra: Intimate Connection to Its Microbial Partners

4.12.1 *Hydra: anatomy and physiology*

Hydra is a simple tube-like organism belonging to the phylum *Cnidaria*. Its slender body has a single opening serving as a mouth and an excretory apparatus. This anatomical feature is surrounded by tentacles armed with tiny specialized cells referred to as nematocysts that are utilized to incapacitate tiny organisms serving as food. The gastric tube links the mouth to the gastric cavity where foods are digested. The aboral end consists of the foot pole made up of the peduncle and the basal disc, an arrangement enabling the animal to stick to the surface where it lives in an aquatic environment. The entire body structure of the hydra from the tip of the tentacles to the basal disc is organized as an epithelial bilayer with an intervening extracellular matrix composed of laminin and collagen. The outermost layer known as the epidermis secretes a protective coating termed the periderm, a carbohydrate-rich component that is a repository of microbial residents. The chemical composition of this outer surface is manipulated to select for the desired microbial community and repel any undesired intruders. The endodermis is the innermost layer that lines the gastrodermis and possesses a variety of cells tasked with diverse functions. For instance, when food enters the mouth enzymes are secreted from the specialized cells into the gastric cavity where it is broken down, a process aided by the mixing ability of the flagella located in this region. The fine particles are engulfed by the digestive cells where they are processed and distributed through the organism. The connective tissue region between the epidermis and the gastrodermis is the collagen-rich mesoglea. It extends through the organism except for the mouth and the aboral region and plays an important role in the physical support, mobility, regeneration and reproduction.

4.12.2 *Hydra and its specialized cells*

Hydra has multiple cell types that are representative of the building blocks of all multicellular organisms and are involved in a variety of functions. The interstitial cells are pluripotent cells like stem cells capable of proliferation, self-renewal and differentiation. They can differentiate into nematocytes, nerve cells, germ-line cells (sperm and egg), gland cells and mucous cells. Although the differentiation process is initiated in the column of the body, at late stages these differentiated cells migrate towards the mouth and the basal disc depending on their biological functions. Hydra has a diffuse nerve arrangement extending through the body with significant concentration in the head and foot regions. It is composed of approximately 6,000 nerve cells consisting of the sensory nerve cells and the ganglion cells. These cells are arranged as net-like assemblage intercalated into the ectodermal and endodermal epithelial layers. They respond to mechanical, light and chemical stimuli. They coordinate feeding behaviour, control morphogenesis and modulate locomotion activity. The ability of these cells to synthesize neuropeptides helps shape the microbial community residing within the holobiont.

The hydra does not have an adaptive immune system but possesses an innate immune defence system that enables the organism to detect, intercept and phagocytise microbial invaders. This surveillance system is finely calibrated with a range of molecular constituents aimed at recognizing and promoting the establishment of microbial symbionts. Antimicrobial peptides (AMPs), protease inhibitors, pattern recognition receptors, mucosal guards and microbe-associated molecular patterns all contribute to maintaining microbial homeostasis dedicated to the normal functioning of the holoorganism. The reproductive process takes place either via the asexual or sexual route. As hydra can readily regenerate, cells divide on an ongoing basis. In order to manage its shape and size, these cells need to be eliminated especially when food is plentiful. They are packaged as buds on the body wall where they develop as miniature adults and break off. Sexual reproduction is initiated with the differentiation of the interstitial cells into sperm and egg cells. The former occurs towards the upper part of the body while the latter is localized nearer to the aboral region. Following the release of the sperm

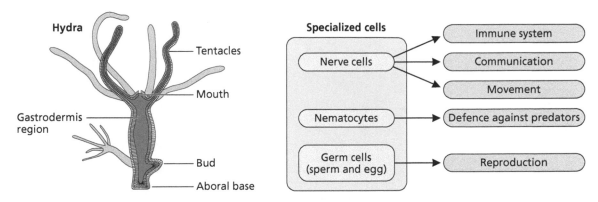

Figure 4.13. The anatomy and physiology of the hydra: functions of specialized cells. The hydra contains specialized cells, all contributing to specific functions the animal needs to survive. The nematocytes defend against predators while the nerve cells are involved in the immune response, communication and movement. Sexual reproduction is mediated by germ cells. Hydra possesses an aboral base and a mouth to intake nutrients and expel wastes.

in the water, it can fertilize the egg where a zygote is formed within the hydra. The zygote forms a thick protective coating that drops on the sediments from where a hydra emerges depending on a variety of environmental factors (**Figure 4.13**).

4.13 The Microbial Community and Its Contribution to the Holobiont

4.13.1 Shaping of the microbial population

The microbial colonization of the hydra is not only actively calibrated by the immune system and the nerve cells but is also dependent on the genetic make-up of the host and on the inter-microbial communication. The glycocalyx, a glycoprotein and glycolipid-rich lining of the ectoderm, is an important protective barrier dedicated to select symbiotic microbes and thwart the invasion by pathogens. The nature of this carbohydrate enables the selection of specific microbes as only microorganisms capable of utilizing this glycan polymer are able to access the ectoderm. These bacteria tend to have complex carbohydrate-degrading enzymes and saccharide transporters. The innate immune network with a range of effectors like AMPs, pattern recognition molecules and lectin-carbohydrate selectivity module adds another layer of control on the microbial residents. The hydra has also the ability to modify the quorum-sensing signals that play a critical role in assembling the appropriate microbial community. Lactonases, oxido-reductases and acylases are involved in enhancing or mitigating the AHL communicating moieties. These enzymes also interfere with the signalling mechanisms of competing pathogenic and opportunistic bacteria. The chemical modification of AHL can help switch the phenotype of *Curvibacter* spp., a prominent microbe in hydra. The microbes housed in the outer layer contribute to this defence strategy. Elimination of these microbes results in fungal infection.

The selection and establishment of the microbiome are also aided by the nerve cells known to produce antibacterial factors aimed at favouring a specific group of microbial residents. These biomolecules (peptides) secreted by these specialized cells modulate the immune system so that the desired microbes can escape its phagocytotic activity. This strategy favours the concentration of *Curvibacter* spp. predominantly around the hypostome and the aboral region. Removal of the nerve net leads to a decrease in *Proteobacteria* and an increase in *Bacteroidetes*, followed by an augmentation of antimicrobial activity, factors not conducive to the lifestyle of the hydra and exposing the holoorganism to mortal danger. Hence, the

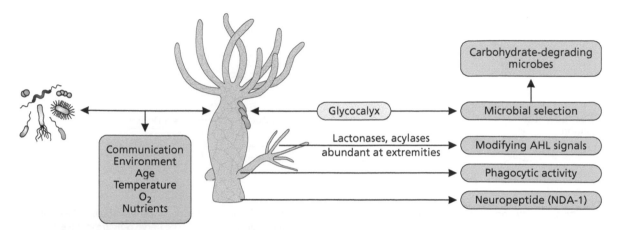

Figure 4.14. Factors modulating microbial colonization of hydra. The assembly of the microbial community is dependent on numerous factors including temperature, pH, O₂, nutrients and communication between the host and the microbes. The neuropeptide (NDA-1), a cysteine-rich antibiotic, helps in the spatial distribution of *Bacillus* spp. and *Pseudomonas* spp. The presence of glycocalyx on the epidermis selects for microbes with the ability to degrade carbohydrates while the modulation of AHL (acyl homoserinelactones) controls the microbial population. Phagocytic activity of the host is also an important selector of the microbiome.

host-associated microbial community is in constant communication with the host nervous system in order to maintain a homeostatic condition, essential for an effective biological activity. The neuropeptide NDA-1, a biomolecule with a conserved 10-cysteine residue synthesized by the sensory and ganglion cells, is abundant at the extremities and less common along the body column. This antibiotic that is more toxic to *Bacillus* spp. than *Pseudomonas* spp. helps orchestrate the spatial distribution of these microbes. The microbial community is dynamic and varies as the organism develops. It attains a stable organization during adulthood. The adult has preponderance of *Curvibacter* spp. and *Duganella* spp. (*β-Preoteobacteria*), while the hatchlings tend to have a higher amount of *Bacteroidetes* (**Figure 4.14**).

4.13.1.1 Microbial input and lifestyle of the hydra

This cross-talk between the nerve cells, the immune components and the bacterial cells clearly illustrates the importance of the microbial community for proper functioning of the host. In fact, the evolution of the hydra has been prompted by microbial input. Hence, the hydra is a holoorganism comprising different organisms working for the upkeep of the whole community. It is clear that the resident bacteria contribute to numerous pivotal functions as if they are specialized tissues assigned to specific tasks but with their own genetic information. They play an important role in mobility, reproduction and defence mechanisms. Mobility is critical for respiration, digestion, morphogenesis, somersaulting and reproduction. The microbiome modulates body contractions, activities pivotal to the host. Spontaneous contractions triggered by pacer cells are common in organs such as the heart and the digestive tract. For instance, during digestion in humans, the movement of the food once it leaves the mouth starting from the oesophagus till the excretion process is propelled by rhythmic electrical pulses generated without any external stimuli. This motion spreads throughout the gastrointestinal system despite no direct input from the nervous system. The nerve cells are involved in the motion generated in the hydra. The removal of the nerve cells abolishes this movement. The hydra devoid of bacteria is severely handicapped in its mobility and spontaneous contractions are altered. Microbes play a role in aiding the pacemaker directly or helping transmit this information to the epithelial and muscular cells. When the germ-free hydra is recolonized by its microbiome comprising *Clavibacter, Duganella, Undibacterium, Acidovorax* and *Pelomonas* genera, contractions are re-established. Hence,

BOX 4.5 BODY MOVEMENT IN HYDRA AND MICROBIOME

Spontaneous involuntary contractile movement in the gastrointestinal tract is facilitated by pacemaker cells known as the interstitial cells of Cajal. These cells generate slow electrical waves and trigger Ca^{2+}-induced contraction that is modulated by the nervous system and the host microbiome. These movements are severely curtailed during microbial imbalance in the digestive canal. Hydra also possesses the ability to undergo spontaneous body contraction, an event mediated by symbiotic bacteria. Gnotobiotic hydra displays reduced and abnormal contractions. These conditions can be readily restored by the inoculation with native microbes. Essential functions such as morphogenesis, regeneration, osmoregulation, respiration and mobility are dependent on the ability of the hydra to undergo these organized contractions. Bacteria lodged within the epithelial cells are pivotal to this contractile activity. The metabolites they produce appear to initiate these movements and control both their frequency and regularity. Hence, the microbial population the hydra possesses plays a critical role in the movement within the host, an attribute that fulfils numerous biological functions. The holobiont needs all its members to be viable (in Box Figure Contraction/Movement in Hydra and Microbiome).

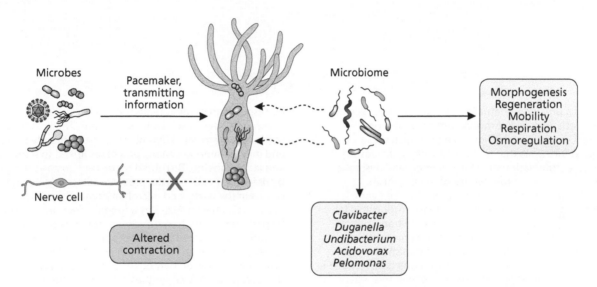

Box Figure 5. **Microbial input in the movement of hydra, an activity vital for numerous biological activities.**

the microbes are part of the hydra inner workings and may facilitate the relaying of the signals of the nerve net to the effector cells that respond to these cues.

Cellular differentiation in the formation of gamete synthesizing protrusions depends on such molecular movements. There tends to be an increase of *Pseudomonadaceae* in budding polyps compared to the non-budding ones where an augmentation in *Rhizobiales* is seen. During embryogenesis, a reconfiguration of the microbial population is observed. The ability of these microbes to express such enzymes as lipoxygenases may have a role in morphogenesis. Lipoxygenases are known to generate morphogenic metabolites from arachidonic acid that contribute to this process. The transcription factor forkhead box O (FoxO) is associated with longevity owing to its ability to orchestrate stem cell differentiation and expression of antimicrobial proteins. These biomolecules are involved in shaping the microbiome of the hydra. Abnormal FoxO signalling results in defective AMP production, an event that results in dysbiosis and leads to the invasion by pathogens. This change in microbiome is associated with the inability of the holobiont to renew its cells and contributes to the demise of the hydra. Thus, a stable microbial population promoted by FoxO is key to the renewal of the host and may contribute to its longevity (**Figure 4.15**).

4.14 Conclusions

Like most animals, corals, tunicates, sponges and hydras need their microbes. They have evolved with their microbial constituents in such a

Hydra and the Microbiome

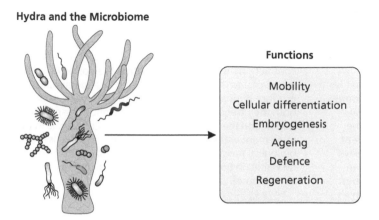

Functions

Mobility

Cellular differentiation

Embryogenesis

Ageing

Defence

Regeneration

Figure 4.15. Functions of the microbiome in hydra. The microbes residing within the hydra are involved in mobility, embryogenesis, ageing, regeneration and defence. Lipoxygenases expressed by the microbes produce metabolites critical during morphogenesis.

BOX 4.6 MICROBIOME AND THE AGELESS HYDRA

Ageing is a process punctuated by a decrease in cellular homeostasis, immune response, antioxidants and microbial diversity in the host. The decline in these biological functions that is coupled with an increase in oxidants, accumulation of cellular defects and dysbiosis usually culminates in the demise of the organism. Molecular markers like apolipoprotein E tasked with lipid homeostasis, cytochrome C oxidase responsible for the aerobic production of ATP and FoxO, a transcription factor linked to cellular proliferation are amongst some of the biomolecules that are sharply diminished as age progresses. The non-senescent polyp Hydra is not known to undergo any ageing process. The lifespan of this organism can be as long as 1,400 years. It has three immortal stem cells with the responsibility to regenerate twenty cell types. This organism continuously reproduces asexually by budding with the assistance of these stem cells and is known to express a homolog of the foxo gene. The product of this gene, the FoxO transcription factor and its

corresponding binding regions in various genes are known to markedly decrease in ageing humans. The FoxO protein modulates stem cell differentiation, tissue maintenance and innate immune system, processes linked to longevity and lack of ageing. Disrupted tissue homeostasis triggered by defective FoxO in hydra provokes altered AMPs and results in the perturbed microbial population. This dysbiosis is characterized by a fungal infection, invasion by pathogens and dysregulation of the microbiome. Such a condition triggered by FoxO that provokes a decline in survival and an increase in death as the host is unable to distinguish between pathogens and commensals due to the disruption in AMP synthesis. Microbial transplantation with native microbes can remedy this situation. Hence, the ability of FoxO to maintain tissue homeostasis that then generates the proper AMPs leads to a viable holoorganism with its appropriate microbes that are crucial for survival and longevity. Hence, the longevity in hydra is linked to its microbiome via the FoxO protein.

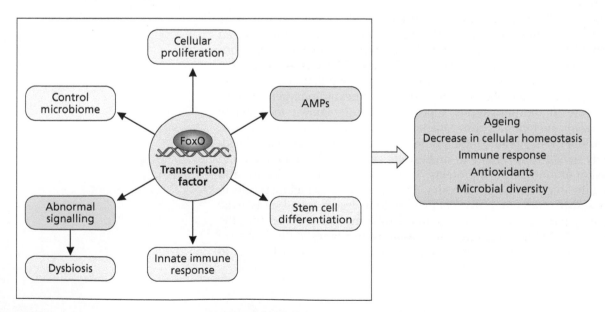

Box Figure 6. **FoxO protein regulates the microbial population and numerous other functions in hydra.**

fashion that numerous pivotal functions are accomplished with the assistance of their microbes. This arrangement requires the acceptance of the microbial residents as endosymbionts or/and as exosymbionts. They perform key tasks that obviate the need for specialized cells, tissues or organs and extend the adaptive ability of the host. The same host can acquire different microbes to colonize or survive in a disparate environment. The occurrence of sponges in such diverse ecosystems as ocean oil vent and in sunlight-rich habitats is prompted by genetic information in the microbes capable to metabolize hydrocarbon and produce photo-protective molecules. The nutritional diversity is aided by microbes with enzymes to readily degrade such polymers as chitin. Hence, it stands to reason that these aquatic animals with limited locomotion can settle in any ecological niche they find themselves in only due to microbes they harbour. Devoid of their microbes, these organisms would survive or would exist in forms that have yet to be elaborated by nature. Hence, microbes are integral components of these aquatic holobionts.

SUGGESTED READINGS

Augustin, R. et al., (2017). A secreted antibacterial neuropeptide shapes the microbiome of Hydra. *Nature Communications, 8*(1), 698–707. doi:10.1038/s41467-017-00625-1

Beyersmann, P. G. et al., (2017). Dual function of tropodithietic acid as antibiotic and signaling molecule in global gene regulation of the probiotic bacterium *Phaeobacter inhibens. Scientific Reports, 7*(1), 730–739. doi:10.1038/s41598-017-00784-7

Cheng, L. K. et al., (2010). Gastrointestinal system. *Wiley Interdisciplinary Reviews: Systems Biology and Medicine, 2*(1), 65–79. doi:10.1002/wsbm.19

Davy, S. K., et al., (2012). Cell biology of cnidarian-dinoflagellate symbiosis. *Microbiology and Molecular Biology Reviews, 76*(2), 229–261. doi:10.1128/mmbr.05014–11

Gao, Z. et al., (2014). Symbiotic Adaptation Drives Genome Streamlining of the Cyanobacterial Sponge Symbiont. *Candidatus Synechococcus spongiarum. MBio, 5*(2)., 79–90. doi:10.1128/mbio.00079–14

Glauber, K. M. et al., (2015). A small molecule screen identifies a novel compound that induces a homeotic transformation in Hydra. *Development, 142*(11), 2081-2081. doi:10.1242/dev.126235

Hernandez-Agreda, A., et al., (2017). Defining the core microbiome in corals' microbial soup. *Trends in Microbiology, 25*(2), 125–140. doi:10.1016/j.tim.2016.11.003

Hernandez-Agreda, A., et al., (2018). Rethinking the coral microbiome: simplicity exists within a diverse microbial biosphere. *MBio, 9*(5)., 1–14. doi:10.1128/mbio.00812-18.

Hester, E. R., et al., (2015). Stable and sporadic symbiotic communities of coral and algal holobionts. *The ISME Journal, 10*(5), 1157–1169. doi:10.1038/ismej.2015.190

Hwang, J. S. et al., (2007). The evolutionary emergence of cell type-specific genes inferred from the gene expression analysis of Hydra. *Proceedings of the National Academy of Sciences, 104*(37), 14735–14740. doi:10.1073/pnas.0703331104

Kaur, H., & Greger, M. (2019). A review on Si uptake and transport system. *Plants, 8*(4), 81. doi:10.3390/plants8040081

Maldonado, M. et al., (2020). Cooperation between passive and active silicon transporters clarifies the ecophysiology and evolution of biosilicification in sponges. *Science Advances, 6*(28), eaba9322. doi:10.1126/sciadv.aba9322

Mortzfeld, B. M., et al, (2018). Stem cell transcription factor FoxO controls microbiome resilience in hydra. *Frontiers in Microbiology, 9*, 629–639. doi:10.3389/fmicb.2018.00629

Moya, A et al., (2008). Carbonic anhydrase in the scleractinian coral *Stylophora pistillata. Journal of Biological Chemistry, 283*(37), 25475–25484. doi:10.1074/jbc.m804726200

Muller, E. M., Fine, M., & Ritchie, K. B. (2016). The stable microbiome of inter and sub-tidal anemone species under increasing pCO_2, 1–11. *Scientific Reports, 6*(1). doi:10.1038/srep37387

Murillo-Rincon, A. et al., (2017). Spontaneous body contractions are modulated by the microbiome of Hydra. *Scientific Reports, 7*(1), 1–9. doi:10.1038/s41598-017-16191-x

Obata, Y., & Pachnis, V. (2020). Linking neurons to immunity: Lessons from Hydra. *Proceedings of the National Academy of Sciences, 117*(33), 19624–19626. doi:10.1073/pnas.2011637117

Peixoto, R. S., et al., (2017). Beneficial microorganisms for corals (BMC): Proposed mechanisms for coral health and resilience. *Frontiers in Microbiology, 8*, 341–357. doi:10.3389/fmicb.2017.00341

Thomas, T. R., et al., (2010). Marine drugs from sponge-microbe association—a review. *Marine Drugs, 8*(4), 1417–1468. doi:10.3390/md8041417

Ueki, T., et al., (2019). Symbiotic bacteria associated with ascidian vanadium accumulation identified by 16S rRNA amplicon sequencing. *Marine Genomics, 43*, 33–42. doi:10.1016/j.margen.2018.10.006

Wang, L. et al., (2018). Corals and their microbiomes are differentially affected by exposure to elevated nutrients and a natural thermal anomaly. *Frontiers in Marine Science, 5*, 101–117 doi:10.3389/fmars.2018.00101

Zhang, F., et al., (2019). Microbially mediated nutrient cycles in marine sponges. *FEMS Microbiology Ecology, 95*(11), 155–169. doi:10.1093/femsec/fiz155

INSECT MICROBIOME: MICROBES INSECTS DEPEND ON

5

Contents

Keywords

- Insects
- Microbiome
- Bacteriocytes
- Sexual Manipulation
- Signalling
- Social Behaviour

5.1 Introduction

5.1.1 Insect: anatomy and physiology

Insects belong to the phylum *Arthropoda*, a term derived from Greek, 'Arthro' meaning jointed and 'Poda' meaning foot. This phylum comprises animals with joint feet that live in both terrestrial and aquatic environments. They are distributed in a wide geographical area ranging from arid to arctic regions and are classified depending essentially on body segmentation, appendages, respiratory organs and modes of excretion. The lobster belonging to the subphylum *Crustacea* is mostly aquatic and resorts to O_2 intake through the gills, the millipede representing the *Myriapoda* grouping is land bound and is characterized by a large number of appendages.

Insects, the most diverse and abundant animals on the planet belong to the *Hexapoda* subphylum and are characterized by a body consisting of a head, thorax and abdomen. The body is covered with an exoskeleton known as the cuticle that is made up of the epicuticle, a waxy water-resistant component underneath which lies the chitin-rich procuticle. While the latter contains the polymer constituted of *N*-acetyl glucosamine usually encrusted with calcium carbonate and protein, the major component of the former is lipids or their derivatives. The head possesses one pair of large compound eyes and numerous flexible appendages, referred to as antennae. These anatomical structures enable the organism to sense its

DOI: 10.1201/9781003166481-5

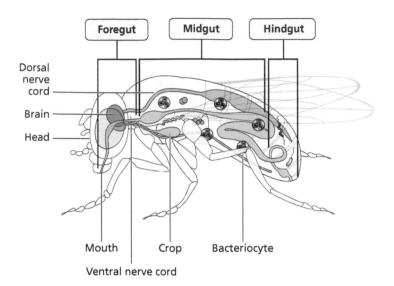

Foregut   Midgut   Hindgut

Dorsal
nerve
cord

Brain

Head

Mouth   Crop   Bacteriocyte

Ventral nerve cord

Figure 5.1. Anatomical features and physiological functions. The major components consisting of the brain and the digestive tract are involved in communication and digestion. The nature of the mouth tends to vary with the nutritional habit of the insects. The pair of wings helps in locomotion while the bacteriocytes lodge microbes essential for numerous functions.

environment. The mouth consists of the maxilla, labium and mandibles, an anatomical arrangement designed to fulfil the organism's feeding need. These are structurally diverse depending on the nutritional habit of the insect. Hence, insects acquiring their nutrients through herbivory, omnivory, carnivory or hemovory tend to have mouth parts dedicated to these nutritional habits. The thorax is composed of three segments, the prothorax, mesothorax and metathorax with the former being the closest to the head. This external body part is home to a nervous system consisting of a brain and a ventral nerve cord. Most insects tend to have wings and legs located in the thoracic region. The abdomen typically contains 12 segments with the terminalia housing the anal–genital organ.

The digestive tract designed to extract nutrients that are consumed is also lodged within. This is where digestion occurs and includes the foregut, the midgut and the hindgut. The O_2 need of insects is fulfilled by internal tubes and air sacs through which gases can diffuse readily. The entry of air is facilitated by the presence of a pore-like structure known as a spiracle. The O_2 is taken in and CO_2 is removed in a continuous cycle. The open circulatory system enables the haemolymph (blood) to flow freely through the cavities where direct contact with the tissues and organs is made. This organization provides a transport network for the organs to communicate, nutrients to flow and wastes to be removed. Hence, the open movement of the haemolymph plays a pivotal role in osmoregulation, storage, skeletal function and the immune process. Females have ovaries to lay eggs, while males produce sperm with the aid of their testes. Most insects are oviparous, and the eggs subsequently go through a series of morphological changes known as metamorphosis. The final stage that culminates in the development of an adult is preceded by the larval and pupal phases. These anatomical features and biological functions they accomplish are to a large extent modulated by the microbial partners residing on and within the insects (**Figure 5.1**).

5.1.2 Ecological and economic significance
Insects are an important component of the ecosystem and contribute significantly to the biodiversity of planet Earth. Their interaction with the fauna and flora is pivotal for the natural renewal process. They represent the majority of the soil fauna where they contribute to the transformation of the soil. These organisms are soil engineers busy with fragmentation, comminution and humidification, activities essential for the fertility of the soil. They also excrete wastes and build subterranean tunnels, undertakings that further help enrich and aerate the soil. Hence, this insect-driven enrichment of the soil is central to the agricultural output globally. Insects also perform a crucial function in the pollination of the flora, an event on which food production relies. Fruits,

BOX 5.1 HISTORY OF THE CARMINE DYE

The carmine dye was discovered by the Aztecs, and it was utilized as early as the 10th century. It was later commercialized upon the colonization of South America by the Spanish. The cochineals growing on pear cacti are the source of this compound. The wingless female insects accumulate the chemical carminic acid and are easily harvested from the shrubs from where the short-lived males can readily be dislodged. Following the drying of the insects, the dye is processed. It is estimated that nearly 70,000 of these tiny creatures (5 mm long) can yield half a kilogram of dye. In order to maintain the monopoly on this trade, the Spanish did not reveal the origin of this chemical and claimed that it was derived from seeds of some plants. During its peak production, this chemical was the second most prized export from South American colonies. It was not almost 200 years later that identity of the dye and its insect source were uncovered following the advent of the microscope. Although synthetic red dyes were able to become a cheap substitute, the negative health impact of these chemicals has provided a boost to

the natural red pigment. Today, it is mainly produced in Peru and generates an important economic activity in rural areas. The export market is valued at several million dollars. The red colour on ice cream, lipstick and textile is essentially insect-derived carminic acid (**Box Figure 1**).

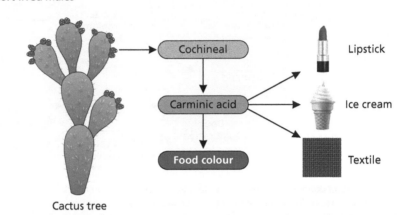

Box Figure 1. History of the carmine dye. A pear cactus tree with the red female cochineals, the source of carminic dye.

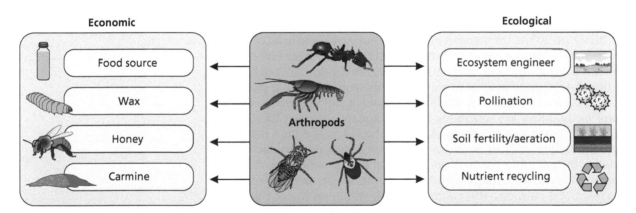

Figure 5.2. Ecological and economic importance of insects. Insects are essential to the survival of planet Earth. They are involved in the recycling of nutrients, contribute to the quality of soils and promote plant life via pollination. They are also important participants in the economy of numerous countries.

flowers and vegetables are dependent on the activity of insect pollinators. For instance, decomposition of dead plants by termites and nutrient recycling by cicadas contribute to the growth of vegetation. In the absence of this essential biological process, ecological disorder will ensue. Anthropods are also an important part of the economy as they are a source of food. Crustaceans are a major component of the aquaculture industry that many countries are reliant on. Other than the well-appreciated products like silk and honey derived from insects, these organisms are a source of bioactive compounds such as antimicrobial peptides (AMPs), proteases and anti-coagulants that are utilized in the health industry (**Figure 5.2**).

5.1.3 Functions of the insect microbiome

Insects like all multicellular organisms will not be able to engage and participate in numerous activities beneficial to the global ecological well-being without the presence of the microbes residing on and within them.

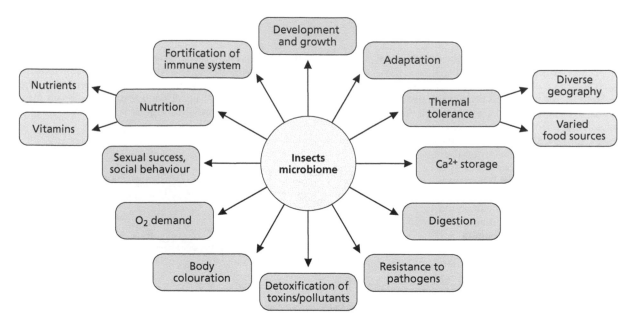

Figure 5.3. Microbial contribution to insect life. Microbes play a major role in the life of insects as they participate in a plethora of functions that the host depends on. The ability of insects to consume a diverse range of foods and populate disparate geographical habitats is dependent on their microbial partners. Social behaviour, sexual success, strengthening the immune system, detoxification of toxins, growth and development are some of the activities insects are reliant on their microbes.

These microbes mostly live in harmony with their hosts and engage in biological processes mutually beneficial or cause no harm. Insects devoid of bacteria are virtually inexistent in nature and are only utilized for laboratory experiments where they have to cope with serious biological defects. These microbes are engaged in nearly all aspects of the insects' life ranging from boosting the immune system to aiding in sexual success. They are a defensive front against diverse environmental challenges and produce a wide array of chemical arsenal to combat toxins secreted by both plants and animals. Social behaviour modulation, moulting and morphogenesis are some of the other key attributes bacteria impart on insect life. The ability of insects to utilize a plethora of food sources like vegetation rich in cellulose and blood with an abundance of heme-containing products is attributed to the microbial partners they harbour. Their prolific presence in most geographical surroundings can be traced to the biochemical versatility of their microbial partners. The communication within an insect colony may also be mediated by the chemical cues originating from the bacterial constituents. Thus, despite the highly diversified organs they possess, insects have evolved in the midst of bacteria and most probably in association with the chemical commands released by the microbes (**Figure 5.3**).

5.2 The Genesis of the Microbiome

5.2.1 Transmission: how insects acquire their microbes

The diverse nature of insects necessitates the presence of a varied microbial population. The acquisition of microbes occurs via vertical or/ and horizontal transmission. The former involves the transfer of microbes from adults to the offspring directly while the latter entails the incorporation of microbes from the environment. The core or primary microbiome usually has a symbiotic relationship with the host, i.e. the microbes are required for survival. The bacterium *Wolbachia* spp. known to be part of 40% of terrestrial anthropods is lodged in a specialized tissue referred to as bacteriocyte. The establishment of this microbial residence within the host necessitates an intricate biochemical conversation between the bacteria and the insect. Coprophagy and trophallaxis, the process of exchanging gut fluids from anus to mouth and mouth to mouth, respectively, may also be operative in the microbial inoculation of the progeny. Maternal smearing of eggs or larvae is another route insects invoke in order to enrich their descendants with a healthy dosage of bacteria. In carpenter ants, *Blochmannia* spp. are vertically transferred following the infection of the ovaries that subsequently impart them to the maturing eggs. *Buchnera* spp., an endosymbiont residing in aphid, is directly transmitted to the eggs, while the vitamin B-supplying microbe of tsetse flies, *Wigglesworthia* spp., is provided in utero to the developing larva via the secretion of the milk gland. Capsule and jelly laden with microbes are deposited on eggs to be consumed by nascent hatchlings, an activity aimed at acquiring select microbial partners. Social interactions these organisms are exposed to in the colony they live in like in a bee hive also provide ample opportunity for the establishment of the microbiome. Worker bees are known to be enriched with such Gram-negative bacteria like *Gilliamella apicola* and *Frischella perra* during social mingling on the surface of the hive.

These well-organized transmission channels are supplemented by dietary intake of microbes found in the different foods insects consume. Thus, insects eating carrion tend to have microbes that are disparate than those partaking in nutrients rich in cellulose or lignin. These stochastic associations can also be dictated by geographical location and by interspecies mingling. The latter is common in bees visiting flowers. For instance, *Riptortus clavatus*, a bug with a characteristic broad head, is infused with the symbiont *Burkholderia* spp. from the soil, while honey bees tend to acquire bacteria present within the flowers they visit. *Lactobacillus* spp. are common flower bacteria that are found in bees. These secondary symbionts are usually located on the surface where they confer biological traits like thermal tolerance and resistance to fungi. Irrespective of the modes of transmission, insects acquire their microbes both the core and non-core types with whom they forge a beneficial partnership critical for their lifestyle (**Figure 5.4**).

5.2.2 Establishment of insect microbial connection: symbiosis

The microbial symbionts housed within insects can be either obligate or facultative. The former usually cannot live independently and has a reduced genome compared to their free-living counterparts. They are harboured in specialized organs like the bacteriocyte, the gut and a variety of other tissues. The host depends on the microbe and numerous physiological activities are performed jointly. The facultative microbes, on

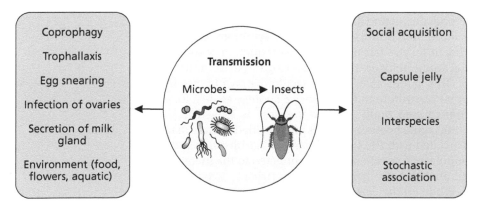

Figure 5.4. Transmission of microbes in insects. Microbes are both vertically, i.e. from parents to offspring, and horizontally, i.e. from the environment. Coprophagy (anus to mouth), trophallaxis (mouth to mouth) and social mingling are some of the routes this microbial transfer occurs. These microbes can also be acquired from the fauna and flora in a specific geographic region.

the other hand, can be accommodated at various locations including the cuticle. They are involved in functions aimed primarily at improving the host's fitness to adapt to a specific environment and contribute to such attributes as thermal tolerance and detoxification of toxins. Hence, it is essential that this symbiotic process is very well-guarded in order to prevent the entry of pathogens or cheaters. A variety of strategies are utilized to accomplish an effective colonization of the hosts by the microbes.

An anatomical feature in the digestive tract termed the proventriculus acts as a micropore filter designed to restrict bacterial movement into the midgut. Only the proper symbionts are admitted. The modulation of cell-wall components such as peptidoglycan and lipopolysaccharide is essential for a successful colonization. The numerous phosphatases that mediate the biosynthesis of cell wall are pivotal as they contribute to the viability of the microbes and provide resistance against lysozyme, osmotic shock and physical resistance. Microbes with flagella have an added advantage as they can travel to the gut to establish a symbiotic partnership. Mutants lacking these motile structures are unable to be symbiotic partners. Insects are also known to modulate their immune system in order to deprive the unwanted microbes to gain access without thwarting the bona fide symbionts. To accomplish this feat, AMPs aimed at the non-symbiotic bacteria are produced. These are toxic to soil-dwelling *Escherichia coli* but not to *Burkholderia* spp. required by the bug *Riptortus*. In the short-honed grasshopper, the cysteine-rich AMPs favour the entry of the desired microbial partner while blocking the entry to adventitious bacteria. Although the insect provides a safe and nutrient-rich environment, the microbes have to be armed with molecular tools that would avert any food constraint. Phasin, a protein associated with polyhydroxyalkanoates, is known to confer resistance to carbon and nutrient depletion. This polymer is an excellent source of carbon and mutant lacking phasin is unable to be involved in this symbiotic relationship. Biofilm formation is another attribute that microbes must have in order to seek a home within insects. This moiety contributes to numerous functions including the initial attachment to the host. Purine biosynthesis is intricately linked to the biosynthesis of biofilms as this metabolic pathway supplies cyclic di-GMP. The latter regulates the production of biofilms.

It is evident that the symbiotic process that seeks to establish a home for the obligate and facultative microbes is fine-tuned by both the host and the microbes. There can be no room for error as it will result in the invasion by pathogens and the demise of the host. This intricate cellular event designed to admit a foreign organism within requires the participation of a series of tightly regulated molecular events as the host has to ensure that pathogens or opportunistic microbes do not find their way into its inner body workings. In pea aphid (*Acyrthosiphon pisum*), the obligate symbiont *Buchnera aphicola* located in bacteriocytes has a 10% genome reduction. It supplies the host with essential amino acids as the latter has to resort to the nutrients in the phloem sap. This aphid also houses facultative symbionts like *Serratia symbiotica* that contribute to the defensive strategy against fungi. *Burkholderia* spp. transmitted orally in *Riptortus* spp. is lodged in the posterior midgut. In the lagnia beetle, microbes are provided a home in the accessory glands associated with the sexual organs. In herbivorous ants, a *Burkholderia* spp. populates a specialized sac connected to the metabolic waste processing tubules and the trachea, organs rich in nitrogen and oxygen respectively. While *Wolbachia*, the sex manipulator, colonizes the germ line tissues in the olive fruit fly, the pharyngeal bulb, located in the head, is inhabited by the microbe *Candidatus Erwinia dacicola*. These tailor-made tissues designed to incorporate these microbes point to an intricate cross-talk between these two participants resulting in a holoorganism (**Figure 5.5**).

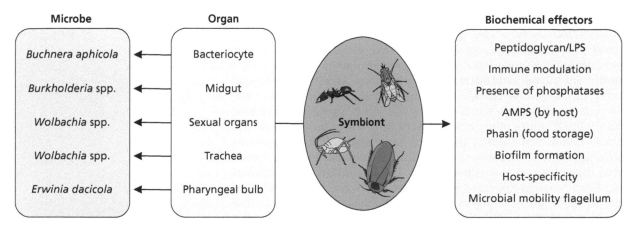

Figure 5.5. Molecular strategies for effective symbiosis and location of symbionts. Microbes that are core to the hosts live in a symbiotic relationship and are acquired in an intricately regulated fashion. This tightly guarded line of communication if decoded results in an invasion by pathogens and an eventual demise of the host. A modulated immune system designed to accommodate the symbionts and fend off the pathogens is crucial. Peptidoglycans and lipopolysaccharides are important as is phasin, a food source microbes depend on to gain access to the host to be lodged in the proper tissue. Biofilm formation and microbial mobility are pivotal to this process.

5.2.3 Insect gut and microbial colonization

Insect is bestowed with a unique gut morphology with its foregut, midgut and hindgut that lends itself to be an ideal location for the microbiome to be housed. It is also replete with food, offers protection against desiccation or ultraviolet radiation and operates at a range of pH and oxygen gradient conductive to bacterial colonization. These conditions are favourable to a wide variety of microbes with diverse nutritional and pH requirements to proliferate. The foregut has a crop region for the temporary transit of food while the hindgut has numerous fermentative chambers before the elimination of the faeces. The bulk of the digestion and absorption occur in the midgut. This digestive structure is home to the majority of the microbial population that ranges from 10^6 cells in adult grasshopper to 10^9 cells in adult honey bee. The highest bacterial biomass to host biomass is observed in wood feeders. It is also important to note that insects have evolved an immune system and physical barriers to limit bacterial load. The midgut secretes a network of chitin-infused proteins and carbohydrates, the peritrophic barrier that selectively allows the entry of nutrients, enzymes and defensive molecules but prevents the direct contact of the microorganisms with the epithelial cell.

The hindgut contains the most bacteria while the midgut is associated with a least amount of microbes. Enzymes such as lysozyme, NADH oxidase and a pH of lower than 3 and in the alkaline range in some regions tend to limit microbial population in the gut. The dominant phyla are *Actinobacteria*, *Bacteroidetes*, *Firmicutes*, *Actinomycetes*, *Spirochetes*, *Verrucomicrobia* and *Proteobacteria*. However, some microbes are common among insects like drosophilae, honey bees and leafhoppers as they have plant-derived diets. The species of Acetobacteraceae are important constituents of these insects. Methanogenic bacteria are typically part of the termite and cockroach microbiome.

The diversity in the phylum, family, genera, strain and species provides the microbial signature that different insects need in order to go through the life cycle in the habitat they operate in and to undergo the developmental stages they are participating in. The solitary lifestyle of the locus is characterized by three main genera *Micrococcus*, *Acinetobacter* and *Staphylococcus* lodged predominantly along the reproductive tract while the onset of the gregarious way of life is punctuated by the *Brevibacterium* bloom, an event common during locust swarm. Praying mantis are solitary predators depending on a diet of smaller insects. Their gut is rich in *Firmicutes* the majority of which consists of the family of *Streptococcaceae* and *Lactobacillaceae*.

In honey bee, the microbiome is dependent on the caste of the colony and the developmental stage. The adult worker has three main phyla (*Firmicutes*, *Proteobacteria* and *Actinobacteria*), while the queen bee is predominantly colonized with *Proteobacteria*. The larval queen has high amounts of enteric bacteria such as *Escherichia* spp. and *Gilliamella* spp. compared to the worker larva, a situation that may be reflective of the protein-rich royal jelly diet it is fed. As the adult queen matures and encounters a different colony environment, it acquires a queen-specific microbial signature including the acetic acid metabolizing bacterium '*Candidatus Parasaccharibacter apium*'. This microbe prefers a microoxic milieu and produces metabolites such as lactate and acetoin. Termites have an abundance of microbes belonging to the signature *Spirochaetes* phylum in their guts in stark difference to their close relative the cockroaches that are populated with the phylum *Firmicutes* with the predominance of the Ruminococcaceae family. In mosquitoes, the aquatic environment where the larval and early adult life begins is an important reservoir of microbes. This aquatic habitat that may be rich or poor in nutrients is known to promote the colonization of different bacteria. While the former promotes the establishment of *Clostridiales*, the latter favours the assembly of *Burkholderia*. The geographical location can also mediate the specificity of microbiome that insects possess. The mosquito *Aedes aegypti* of Brazilian origin is characterized by *Pseudomonas* spp., *Acinetobacter* spp. and *Aeromonas* spp. The counterpart from Burkina Faso has the microbial fingerprint comprising *Wolbachia* spp. and *Acinetobacter* spp. The diversity of the microbiome in insects like in most multicellular organisms is a product of numerous factors including, the host, the microbe, the habitat, geography, nutrient, developmental stage, pathogens and predators. These forces conjugate to form the proper holobiont (insect and the microbiome) that becomes part of the biodiversity it operates in (**Figure 5.6**).

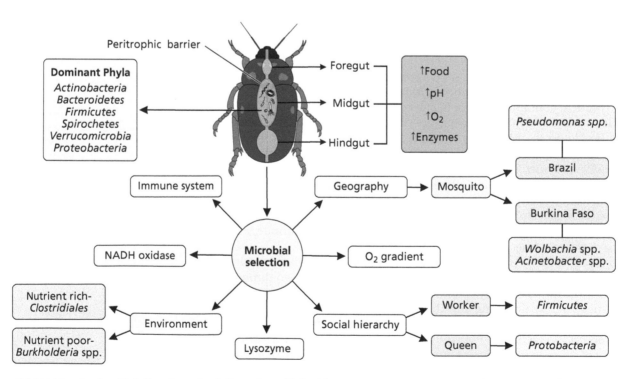

Figure 5.6. Insect gut, microbial diversity, selectivity and specificity. The insect gut harbours the most diverse groups of bacteria. The pH range, the O_2 gradient and the availability of nutrients and the host are the main factors dictating the nature of the microbial population. Microbes are most abundant in the hindgut. Geography, social hierarchy, nutritional requirements and the environment also help fine-tune the specificity of the microbiome insects live with.

5.3 The Tasks Microbes Perform

5.3.1 *Extracting nutrients from disparate food sources*

Microbes perform a variety of functions that is pivotal for the survival of insects. They provide the flexibility that enables insects to live in a wide geographical area and colonize virtually any ecosystem. Acquiring nutrients to fuel daily biological activities is an important task undertaken by the digestive tract, an organ modified anatomical (with microbial input) to suit the diet insects are adapted to. However, these morphological modifications would be redundant if the enzymes needed to hydrolyse the disparate food sources insects are known to consume were not available. In fact, many of these molecular scissors are of microbial origin. The foregut, the midgut and the hindgut are designed to accommodate these bacteria with their large assortment of enzymes. The rich protein diet of carnivorous insects, the complex carbohydrate food sources of termites or the blood-containing meal of the hematophagous insects can only be processed due to the specific microbes residing in the gut.

Honey bee is dependent on the pollen and nectar it acquires from flowering plants. In a highly structured social colony, the worker bee forages for these foods containing mostly carbohydrates and proteins that are processed by such resident microbes as *G. apicola*, *Snodagrasella* spp. and *Lactobacillus* spp. They possess a range of polysaccharide-degrading enzymes (CAZy-carbohydrate-active enzymes) that help release monosaccharides from recalcitrant pollen walls. Polysaccharide lyases, fructosidases and pectinases are some of the enzymes that help hydrolyse the complex carbohydrates. To facilitate the uptake and metabolism of the hydrolysed products, a range of phosphoenol pyruvate transport systems (PTS) are highly expressed in the microbes. Monosaccharides like mannose and xylose that can be toxic are metabolized with the aid of specific PTS and mannose-6 phosphate isomerase. The latter helps the direct entry of mannose into the glycolytic cycle. Microbes are also involved in the degradation of cellulose and other plant glycosides in herbivorous insects. Enzymes such as cellulases, hemicellulases and xylanases participate in these reactions. Insects relying on plants laden with terpene derivatives tend to house microbes with terpenoid metabolic capacity. Mevalonate and methylerythritol degradative pathways are utilized. Aromatic amino acid synthesis via homogentisate is common as tyrosine is incorporated in the lipid layer of the cuticle. Insects feeding on phloem and xylem depend on microbes to provide biotin and riboflavin. These vitamins are essential in the processing of amino acids and monosaccharides present in the sap. *Sodalis* spp. harboured by spittlebug can readily shuttle xylem-derived glutamine to the tricarboxylic acid (TCA) cycle.

Termites rely on wood as their source of food. This natural polymer is rich in phenol and xylose. Although the host helps in the comminution and pre-treatment of the wood, the resident microbes belonging essentially to the phylum *Spirochaete* provide enzymes such as ligninase, xylanases and exoglucanases that are essential in generating nutrients. The catechol and xylose produced are readily shunted via various metabolic pathways. Methane and H_2-metabolizing bacteria are important components of termites as they help in converting these gases into utilizable acetate. Some termites also rear the basidiomycete fungus (*Termitomyces* spp.) with lignin-degrading capacity. Insects depending on vertebrate blood for their nutritional needs have to depend on their bacterial partners as their protein-rich diet is poor in numerous ingredients such as vitamins. The human ectoparasite, the head louse partners with the microbe *Candidatus Riesia pediculicola* to feed on blood. The microbe is housed in the male midgut and the oviducts of females and is a supplier of pantothenic acid (vitamin B_5), essential in numerous biochemical processes including the synthesis of lipids and the TCA cycle. Although female mosquitoes utilize

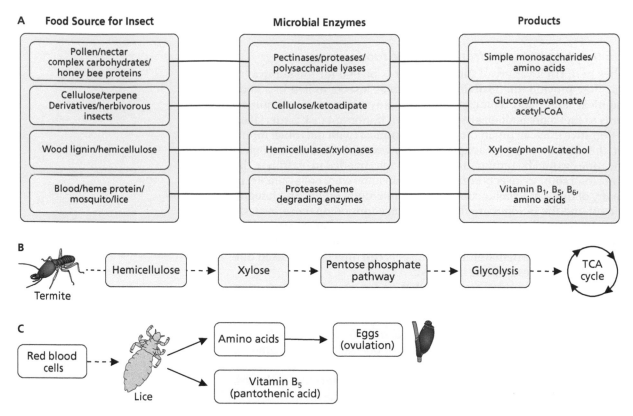

Figure 5.7. Microbes and the diverse nutritional lifestyle of insects. The ability of insects to consume a wide range of foods is dependent on the microbes they house as these microbes possess a plethora of digestive enzymes. (a) Microbial cellulases, pectinases, proteases and lipases enable insects to feed on cellulose, pectin, protein and lipid-containing nutrients. (b) The bacterial residents in termites convert hemicellulose into xylose that can readily be metabolized via the pentose phosphate pathway, the glycolytic cycle and the tricarboxylic acid (TCA) cycle. (c) The red blood cells consumed by lice are degraded into amino acids by microbes that also supply the host with vitamin B_5, an essential nutrient.

blood as the main source of protein for egg production, they depend on their microbes to lyse the red blood cell. This nutritionally restricted diet is also challenging for the tsetse fly, a vector for the African sleeping sickness disease. To overcome this problem, the host has evolved with the integration of the symbiont *Wigglesworthia* spp. in its lifestyle. This obligate symbiont supplies the tsetse fly with vitamins (thiamine B_1, pyridoxine B_6, folate B_9) essential for its survival and proliferation (**Figure 5.7**).

5.3.2 Defensive role of the microbiome

Insects are subjected to numerous environmental changes that they have to respond to. These challenges that may be biotic or abiotic in nature have to be neutralized if these organisms are to survive. Here again, insects invoke the participation of their microbiome to combat parasites, pathogenic microbes, predators, toxins from plants or temperature fluctuation. The resident microbes have a range of strategies including the production of reactive oxygen species (ROS) to the elaboration of proteins aimed at predators. The fruit fly *Drosophila neotestacea* can become easy prey to the infection of parasitic nematodes, an event associated with the sterility of the female insect. To mitigate such an occurrence, the host harbours a *Spiroplasma* spp. with the ability to produce a ribosome-inactivating protein that renders the rRNA devoid of purine bases. This is targeted towards the parasite as the ribosomal machinery of the insect is not affected. The endosymbiont *Hamiltonella defensa* helps the pea aphid suppress the development of a parasitic wasp by arresting the development of eggs to larvae. The synthesis of antibiotics by *Pseudonocardia* spp. is known to rescue fungus-farming ants from pathogens. In *Paederus* beetle, its polyketide-producing symbiont *Pseudomonas* spp. helps defend against wolf spider. *Lagria* beetles living in symbiosis with *Burkholderia gladioli* are saved from fungal infection by lagriamide, an antifungal produced by the microbe. Nutrient-laden eggs and larvae that are immobile can be ready meals for predators. The presence of microbes usually imparted by the female insects is designed to thwart this danger. Brood protection by *Streptomyces* spp. and *Pseudomonas* spp. is a common stratagem utilized. Bioactive molecules like streptochlorin and cycloheximide with activity against both fungi and bacteria are usually involved. Biofilm formation is another characteristic of symbiotic microbes that contribute to defend against the invasion of pathogens. The presence of microbiota is known to improve the immune response in the insect hosts. The holobiont is more resistant to pathogenic attacks compared to the holobiont devoid of its microbes. The ability to produce ROS coupled with the tendency to modulate the synthesis of catalase indicates that the host can also target intruding bacteria and the established microbiome. Microbes may also contribute to the melanization process, an event associated with the defence machinery of the host. Social immunity is an important stratagem employed by insects to fend against intruders. The collective behaviour and action of the colony are invoked to limit infection. In the case of the social aphid *Nipponaphis monzeni*, the soldier ants explode their swollen abdomen laden with tyrosine to protect the colony against aggression. The aromatic amino acid that is derived from the endosymbiont *Buchnera* spp. is regulated by the host and helps seal and protect the colony (**Figure 5.8**).

5.3.3 Microbes help insects counter abiotic stress

The ability of insects to colonize a wide range of ecosystems is to a large extent a result of the microbial partners they harbour. These microbes confer on their hosts' molecular features to navigate temperature variations, water stress and metal pollutants. Insects are small-bodied poikilotherms that are susceptible to temperature change, a physical condition known to not only severely affect biochemical functions but also perturb the intimate relationship between the hosts and the endosymbionts. For instance, an increase in temperature triggers a decrease in microbial population, impedes fertility and retards development. To mitigate this situation, some

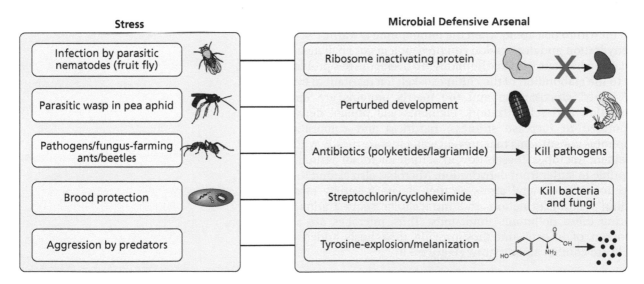

Figure 5.8. Defensive contribution of the microbiome against biotic stress. Insects are confronted with predators and other organisms that they need to fend against. To help them in this task, insects recruit their microbial partners. Ribosome-inactivating proteins designed against nematodes and suppression of development of parasites are some chemical tools elaborated by the microbes. Production of antibiotics mediated the microbiome dissuade pathogens. Microbe-elaborated tyrosine enables soldier ant to create an explosion in order to thwart predators from the colony.

BOX 5.2 FUNGUS-GROWING ANTS AND THEIR MICROBIAL GUARDS

Some insects have evolved to farm fungi that they grow in underground nest chambers as a source of food. In order to be effective farmers, these insects utilize microbes to not only fend against parasites dedicated to destroy the fungal cultivation but to also aid in the extracellular degradation of the biomass that is then ready to be consumed. Leaf-cutter insects gather plant food that is then fed to their fungal cultivars. To deter other fungal pathogens to destroy the farm and the ant colony, antibiotic-producing microbes like *Pseudonocardia* spp. and *Bacillus* spp. are recruited to reside on the exoskeleton where they are provided with nourishment. Antibiotics such as dentigerumycin and polyketides are elaborated to prevent invasion by competing fungi. In an effort to have an effective farming operation, insects like ants, beetles and termites have established an intricate tripartite relationship where the insect farmers, the fungi and the bacterial partners work in harmony. *Enterobacter* spp., *Rahnella* spp. and *Pseudomonas* spp. are some of the microbes common in these fungal farms and are hosted by both insects and fungi. The microbes also act as digestive helpers and contribute to the metabolism of recalcitrant plant biomass into readily utilizable carbon sources destined for the hosts. This external digestive system shaped the evolution of elaborate colonies with millions of insects with distinct task, morphology and behaviour. The microbial contribution is pivotal to the survival of the insect farmers that depend on their fungal farms for nutrition (**Box Figure 2**).

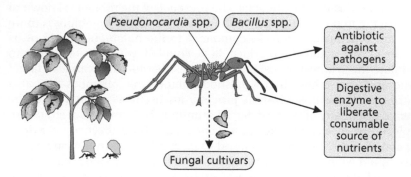

Box Figure 2. Fungus-growing ants and their microbial guards. Microbes help ants farm fungi by supplying antibiotics and degradative enzymes.

of the microbial residents have evolved to adjust to this temperature fluctuation by elaborating heat-shock proteins (HSPs). These molecular chaperones play a key role in stabilizing enzymes and proteins that are prone to denaturation. The pea aphid *A. pisum* houses two obligate *Buchnera* symbionts with different abilities to produce HSPs depending on the temperature of the environment. A single-point mutation in the gene system responsible for HSP production enables one species to be more tolerant to heat stress. This species is more abundant in warmer climates and contributes to the acclimatization of the insect. Furthermore, the colonization by facultative symbionts like *S. symbiotica* dedicated to the synthesis of heat-compatible metabolite such as trehalose also contributes to the adaptation to climate change. In fact, these microbes are more abundant within pea aphid when the temperature is warm.

Dehydration and desiccation can present a major challenge to insects, especially in habitats with low humidity. To overcome this problem, insects resort to the elaboration of the chitin exoskeleton resistant to evaporation. The *N*-acetyl glucosamine and the matrix proteins are modified by melanization and sclerotization. Pigments and phenol derivatives are incorporated to increase stiffness, hardness and colouration. These hydrophobic constituents that prevent water loss are mostly synthesized by the resident microbes. Insects devoid of their microbial partners are more susceptible to suffer from dehydration. Microbes also extend the ability of insects to survive metal stress as they devise a variety of stratagems including biotransformation and intracellular sequestration aimed at rendering these pollutants innocuous. Thus, microbes are an integral component of insects and they provide their hosts added functional flexibility to combat abiotic stress (**Figure 5.9**).

5.3.4 Microbes and detoxification of toxins

Herbivorous insects are subjected to a wide array of toxins that plants produce in order to protect themselves against foraging predators. There is a constant battle between these protagonists to survive. While plants produce deadly chemicals like glucosinolates known to release hydrogen

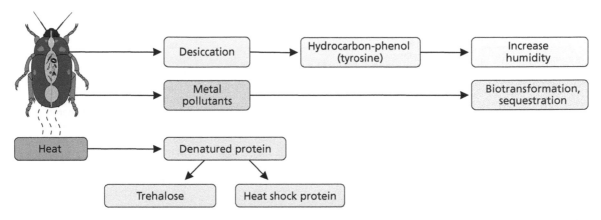

Figure 5.9. Microbes, insects and abiotic stress. Insects can tolerate fluctuating environmental temperature due to the ability of their microbial partners to synthesize heat-shock proteins (HSPs), moieties known to impede the denaturation of proteins. Trehalose, a thermo-tolerant compatible solute is also supplied by microbes. Loss of water can be regulated by the production of chitin decorated with substituents such as tyrosine and hydrocarbon that are of microbial origin. Insects reside in metal polluted areas utilizing the metal detoxification attributes of microbes to adapt to this challenge.

cyanide, insects with the aid of their microbial partners elaborate enzymes dedicated to the neutralization of this moiety. Owing to their immobile lifestyle, plants have evolved intricate chemical strategies to fend dangers posed by insects and other herbivores. A plant diet can be infused with such insecticides as terpenes, caffeine, glycosides and cyanates. Avoiding these toxin-laden meals is a common strategy some insects adopt. However, insects that depend on herbivory harbour a range of microorganisms that enables them to detoxify these chemical compounds. Coniferous plants produce terpenes that are incorporated into the protective resin. Insects living in symbiosis with *Serratia* spp. and *Pseudomonas* spp. metabolize these moieties as these microbes possess the necessary detoxifying enzymes. Oxalate, an insect deterrent, is eliminated with the aid of oxalate decarboxylase expressed by the symbiotic microbes. This enzyme liberates CO_2 and formate. The latter is readily integrated into the primary cellular metabolism. The coffee borer beetle lives in harmony with *Pseudomonas fulva* as the latter provides the biochemical defence against caffeine, a molecule known for its bitter taste and negative impact on numerous physiological processes. This microbe depends on caffeine as the sole source of carbon and nitrogen. It possesses the enzyme caffeine demethylase that mediates the degradation of caffeine. In fact, this beetle, *Hypothenemus hampei*, is the only insect known to complete its entire life cycle feeding exclusively within coffee beans, a feat made possible by the resident microbes.

Unripe fruits are usually nutrient poor and toxin rich as the main goal of these plant products is to provide seeds for propagation, an event that occurs following the ripening of the fruits with the assistance of appropriate vectors. However, a fruit fly is known to lay its eggs on unripe olive fruits containing a high amount of the phenolic glycoside, oleuropein, a pivotal defensive metabolite in the olive plant. The larva is only able to develop in this environment due to the presence of the microbe *Candidatus Erwinia dacicola* it harbours. This symbiont is vertically transmitted via eggs that develop into larvae. The microbe expresses the enzymes like catechol 1,2, dioxygenase responsible for the degradation of the glycoside. Some insect-residing bacteria can also metabolize glucosinolates that have an insecticidal attribute. Following the release of these toxins upon an insect bite, the plant also secretes enzymes such as myrosinase that orchestrate the release of cyanide and thiocyanate, two deadly products known to interfere with the energy-generating machinery of the organism. The presence of symbiotic bacteria helps neutralize these toxins into NH_4^+, CO_2 and SO_4. Some insects are known to become resistant to pesticides such as

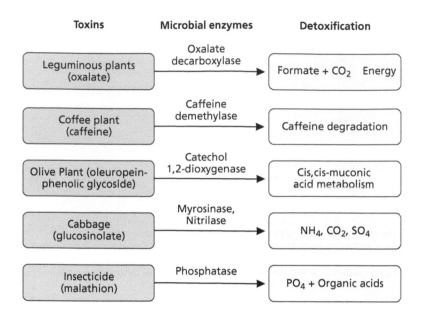

Toxins	Microbial enzymes	Detoxification

Figure 5.10. **Insect and microbial symbiosis in the detoxification of toxins.** Insects are challenged by toxins in nature and man-made pesticides. To counter these, microbial metabolism with its wide array of degradative networks is invoked. Plants produce oxalate, glucosinolates, caffeine and phenol derivatives to protect against herbivores. Enzymes like oxalate decarboxylase, caffeine demethylase, catechol dioxygenases and nitrilases of microbial origin are deployed to detoxify these toxins so that hosts (insects) would not succumb. Chemicals like organophosphates are neutralized by bacterial phosphatases.

organophosphates. The ability to acquire *Pseudomonas* spp. with organo-hydrolases tends to give insects an edge over man-made pesticides like malathion, a competition fuelling resistant insect pests (**Figure 5.10**).

5.4 Microbe–Insect Cross-Talk: The Molecular Language

Microbes residing in these multicellular organisms are an integral part of these supra organisms as they also constitute an important aspect of the communication network utilized to perform a variety of functions. These chemical signals emanating from the resident microbes can signal danger, can promote mating, can inform the social integrity of the colony and can help congregate the hosts in an effort to perform a specific task. A *Bacteroidetes* symbiont in saw-toothed grain beetle supports the synthesis of the cuticle by supplying lipid derivatives, an important component aimed at attracting members of the same colony. The cuticular hydrocarbon fingerprint can help discriminate between a foe and friend and allow the identification of the proper nest mate. In termites, the addition of succinate, acetate and methyl groups in these exocellular moieties mediated by the microbial residents is critical in locating mates, in the choice of partners and social interactions. Pheromones and organic acids are also elaborated to modulate kin recognition, recruitment and social hierarchy. These signalling molecules are synthesized de novo from fatty acids, isoprenoids, monosaccharides and amino acids. They guide the adaptive behaviour of the host.

The grass grub beetle partners with the bacterium *Morganella morganii* to convert the aromatic amino acid tyrosine into phenol, a pheromone known to be an attractant. Locust aggregation is triggered by phenol and 2-methoxy phenol, molecules microbially elaborated. The gregarious lifestyle of these pests is known to cause major devastation to agriculture and contributes to food insecurity in some parts of the world. Antibiotic treatment of locusts results in a sharp reduction of these metabolites, an observation further confirming the critical role the resident microbes play in this process. Insects feeding on conifers utilize these chemical cues to coordinate this aggregation behaviour in order to mass attack these plants. Repellent feature of the insect–microbe communication can be on display in an attempt to limit competition when vegetation is poor. For instance, the symbiotic microbes associated with the bark beetles convert α-pinene into verbenol and verbenone, chemicals known to display attractant and repellent signalling characteristics. Some insects use these chemicals as

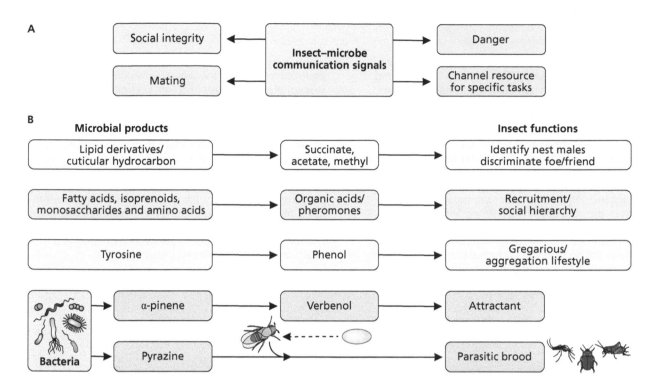

Figure 5.11. Microbe–insect cross-talk: the molecular language. (a) Microbes are the origin of numerous signals responsible for guiding the host to perform various tasks including mating and alerting dangers. The presence of succinyl, acetyl and methyl derivatives on the exocellular cuticle help distinguish between foe and friend. (b) Social order in various colonies is maintained with the assistance of chemicals elaborated by the resident microbes. Microbial pyrazine enables the parasitic brooding in some species. Phenol derived from the microbiome plays an important role in the gregarious lifestyle of some insects.

camouflage, a strategy that enables their eggs to be nurtured and fed by another insect. This is possible due to the presence of symbiotic bacteria that produce these pyrazine-derived, information-laden molecules. Hence, these resident microbes serve almost like specific organs in higher organisms designed to communicate with other parts of the holoorganism. Without the input of the microbial partners, insects would have to evolve other onerous tissue specialization to fit their lifestyle (**Figure 5.11**).

5.5　Microbe-Induced Feminization of Insects: Sexual Manipulation

Numerous insects house a variety of microbes that helps them in an array of biological functions like acquiring vitamins and essential amino acids. Some of these bacteria can be obligate or facultative and to ensure their own survival they need to be transmitted in a secure manner. In many cases, the transmission channel is via the female eggs. *Wolbachia*, an obligate endosymbiont in numerous anthropods, engages in a plethora of molecular manipulations to promote females at the expense of male counterparts. To achieve this objective, *Wolbachia* exhibits essentially four distinctive phenotypes aimed at limiting male progeny and increasing female insects. Cytoplasmic incompatibility, male killing, feminization and parthenogenesis are the strategies invoked. During male killing, the infected males die at the embryonic or larval stages. The development of genetic male is sabotaged by bacterial infection to become female. Parthenogenesis results in virgin females producing daughters, an altered cellular division event commandeered by the microbe. Testis development in the pupal stage mediated by the enhanced synthesis of female pheromones also contributes to arresting the male population.

BOX 5.3 THE WEAK LINK: PLANTS TARGET INSECT GUT MICROBES TO KILL THE HOSTS

Plants are constantly under attack by various herbivorous organisms including insects. This ongoing challenge has compelled plants to devise elaborate stratagems in an effort to thwart the assault by folivorous insects. Various toxins aimed directly at the insects provide some relief to plants even though the recruitment of microbes with the ability to detoxify these insecticides enables some leaf-seeking predators to survive. To further fortify their defence system, plants are also known to enlist the assistance of the gut microbiome to debilitate insect intruders. In this instance, these microbes that impart numerous benefits to the host become a liability. Some plants trigger perturbation in the digestive system that creates a favourable environment for the microbes to proliferate and invade the body cavity of the hosts resulting in their demise. Physical disruption evoked by trichomes (thorny outgrowth), proteases and chemicals produced by the plants increases the permeability of the protective peritrophic matrix of the digestive tract rendering it susceptible to microbial invasion. Lectins and peptides are also deployed to accomplish this task. While the former binds directly to the matrix increasing its permeability to microbes, the latter disrupts the different components of the gut releasing bacteria into the haemolymph. This strategy of perforating the outer lining of the digestive tract and providing access to benign microbes to invade the body cavity triggers sepsis and leads to the death of the predator. Thus, plants utilize the insects' own microbes to inflict maximum damage. Plants have exploited this weak link in order to create a leaky gut that becomes a fertile environment for microbial proliferation resulting in the death of the predatory insects (**Box Figure 3**).

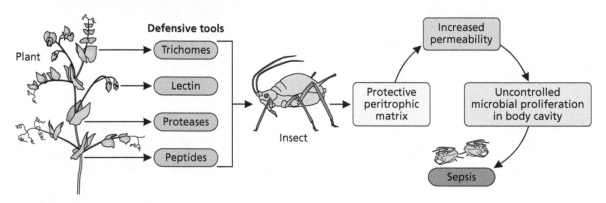

Box Figure 3. The weak link: plants target gut microbes to kill the insect hosts. Microbes become a liability as plants disrupt the gut with diverse strategies to assist bacteria in infecting their own host.

Uninfected females are sterile and can reproduce only upon infection by the microbe when the normal ovaries are restored. While infection is crucial for oogenesis, it also inhibits the expression of masculinizing genes. Cytoplasmic incompatibility is another strategy *Wolbachia* adopts in order to promote its sexual regulatory power by tilting the balance in favour of female insects. In this instance, male gamete chromatin packaging is perturbed and prevents mating between an infected male and an uninfected female. The delayed removal of protamine alters sperm chromatin and renders the reproductive process ineffective. Even females associated with phenotypes devoid of the genetic information to manipulate the sex ratio of the insect population are unable to reproduce. Although the insect microbiome evolves with the host and contributes to essential functions of the holobiont, the seemingly manipulative attribute of *Wolbachia* may be designed to promote its own survival for the betterment of the holoorganism. It is also important to note that insects without and/or limited amount of microbes have a high mortality rate as numerous microbial cues like the hypoxia induced by the midgut microorganisms help shepherd the development of the host (**Figure 5.12**).

5.6 Microbiome in the Moulting Process: A Growing Pain That Needs Microbial Assistance

Anthropods possess a characteristic exoskeleton that serves as a protection against predation, desiccation and mechanical injury. It does also impart

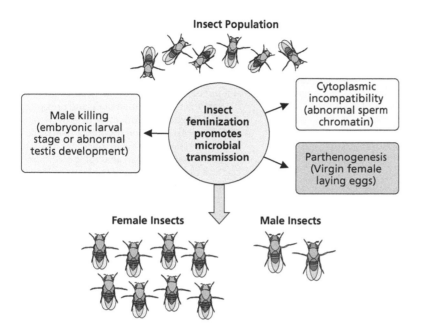

Figure 5.12. **Microbes dictate sexual pheno-types.** *Wolbachia* spp. are obligate symbionts in many insects. They are usually transmitted by the females via eggs. To promote their survival, these microbes manipulate the sexuality of insects in such a manner to favour females. Male killing, feminization, cytoplasmic incompatibility and partheno-genesis are some of the strategies invoked to ensure a healthy female population as the latter is responsible for the transmission of the microbes.

shape and colour to the organism and acts as an anchor where the body muscles are attached. This armour-like anatomical structure is a critical survival shield and has enabled insects to successfully colonize most ecosystems on this planet. This rigid component referred to as the cuticle is essentially composed of an outer layer, the epicuticle and an inner layer, the endocuticle. These layers composed of proteins, chitin, lipids, Ca and phosphate need to be replaced on a regular basis to allow for growth. The process of shedding the old cuticle known as ecdysis is under hormonal control. The release of ecdysteroids in a pulsated manner is accompanied by a network of biochemical transformations resulting in the replacement of an old cuticle with a new one.

The amounts of Ca and phosphate required to regenerate this body part are manipulated by the symbiotic bacteria residing in the calcium bodies lodged in the epithelial sacs located along the digestive tract. The ability of the microbes in producing polyphosphate plays a key role in maintaining the homeostasis of these two critical ingredients. The digestion and resorption of the old tissue coupled with the storage of Ca is an important biological event before the exuviation of the cuticle. The polyphosphate-producing microbes within the calcium bodies are endowed with polyphosphate kinases and phosphatases that mediate the sequestration and release of Ca. Furthermore, the phosphate polymer may also act as a source of energy during nutrient-deficient conditions prevailing as the organism moults. This symbiotic utilization of Ca and phosphate is an important aspect of this communal life aimed at the effectiveness of the morphological renewal in the host. The associated microbes do not only assist in this biosynthetic process but also contribute to its regulation. In fact, the hypoxic environment orchestrated by the bacteria induces the release of the moulting hormone. The enzyme cytochrome bd oxidase modulates the O_2 levels in the gut, a situation conducive to anaerobiosis and subsequent slowing of metabolic activity. Here, the role of HIF1-α (hypoxia-inducible factor) has been reported. The inhibition of the prolyl hydroxylase in these conditions leads to the stabilization of the transcription factor promoting the expression of the biochemical machinery dedicated to a lifestyle with reduced O_2. In this instance, decreased ATP production observed during moulting would result in diminished mobility, a condition beneficial for the renewal of the exoskeleton. Thus, the initial signal to prompt this developmental process is triggered by the resident microbes resulting in the decreased mobility of the host. The Ca and phosphate, two

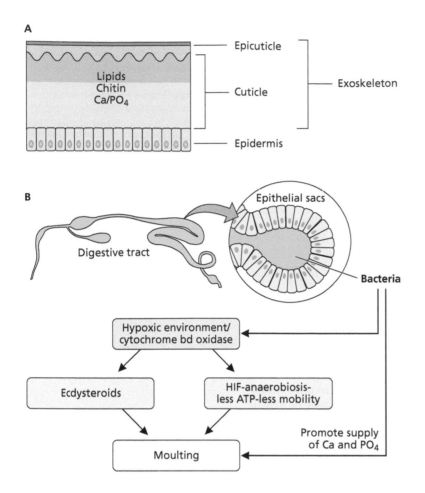

Figure 5.13. Microbial contribution to insect moulting. The exoskeleton composed of numerous compounds including Ca and phosphate plays a central role in anthropods and is periodically renewed. Microbes are directly involved in this process as they regulate the concentration of Ca and phosphate and trigger the production of ecdysteroids, chemicals signalling moulting. To accomplish this physiological event, microbes mediate the generation of an anaerobic environment and the stabilization of the hypoxia-inducible factor (HIF). The decrease in metabolic associated with this event signals moulting.

ingredients essential in the synthesis of the exoskeleton are supplied by the microbial partners. In herbivorous turtle ants, the gut microbiome is known to convert nitrogen wastes into amino acids that fuel the formation of the cuticle. In this instance, proteins, catecholamine cross-linkers and chitin are supplied by microbial residents to the ant in order to build the external structure. Microbes, integral constituents of insects, are essential for the rejuvenation and synthesis of the exoskeleton of the hosts (**Figure 5.13**).

5.7 Microbes and Colour Formation in Insects

Colour is a multi-faceted feature that permeates most living organisms and serves a multitude of functions essential for the survival of the hosts. Pigments on an organism provide protection against predators, help attract sexual partners, warn potential aggressors (aposematic), act as a camouflage, shield against UV light and contribute to thermoregulation. Numerous optical mechanisms that involve iridescence, luminescence, reflectance and photonic crystals participate in the array of visible manifestations exhibited by living systems. The contribution of chemical moieties in the colourful display in nature is very important. Biomolecules possessing aromatic, azo and polyene functionalities coupled with enhanced steric hindrance tend to have the characteristic colour-emitting property. Polyphenol and carotenoid derivatives are two important pigments insects have. These colour displays are essentially associated with the body or located on eggs.

The pea aphid (*A. pisum*) usually manifests itself as red and green phenotypes. The red tinge imparted to the body is predominantly derived from carotenoids, while the green hue is attributed to polycyclic quinone-containing compounds. The biosynthesis of these pigments necessitates the participation of a wide variety of precursors and enzymes like fatty acid synthase and polyketide synthase. The host and the microbial partners work in tandem to elaborate these colourful body covers essential for the holobiont. This aphid has an obligate symbiont *Buchnera aphidicola* and a facultative partner '*Candidatus Rickettsiella viridis*'. Upon infection by the latter, the red aphid turns green, an event punctuated by the increase in the production of polycyclic quinone. This phenomenon occurs as a result of the microbe provisioning the host or other symbionts with acetyl CoA and malonyl CoA, ingredients that activate and propel the synthesis of the green pigment. This symbiotic biosynthetic process may confer an evolutionary advantage to the holobiont. Migrating locusts exhibit different outer colouration depending on their solitary or gregarious lifestyle. They tend to be green when living alone. A black-brown hue is the colour of choice when these insects live in groups. This change in colour is characterized by behavioural modulation and is influenced by a variety of factors including hormonal secretions. The increased expression of the β-carotene binding protein results in the accumulation of a red pigment that interacts with green colour common during the solitary lifestyle to give rise to the black-brown tinge associated with the sociable nature of the locusts.

The commercial production of carmine, a red phenolic dye from cochineal, a scale insect (*Dactylopius coccus*) has been going on for more than 500 years. This dye is utilized in a variety of products including foods, cosmetics and textiles. The cochineal feeds essentially on cactus sap. This nutrient-poor diet compels the insect to forge a symbiotic relationship with β-proteobacterium, *Candidatus Dactylopibacterium carminicum* housed in the ovaries. Nearly 10% of the biomass of the insect is microbial. This arrangement allows the microbe to fix nitrogen, an anaerobic process, to supply essential amino acids and to recycle uric acid. In return, bacterium gets a secure environment and carbohydrate-replete fuel from the host. The carminic acid that constitutes almost 25% of the dry weight of the female is also present in eggs. This pigment is postulated to deter predators as the female are wingless and to protect the egg from UV radiation. The males are short-lived, can fly and do not tend to have the very visible pigment the females accumulate. Microbes are again at work contributing to the well-being of their insect hosts (**Figure 5.14**).

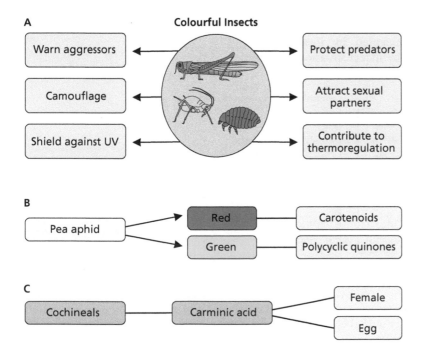

Figure 5.14. Colourful insects, their microbes and the synthesis of pigments.
(a) Colour fulfils numerous tasks in insects. It is a camouflage, an attraction for sexual partners, protects against ultraviolet light and warns aggressors. Both the host and microbes contribute to the synthesis of these pigments. (b) In pea aphids, carotenoids with their extended conjugated chains that are derived from bacteria give a red colouration while the green pigment is a reflection of the polycyclic quinones elaborated with microbial input. (c) Carminic acid that imparts a red colour to female cochineals constitutes nearly 25% of the dry weight of the insect. It helps protect eggs against predation.

5.8 Social Insect Cohesion Is Dictated by Microbial Chemical Signals

Several insects including bees, ants and termites are known to live in social groups where each group is assigned specific tasks aimed at the proper functioning of the social network. For instance, in a bee society, the queen is responsible for reproduction, a process aided by male drones. For this activity to be successful the queen is constantly fed royal jelly by the worker bees. In order to operate properly, the members of disparate social groups need to be recognized and have to exhibit cues of a specific group membership. These members have first to discern each other and differentiate their kinship from intruders. Such identity recognition is not just important for the colony but is also essential in preventing intruders and parasites from invading their organization. To maintain and regulate colony membership and integrity, recognition cues that are independent of specific genetic variants but are colony specific are established. These signals are dependent on the common factors prevalent in the colony and on the shared social environment. The cuticular hydrocarbon and the chemical profile of each individual insect that are shaped by the community-level variation and similarity of the microbial population play a central role in the organization of the colony. The resident microbes regulate the composition of the cuticle by supplying or depleting the precursors involved in the synthesis of this external structure. The chemical moieties within this epicellular layer can attract members and dissuade intruders. They are modulators of sociability and behaviour. The presence of symbionts like *G. apicola* directs the synthesis of cuticular hydrocarbon with a disparate chemical profile compared to the structures of opportunistic microbes acquired from the environment. Hence, microbes are important guardians of the social structure that numerous insect colonies have come to depend on (**Figure 5.15**).

5.9 Conclusions

Microbes are essential to the contribution insects bring to the ecosystem. The intricate relationship bacteria forge with their insect hosts is critical to the biological activity of these holoorganisms. The plethora of functions that microbes perform is deeply rooted in the very existence of these organisms with very prolific adaptive attributes. The understanding of this intimate interaction of these disparate biological entities will pave the way to generate improved agricultural yield, enhanced food production and mitigation of various diseases that are propagated by insects. The decimation of agricultural products provoked by insect invasion can be

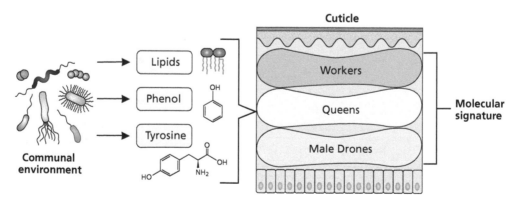

Figure 5.15. Sociability in insect colony dictated by microbes. The specific chemical organization of the exoskeletal layer contributes to the social structure in numerous insect colonies including those found in termites, bees and ants. These chemicals elaborated with the input of the microbiome allow for an intricate society with each member group dedicated to an assigned duty to operate. The presence of lipids, phenol and tyrosine are characteristic of the worker bee, the queen bee and the male drone, respectively. This well-defined hierarchy exists only with the interaction of the host and the resident microbes.

Maculinea phengaris is a grassland butterfly that has a unique life cycle. This blue-colour butterfly depends on a specific ant colony to complete its metamorphosis where the larva becomes a pupa and eventually emerges as an adult, all nurturing done by the members of the *Myrmica* ants. To achieve this feat, the different entities of the developing butterfly use appeasement and mimicry to be amenable to the host ant colony. Acoustic signals and chemical cues are central in this strategy. Microbes play an important role in elaborating the metabolites laden with deceptive information that enables the social parasite to be integrated within the colony. The butterfly first lays its eggs on select plants belonging to the *Thymus* spp. and after 2–3 weeks of feeding on the flowers, the larva in the final stage drops to the ground from where it awaits collection by ant workers of the *Myrmica* spp. These ants rear the caterpillar in the nest where the parasite's nutritional habit changes from phytophagy to carnivory as it preys on the brood of the colony. To succeed in this social parasitic behaviour, the caterpillar seeks the assistance of such microbes as *Serratia marcescens* and *Serratia entomophila*. These microbes contribute to the production of surface hydrocarbon moieties that are similar to the *Myrmica* ant workers. This chemical mimicry allows the adoption and integration of the caterpillar in the colony where it is treated as a native brood and fed by trophallaxis. This behaviour coupled with the co-mingling within

the ant colony further promotes microbial exchange and fortifies the incorporation of this parasitic larva as a member of the nest. The synthesis of fatty acids and volatile pyrazines orchestrated by the microbes serve as deceptive signals that enable the butterfly to have its eggs reared into an adult within the colony of the *Myrmica* ants (**Box Figure 4**).

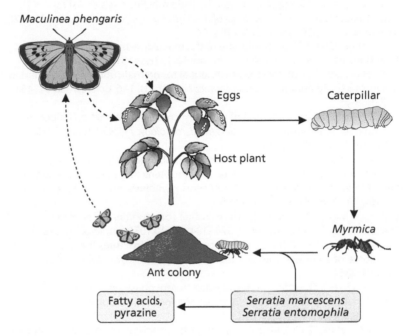

Box Figure 4. Microbes assisting insect behaviour resulting in social parasitism. The butterfly larva is able to use camouflage signals produced by microbes to integrate an ant brooding nest.

effectively alleviated by manipulating the microbial residents that are essential for the survival of the pest. For instance, locusts known to destroy crops in tropical countries can be impeded by exploiting the microbial residents responsible for their feeding-frenzy gregarious lifestyle. Perturbing the reproductive success of insect vectors mediating the transmission of diseases by exploiting bacteria within will help improve global health. Hence, the microbiome, an integral constituent of the insect holobiont can provide interesting possibilities to ameliorate the biological roles that insects play in the global ecological framework and to benefit humans in combatting diseases and hunger.

SUGGESTED READINGS

Aylward, F. O. et al., (2014). Convergent bacterial microbiotas in the fungal agricultural systems of insects. *MBio, 5*(6), e02077–14. doi:10.1128/mbio.02077-14.

Bratburd, J. R., et al., (2020). Defensive symbioses in social insects can inform human health and agriculture. *Frontiers in Microbiology, 11*, 1–8. doi:10.3389/fmicb.2020.00076

Brune, A., & Dietrich, C. (2015). The gut microbiota of termites: Digesting the diversity in the light of ecology and evolution. *Annual Review of Microbiology, 69*(1), 145–166. doi:10.1146/annurev-micro-092412-155715

Bustamante-Brito, R. et al., (2019). Metatranscriptomic analysis of the bacterial symbiont *Dactylopiibacterium carminicum* from the carmine cochineal *Dactylopius coccus* (Hemiptera: Coccoidea: Dactylopiidae). *Life, 9*(1), 4. doi:10.3390/life9010004

Calcagnile, M., et al., (2019). Bacterial semiochemicals and transkingdom interactions with insects and plants. *Insects, 10*(12), 441. doi:10.3390/insects10120441

Ceja-Navarro, J. A. et al., (2015). Gut microbiota mediate caffeine detoxification in the primary insect pest of coffee. *Nature Communications, 6*(1), 1–9. doi:10.1038/ncomms8618

Coon, K. L. et al., (2017). Bacteria-mediated hypoxia functions as a signal for mosquito development. *Proceedings of the National Academy of Sciences, 114*(27), 5362–5369. doi:10.1073/pnas.1702983114

Di Salvo, M. et al., (2019). The microbiome of the maculinea-myrmica host-parasite interaction. *Scientific Reports, 9*(1), 1–11. doi:10.1038/s41598-019-44514-7

Duplais, C. et al., (2021). Gut bacteria are essential for normal cuticle development in herbivorous turtle ants. *Nature Communications, 12*(1), 1–6. doi:10.1038/s41467-021-21065-y

Gupta, A., & Nair, S. (2020). Dynamics of insect–microbiome interaction influence host and microbial symbiont. *Frontiers in Microbiology, 11*, eCollection 2020. doi:10.3389/fmicb.2020.01357

Kaczmarczyk, A. et al., (2018). First insight into microbiome profile of fungivorous thrips *Hoplothrips carpathicus* (Insecta: Thysanoptera) at different developmental stages: Molecular evidence of *Wolbachia* endosymbiosis. *Scientific Reports, 8*(1), 1–13. doi:10.1038/s41598-018-327470-x

Kwong, W. K., & Moran, N. A. (2016). Gut microbial communities of social bees. *Nature Reviews Microbiology, 14*(6), 374–384. doi:10.1038/nrmicro.2016.43

Lanan, M. C., et al., (2016). A bacterial filter protects and structures the gut microbiome of an insect. *The ISME Journal, 10*(8), 1866–1876. doi:10.1038/ismej.2015.264

Lavy, O., et al., (2019). The effect of density-dependent phase on the locust gut bacterial composition. *Frontiers in Microbiology, 9*, 1–8. doi:10.3389/fmicb.2018.03020

Lemoine, M. M., et al., (2020). Microbial symbionts expanding or constraining abiotic niche space in insects. *Current Opinion in Insect Science, 39*, 14–20. doi:10.1016/j.cois.2020.01.003

Lenaerts, C. et al., (2017). The ecdysis triggering hormone system is essential for successful moulting of a major hemimetabolous pest insect, *Schistocerca gregaria. Scientific Reports, 7*(1), 1–14. doi:10.1038/srep46502

Mason, C. J., et al., (2019). Plant defenses interact with insect enteric bacteria by initiating a leaky gut syndrome. *Proceedings of the National Academy of Sciences, 116*(32), 15991–15996. doi:10.1073/pnas.1908748116

Perlmutter, J. I., & Bordenstein, S. R. (2020). Microorganisms in the reproductive tissues of arthropods. *Nature Reviews Microbiology, 18*(2), 97–111. doi:10.1038/s41579-019-0309-z

Rennison, D. J., et al., (2019). Parallel changes in gut microbiome composition and function in parallel local adaptation and speciation. *Proceedings of the Royal Society B, 286*, 1–9. doi:10.1101/736843

Rio, R. V., et al., (2016). Grandeur alliances: Symbiont metabolic integration and obligate arthropod hematophagy. *Trends in Parasitology, 32*(9), 739–749. doi:10.1016/j.pt.2016.05.002

Salem, H., et al., (2015). An out-of-body experience: The extracellular dimension for the transmission of mutualistic bacteria in insects. *Proceedings of the Royal Society B: Biological Sciences, 282*(1804), 1–10. doi:10.1098/rspb.2014.2957

Van Arnam, E. B., et al., (2018). Defense contracts: Molecular protection in insect-microbe symbioses. *Chemical Society Reviews, 47*(5), 1638–1651. doi:10.1039/c7cs00340d

Vernier, C. L. et al., (2020). The gut microbiome defines social group membership in honey bee colonies. *Science Advances, 6*(42), 1–9. doi:10.1126/sciadv.abd3431

Yang, M., et al. (2019). A β-carotene-binding protein carrying a red pigment regulates body-color transition between green and black in locusts. *ELife, 8*, e41362. doi:10.7554/elife.41362

FISH MICROBIOME: FUNCTIONAL PERSPECTIVES

6

Contents

Keywords

- Fish
- Microbiome
- Mobility
- Digestion
- Ageing
- Signalling
- Defence

6.1 Introduction: Classification, General Anatomical and Physiological Features

Fishes belong to the phylum of *Chordata* and are part of the subphylum of *Vertebrata*, a grouping also shared by amphibians, reptiles, birds and mammals. These organisms are bestowed with a spinal cord that is supported by a backbone. During development of the embryo, the first opening evolves as the anus while the mouth appears later. This phenomenon helps further categorize fishes as deuterostomes that are organisms where the anus is formed before the mouth. Depending on various anatomical features and physiological characteristics, fishes are further classified into three groups: Agnatha – fish devoid of jaws (e.g. hagfish), Chondrichthyes – fish comprising of cartilage, a biomaterial constituting the skeleton (e.g. shark) and Osteichthyes – fish with a calcified bone structure and punctuated with protective scales (e.g. salmon). These aquatic organisms live either in marine environment or in freshwater and occupy a vast geographical area. They are found in tropical, temperate, Antarctic and Arctic regions. They also live at the bottom of oceans or lakes and in water bodies located at high altitude. The diverse ecological niche fish occupies is facilitated by the physiological, morphological and microbial attributes the organism has evolved with in order to adjust to its natural surroundings.

For instance, most fishes have a pair of round eyes that allow them to see clearly despite the refractive strength of the water they live in; however,

DOI: 10.1201/9781003166481-6

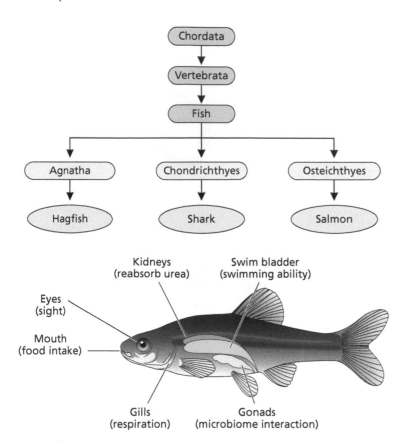

Figure 6.1. Classification, general anatomical and physiological features of fishes. Fishes are classified into three main groups which belong the hagfish, shark and salmon respectively. The anatomical characteristics are reflective of the environment and nutritional habits of the fish. For instance, some fishes have jaws while others are jawless. Gills play an important role in respiration and excretion is mediated by the kidneys. The swim bladder aids in swimming and vision is promoted by the eyes even though cave-dwelling fishes tend to be blind. Many of these discrete organs including the gonads are assisted in their functions by the resident microbes.

cave-dwelling fish (*Astyanax mexicanus*) does not possess eyes as the need for vision is completely obviated in this habitat. On the other hand, fishes living above-ground in rivers have eyes with excellent vision. The loss of eyes may have compelled cave fishes to limited mobility as food or even predators are scarce in this ecosystem, thus enabling them to save energy. This lack of vision is compensated with a highly specialized sense of taste and smell, attributes that are essential to satisfy the omnivorous lifestyle of these fishes. The skin that serves as a protective layer has scales and fins usually covered in mucus. While the fins are pivotal in movement and steering, the mucus reduces the frictional force of water that is nearly 800 times denser than air. The gills allow the uptake of O_2-dissolved water, discharge NH_3 and are equipped with ion pumps aimed at regulating the concentration of NaCl. Tiny capillaries provide the blood stream easy access to O_2. The air/swim bladder is a buoyancy-modulating organ filled with air that plays a critical role in the swimming ability of the fish. The structural feature of the mouth is usually reflective of the nutritional habit the fish is accustomed to. The brain, liver, digestive tract, heart, kidneys, gonads and muscles are important components of this vertebrate and function in partnership with the microbiome to perform a variety of biological processes vital for the survival of the fish in the ecosystem it has elected to reside in. For instance, the kidneys have elaborate structural features aimed at reabsorbing urea, an important osmoregulatory strategy to maintain a proper intracellular ionic concentration (**Figure 6.1**).

6.2 Ecological and Economic Significance

6.2.1 Nutrient recyclers and ecosystem engineers

Fish is an important component of the ecosystem and its contribution extends far beyond the marine and freshwater environment it thrives in. Its foraging activity of coral reefs alters the physical habitat on which a range of diverse organisms depends on. Parrot fishes are referred to as the biological bulldozers shaping and building the ocean floor with the sediment they deposit. These marine engineers are herbivores and feed essentially on algae living in symbiosis with corals. They help control algal population and thus support the biodiversity in a given ecological niche. Both in the marine and freshwater habitats, fish contributes to the food web as it is a source of food and also controls the rampant proliferation of other species. Thus, this organism is a crucial link in the functional ecology in water bodies that has a tremendous impact even on terrestrial organisms. The high biodiversity and abundance of fish afford the aquatic ecosystem a rich supply of protein that fuels other organisms and play a key role in the biogeochemical fluxes of C, P and N. The excretion of NH_3 and urea, a routine activity fishes engage in is pivotal to sustain life, especially in organisms devoid of nitrogen-fixing characteristics. Fish is an important producer of NO_2 and NO_3 as these nitrogen moieties are generated when nitric oxide (NO) is involved either as an intracellular signalling molecule or in the formation of reactive nitrogen species. In fact, NO participates in such physiological functions as reproduction or the pumping of blood. NO_2 and NO_3 can become a ready source of N. Calcium carbonate, a moiety where CO_2 is sequestered, is also deposited by fish. In fact, several tonnes of this mineral originate from fish as part of its cellular metabolism. The excess of calcium ingested from seawater is converted in the gut, a process mediated by metabolic CO_2 and eliminated as $CaCO_3$. It was initially thought that other marine organisms were involved in this process while fish was not a participant in this carbon cycle. The finding that fish secretes copious amount of $CaCO_3$ further confirms the Ca recycling role of this organism in marine ecosystems. It is important to note that this mineral is an important modulator of pH, a physico-chemical property dictating the flow and success of life in all environments. These mineral nutrients are not only produced by this organism but it also helps in their movement across the ocean. The bioturbation induced by fish during their foraging and swimming activity increases the exchange between sediment and water. This activity further aids in the distribution of these nutrients via water column processes. Hence, the translocation of foods and cycling of nutrients are vital activities fishes are engaged in in order to maintain the vast biodiversity that exists on planet Earth. These ecosystem services ranging from nutrient cycling to pH modulation provided by fish are delicately intertwined with the well-being of all organisms.

6.2.2 Economic benefits

Fish contributes immensely to the economic well-being of the globe. It is estimated that approximately 250 million people worldwide depend on the fishing and aquaculture industries for their livelihood. Thus, this aquatic vertebrate provides food security, economic benefit and empowerment. Fish products are an important source of proteins and essential oils. In many countries, fish is the primary source of animal protein and contributes to a good portion of the gross domestic product. Furthermore, its content in vitamins (A, B, D) and minerals (Ca, Zn) is critical in diminishing nutrient deficiency in numerous countries. Readily dried fish can be a source of excellent nutrients all year long. Fish scales commercialized as artificial pearls and decorative beads are also an income earner. The ornamental fish industry is another component of economic activity generated by this vertebrate. Furthermore, recreational fishing and the colourful marine life are also part of the social fabric with significant monetary impact. Fish feeding on insect larvae is a crucial component of the public health strategy aimed at controlling diseases transmitted by insect vectors. The multifaceted influence of fish on our daily life undoubtedly makes this organism an important subject to understand and to learn how it operates in nature. Although its visible anatomical features have been

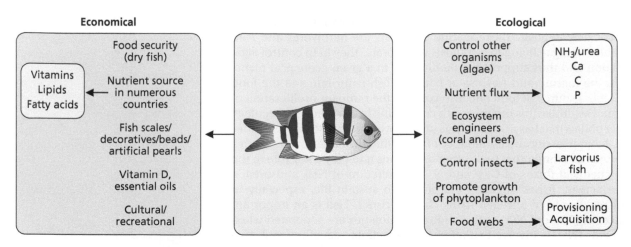

Figure 6.2. Ecological and economic significance of fishes. Fishes contribute to the well-being of the planet as they are essential components of the ecosystem. They deposit CaCO₃, recycle nutrients and control insect population. Their activity is pivotal to the coral reef, a habitat where a wide biodiversity thrives. Economic activity in numerous countries is reliant on the fish industry as these organisms are an important source of proteins, vitamins and essential oils for humans. Recreational activity centred around the fish is also a sizable income generator. Healthy fishes largely maintained by their microbial partners are key to these activities.

delineated, the functions of the microbial partners residing within and on fishes have yet to be fully elucidated. These microbial constituents are an important component of fishes as they participate in a variety of biological functions essential for the growth and survival of these vertebrates (**Figure 6.2**).

6.3 Fish Microbial Communities

6.3.1 Factors dictating the establishment of microbial partners

Like all multicellular organisms, fishes harbour a variety of microbes that makes these aquatic organisms the way they are. Numerous biological activities performed by fish are aided and assisted by the assemblages of bacteria residing on the skin, within the gills, the digestive tract and almost all of the other organs. These microbes are mostly invisible, except for the bacteria that make themselves evident by the light emitted by fish or the mucus they produce on the skin. The latter is an exopolymer known not only for its antibiotic properties but also for its ability to enable the host to swim efficiently. These micro-organisms participate in a plethora of other biological tasks like reproduction, fortification of the immune system, digestion and elimination of toxins.

The microbial communities within the fish that constitute the hologenome may be divided into three main groups:

1. A 'core' microbiome that is adapted to the host and determined by numerous factors including genetics.

2. A residential group that is flexible in its assembly and is dependent on environmental diversity and external parameters.

3. A transient assemblage of microbes depending on where the host is located.

Thus, the microbial composition of a fish is dictated by such factors as host phylogeny, geography, age, sex, genetics, feeding habit, water temperature, salinity, weight and environment (in captivity, in aquaculture or in natural milieu). In Atlantic salmon, Mycoplasmataceae phylotypes are known to be harboured during all stages of the lifecycle. Regardless of whether the Zebra fish is domesticated or lives in the wild its gut microbiota is populated with a high abundance of *Proteobacteria*. The genera belonging to *Shewanella* and *Aeromonas* are more prominent microbes. *Aeromonas* spp. is a characteristic of freshwater fish, the marine counterpart is more likely to house *Vibrio* spp. While the latter depends on its microbial partners residing in the hindgut to process its nutrients acquired through a herbivorous diet, the freshwater herbivores are not known to rely on hindgut bacterial fermentation. The algal-rich food that marine fishes consume contains diverse complex and chemically modified carbohydrates. These liberate monosaccharide-like mannitol and uronic acids that need to be further processed, a metabolic event mediated by the resident fermenters in the hindgut.

These bacteria can readily convert the glycosidic-rich monomers into short-chain fatty acids (SCFAs). The primarily cellulose-based nutrition that the freshwater fishes are accustomed to does not necessitate such a fermentative treatment. In fact, freshwater fishes display a shorter gut transit than their marine relatives.

Cetobacterium somerae, an obligate anaerobe that is involved in the synthesis of vitamins, is an important constituent of such diverse fishes as channel catfish (*Ictalurus punctatus*), largemouth bass (*Micropterus salmoides*) and blue gill (*Lepomis macrochirus*). Rainbow trout also tends to have a core microbiome as demonstrated by minor changes in microbial residents despite living in high or low-density environment or being fed disparate diets. The nature of the microbiome is dependent on the nutritional habit of these vertebrates, a phenomenon common in most multicellular organisms. Species engaged in carnivorous, herbivorous, omnivorous and filter-feeding behaviours possess microbial assemblages that are reflective of these dietary characteristics. *Clostridium* spp. and *Citrobacter* spp., microbes expressing cellulase activity, are more prominent in herbivores while carnivores are more prone to house *Cetobacterium* spp. and *Halomonas* spp. with their enhanced proteolytic enzymes like trypsin. Omnivores and filter-feeders vary in their content of *Planctomycetes* spp. These microbial features may be attributable to a core microbiome that these aquatic organisms have become reliant on in order to be a full-fledged multicellular organism, to extend their adaptive attribute and to limit further cellular specialization (**Figure 6.3**).

6.3.2 Influence of habitats on fish microbiome

The habitats fishes live in are known to be subjected to constant fluctuation in temperature, hydrostatic pressure, salinity, availability of foods and parasitic microbes. These highly variable factors also help calibrate the nature of the microbiome these aquatic organisms possess. However, the microbial assemblage acquired to adapt to these conditions may be transient and is remodelled on an ongoing basis depending on the situation these vertebrates have to face. The intestine is equipped with elaborate osmoregulatory mechanisms designed to adjust to the oscillating osmolality of the water body. The increased expression of protein pumps such as Na^+/K^+ ATPase dedicated to modulate the concentration of Na^+ and Cl^- coupled with the constant intake of water is part of the elaborate stratagem to thwart the toxicity of elevated amounts of these ions. The presence of appropriate microbial partners also contributes to this task. Microbes like *Clostridium* spp. and *Pseudomonas* spp. on the skin with the ability to produce carbohydrate-rich mucus is known to play a role in osmoregulation in water bodies with disparate NaCl levels. In cold water, the psychrophilic microbes with diminished-energy metabolism and anti-freeze producing ability are important constituents of fish microbiome. Seasonal change can also shape fish microbial landscape. In bluegill (*L. macrochirus*), *Clostridium* spp. is dominant during fall while in spring a shift towards *Cetobacterium* spp. is observed. The microbiome of tilapia is characterized by an increase in *Micrococcus* spp. and *Flavobacteria* spp. during winter months.

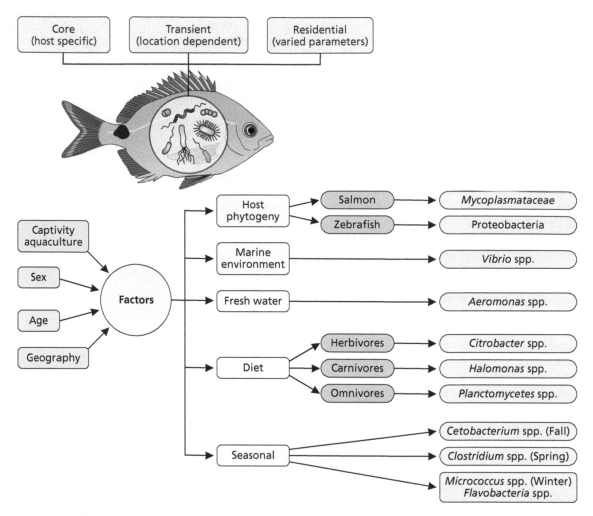

Figure 6.3. Factors shaping fish microbiome diet and seasonal habitat. Microbes residing within fishes can be grouped as core, residential and transient. The core microbes are essential for the survival of the host while the residential and transient members of the microbiome vary with numerous factors including climate and habitat. Age, diet and captivity are important modulators of the microbial population. Freshwater fishes possess *Aeromonas* spp., while *Vibrio* spp. is predominantly found in marine fishes. Herbivores harbour *Citrobacter* spp. and carnivores *Halomonas* spp. respectively.

6.3.3 Diet and microbiome

Diet is a major modulator of the microbiome. A protein-rich diet favours a higher ratio of *Firmicutes* to *Proteobacteria*. Plant oil nutrition triggers colonization by *Firmicutes*, while fish oil induces the assembly of microbial communities abundant in *Proteobacteria*. The ability of these microbes to produce enzymes like amylase, cellulase, lipase, chitinase, protease and phytase is a key factor in the selection of the most conducive microbiome to extract maximum nutrients from the diet. The Red sea surgeonfish relies on a diet consisting mainly of different types of algae. To help in the task of digesting these photosynthetic organisms, the host recruits giant bacteria referred to as *Epulopiscium*, an important component of the microbiome. In aquacultures, fishes feeding on prebiotics (fucosylated oligosaccharides, FOS) and probiotics (*Lactobacillus* spp.) display an increase in microbes exhibiting an aerobic lifestyle. A lack of food or an unpredictable food source can also provoke a reprogramming of the microbial population inhabiting fishes. Starvation induces the assembly of a microbiome composed of bacteria with a range of metabolic activities suited for utilizing diverse nutrients and requiring less energy to function. Such a situation is punctuated with a dramatic enrichment of *Bacteroidetes* followed by a reduction in *Proteobacteria*. The genus *Bacteroides* known for its enzyme systems to

BOX 6.1 TEMPERATURE AND SALINITY: MODULATORS OF MICROBIAL ASSEMBLY

Fishes are poikilothermic organisms, i.e. they are unable to control their body temperature and their metabolic processes are adapted to the temperature of the environment. Thus, it is not surprising that numerous fishes migrate to warmer climate when the temperature is below freezing. Although they are known to produce anti-freezing agents like glycoproteins and polyol to survive, they also tend to house cold-tolerant bacteria in order to adapt to low temperatures. The microbiome constituted of cold-resilient microbes is able to shape the tolerance of the host to any temperature change. For instance, fishes living in colder waters tend to possess microbes more adapted to such an environment compared to microbes residing in a host from warmer surroundings. For instance, the arctic char (*Salvelinus alpinus*) has psychrophilic microbes composed of the genera *Psychrobacter* and *Flavobacterium*, while cold-adapted tilapia has more microbes resilient to cold than tilapia living in a warmer water body (**Box Figure 1**).

Anadromous fishes are accustomed to their annual migration to warm freshwater in winter to escape freezing and return to the marine environment in summer when nutrients abound. The change in salinity has also to be properly dealt with if both the host and the microbiome are to make this transition without any complication. The Atlantic salmon undergoes physiological changes in the gills, kidney and intestine to adjust to the osmotic change encountered during the migration from the freshwater body to a marine milieu. Marine fishes have to intake water constantly in order to combat the hyperosmotic environment, a situation that results in the alkalinization of the intestine. The gills are equipped with an abundance of Na^+/K^+ATPases in order to maintain proper osmolality. These adaptations are further fortified by the acquisition of microbial partners that are tolerant to the increased salinity due to the change of habitat to seawater. Although there are some core species like

Lactobacillus spp. and *Streptococcus* spp. that are common to the salmon from both water bodies, the seawater ones are characterized by *Shigella* spp. and *Mycoplasma* spp. On the other hand, the freshwater fishes house more *Propionibacterium* spp. and have a wider diversity of microbial communities. In the arctic char, the *Deinococcus* spp. known for its adaptability to environmental stresses is increased in saline condition. The metabolism of fatty acid, arachidonic acid and cholesterol is more pronounced in the microbes residing in the intestine of the salmon from the sea. This biochemical adaptation may be reflective of change in the membrane lipid composition necessary to accommodate the osmotic tension associated with seawater. Hence, a change in salinity in the environment is mirrored in the hologenome as a whole where both the host and the microbiota have to respond to the ionic challenge.

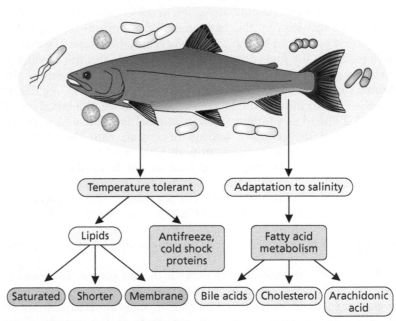

Box Figure 1. Microbes help fishes adapt to temperature and osmotic fluctuations. Modulation of anti-freeze molecules and lipid metabolism.

harvest energy from a diverse array of complex carbohydrates and to strengthen the immune system of the host is favoured. Even the increase in microbial remodelling is compartmentalized in the hindgut with an increase in the colon concomitant with a decrease in the caecum. This arrangement enables the microbes to extract simple nutrients that can be utilized by the host.

6.4 Fish Microbiome: Genesis

6.4.1 Egg and larva

To understand the genesis of microbiome in fish, it is important to have an appreciation of the different stages of the lifecycle of this organism. Each of these phases is associated with bacteria that contribute to the development of the fish and become an integral part of the host. Even though there is remodelling of the microbial residents in the same way there is reprogramming of the visible organs, fishes are composed of a microbial population that manifests itself in different specificity and abundance throughout the lifecycle of the host. Following the fertilization of the eggs that are laid in open water bodies, these developing entities come in contact with the microbes residing in the ecosystem. Both stochastic and deterministic factors come into play in the establishment of the microbial community within and on the eggs. Although the presence of microbes in a given environment acts as the pool of microbial population from which the microbiome is assembled, there are specific host-driven cues that mediate the selection of microbial cells associated with the eggs and the sperms.

The biopolymers within these reproductive cells may promote certain select microbes. The fertilized eggs are quickly colonized by microbes, a phenomenon that continues on during the embryonic stage and beyond until a relatively stable microbiome is established at the adult stage. There is an intricate link between the modulation of the microbial community and the various developmental processes the host undergoes that has yet to be fully deciphered. The host-microbe and microbe-microbe communications are pivotal in the genesis of the supra-organism. The larval phase begins with its hatchment from the embryonic chorion that may have programmed receptors designed to select only certain bacteria and thwart the invasion by the pathogenic or opportunistic ones. The larva is quickly colonized by the ovum debris and the intake of water by the larva results in the ingestion of microbes that initiates the seeding of the invisible organ. This then becomes an integral component of the developing organism. Thus, this initial passage of water and egg debris enables the colonization of the larval gut. For instance, the larval gut of the farmed gillhead sea bream (*Sparus aurata*) is populated with the anaerobic *Ruegeria mobilis*, a marine bacterium with the ability to produce biofilms and the antibiotic tropodithietic acid, two moieties that may help promote the developmental process. The larva of sea bass has *Vibrio anguillarum* when fed with rotifers, while *Vibrio alginolyticus* is an important microbial partner in the larva nourished with shrimp brine. The dominant microbes in larva of silver carp are *Proteobacteria*, *Actinobacteria*, *Bacteroidetes*, *Firmicutes* and *Cyanobacteria*. However, as the larva proceeds to juvenile stage reconfiguration of the microbial composition occurs.

The fries and the juvenile acquire bacteria that are purposely chosen to be part of the microbiome. In marine juvenile fishes, *Vibrio* spp., *Pseudomonas* spp. and *Flavobacterium* spp. are most abundant. Although the microbes in the environment do contribute to the genesis of the microbiome, bacteria distinct from the environment are also involved. For instance, in Coho salmon, *Pseudomonas* spp. located in the eggs and juveniles does not have its origin in the environment. A vertical transmission may be at play as in the case of insects. In rainbow trout, gut microbes are detected before the first feeding. The further development of the juvenile and the formation of organs are punctuated by the reconfiguration of the microbial population of the organism. Ontogenetic factors, environment and nutritional habits contribute to generate the microbiome of the fish that is located in most organs the organism possesses. The gut, skin, gills, reproductive organs and specialized bioluminescent organ are the main locations where the microbes are housed. This symbiotic arrangement that is mutually beneficial necessitates an intricate molecular cross-talk resulting in the proper density of microbial cells and the right anatomical features to establish a functional holoorganism. The gills of freshwater fishes have *Cytophaga* spp. and *Acinetobacter* spp. while those of marine fishes are abundant in *Achromobacter* spp. and *Alcaligenes* spp. The skin microbiome consists primarily of *Gammaproteobacteria* and an archaeal microbial lineage termed *Thaumarchaeota*, a phylum analogous to some microbes in the human skin (**Figure 6.4**).

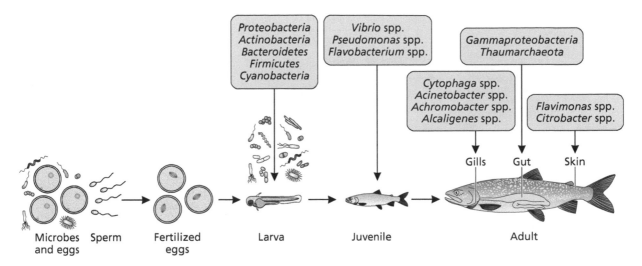

Figure 6.4. Genesis of the microbiome and lifecycle of fish. The microbiome of a fish varies with its lifecycle reaching a relatively stable population during adulthood. The egg and sperm attract microbes in the open body of water where they are released. Following fertilization and development of the larva, microbes are further selected. The production of biofilm and antibiotics like tropodithietic acid helps in the genesis of the microbiome. The juvenile and the adult stages also contribute to further reconfigure the assembly of the microbial residents. The microbes tend to be specific to the organs where they reside.

6.5 Microbial Colonization of the Gut

The gut contains the most microbes, a situation common in a wide range of vertebrates. The digestive tract, made up of the foregut, midgut and hindgut is home to the highest amount of microbes. Approximately 10^8 cells have been reported to populate this organ that begins from the posterior edge of the gills with an oesophagus. In the European cuttlefish, *Vibrio* spp. is the dominant microbe in this organ that is preceded by the stomach and the midgut. The latter region harbours microbes which have enzymes known to contribute to the digestion of the foods the fish consumes. The hindgut is the longest part of the digestive tract and plays an important role in the harvesting of energy from food sources known to be recalcitrant to the regular digestive enzymes. In fact, like in mammals, the length of the digestive tract is indicative of the nutritional habit of the fishes. Herbivorous fishes possess the longest alimentary canal usually three times the length of the organisms, while carnivores have the shortest system. This organ presents an environment with disparate pH and O_2 and tends to attract microbes reflective of the ecological niches they encounter. There is a wide diversity of species ranging from aerobes, facultative anaerobes, obligate anaerobes, acidophiles and alkaliphiles. Although nearly 90% of the microbes belong to the phyla *Proteobacteria, Firmicutes* and *Bacteroidetes*, bacteria from *Fusobacteria, Actinobacteria, Actinobacteria, Tenericutes* and *Verrucomicrobia* phyla are also present. The microbial assemblages depend on a variety of factors including genetics, feeding habits, environment and age. *Silurus meridionalis* (Chinese catfish) have species representing *Fusobacteria, Firmicutes, Proteobacteria* and *Bacteroidetes* phyla. The increased presence of *Firmicutes* confers a faster growth rate than *Bacteroidetes*. In tilapia (*Oreochromis niloticus*), the stomach contains the least amount of microbes (0.6%) and harbours predominantly *Flavimonas* spp. while the hindgut has a microbial population of *Citrobacter* spp. and *Burkholderia* spp. as its main residents.

The ability of the microbes to attach to the mucosal epithelium of the intestine is central to this symbiotic relationship. The surface receptors on this tissue enable the proper microbes to create an appropriate ecological niche where they can establish and contribute to the well-being of the host. They become permanent (autochthonous) dwellers or part of the core microbiome. Some allochthonous (temporary) residents may also find a home here. However, parasites are rapidly eliminated, a

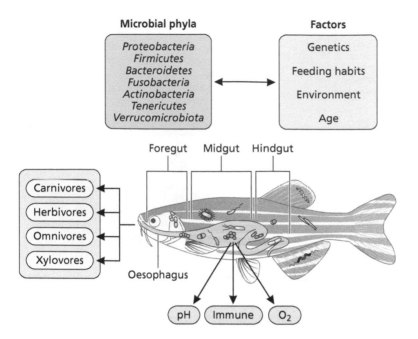

Figure 6.5. The microbiome in the fish gut. The fish gut houses the most bacteria with the hindgut being the component lodging the highest amount of microbes. Carnivores have the shortest gut, while herbivores have the longest. The microbial selection is guided by genetics, the pH, O_2 and modulation of the immune system. Enzymes like cellulases, hemicellulases, ligninases, proteases and chitinases found within the microbial residents contribute to the nutritional habits of fishes. These aquatic organisms possess more than the seven phyla of bacteria. *Firmicutes* tends to promote a faster growth rate than *Bacteroidetes*.

task also aided by the resident microbes. The differential microbial composition of the intestine helps train the host immune system to guard against opportunistic bacteria. It is also becoming evident that the genesis of the microbiome and the development of the host operate in tandem, i.e. each process taking cues from the other. The formation of the various organs exhibiting disparate structures with a well-defined chemical and physical environment that can accommodate a very select group of microbial community will indeed necessitate input from all the partners before such an event can ensue. These symbiotic pockets within the organism indicate that the microbiome and the fish evolved concurrently as a holoorganism. The size of the intestine geared to the feeding habit of the fishes undoubtedly provides further evidence for such an occurrence. In fact, the imaging of the genesis of the microbiome within Zebra fish provides tantalizing visual evidence that the microbiome is an important driver in the development of fish and the associated physiological processes. Intestinal maturation, education of the host's immune system, nutrient absorption and metabolism are guided by an intricate communication network between the microbial partners and the host. Hence, the presence of *Cytophaga* spp. and *Flexibacter* spp. in Atlantic halibut compared to the association of *Pseudomonas* spp. and *Aeromonas* spp. with herring illustrates the specific nature of the community of microbes that fishes acquire as they develop. The biogenesis of specialized organs like the air bladder and gonads is also accompanied by the assemblage of specific microbial communities that play an important role in the physiological functions these components are dedicated to (**Figure 6.5**).

6.6 Functions of the Microbial Residents in Fish

6.6.1 Skin microbiome

The microbiome is an integral component of all multicellular organisms and contributes to numerous activities the host engages in order to proliferate and survive. The fish skin is the most exterior part of the organism and thus is in direct contact with the natural environment. Microbes residing within this expansive space are usually housed in the

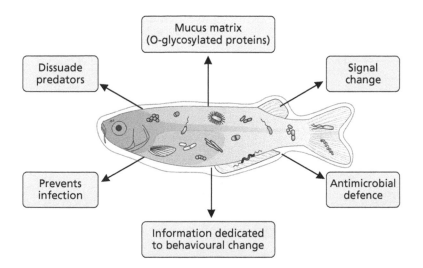

Figure 6.6. Functions of microbial constituents of fish skin. The skin is usually covered by a slimy mucus elaborated jointly by the host and its microbes. This epicellular layer plays an important role in assembling the proper microbiome and thwarting pathogenic invasion. The microbes residing within are pivotal for signalling, alerting dangers and fortifying the host immune. This mucus layer is infused with enzymes like phosphatases and esterases aimed to prevent infection.

mucus matrix consisting of O-glycosylated proteins that are elaborated by the epithelial cells with the assistance of the microbial partners. In fact, this biopolymer usually acts as an attractant to select the designated microbial community and as a repellent against opportunistic microbial invaders. It does not only play a pivotal role in the establishment of the symbionts but also participates in a variety of other functions. This nutrient-rich viscous layer dissuades predators, prevents infection, signals danger and relays information regarding shoaling, feeding, resting and spawning behaviours. It also possesses antimicrobial properties and helps diminish friction during swimming. The viscosity of the mucus impedes pathogen mobility and thwarts attacks by predators. It is infused with lysozyme, immunoglobulins, antimicrobial peptides, phosphatases, esterases, transferrin and phospholipids. These moieties are in fact involved in numerous biological responses. While lysosome and antimicrobial peptides help combat pathogens, transferrin is known to starve organisms of Fe, an essential nutrient. In rainbow trout, the microbe *Flectobacillus major* secretes sphingolipids that elicit the synthesis of IgT (immunoglobulin specific to teleosts). This response prompts the priming of the host immune system. The IgT covers the surface of the symbionts and regulates the microbial population within the fish. Depletion of this immunoglobulin results in dysbiosis and diseases. The presence of glycosaminoglycan is indicative of an alarm signal and the release of phosphatidylcholine serves as a deterrent. Hence the mucus, home to the skin microbes is replete with molecular information that signals the execution of numerous tasks that ensure the well-being of the supra-organism (**Figure 6.6**).

6.6.2 Microbial Residents within the Gill

Like the skin, the gill an organ in contact with the external milieu harbours a variety of microbes. This organ known for its respiratory attribute is covered with mucus abundant in bioactive molecules where a select microbial community is engaged in diverse tasks. Recycling and removing wastes such as ammonia and urea are important communal functions these microbes execute. The gill is high in oxygen and nutrients, conditions conducive for microbial colonization. *Nitrosomonas* spp., *Ferrovum* spp. and *Shewanella* spp. are known to occupy this area and invasion by pathogens is strongly dissuaded. These organisms are ammonia oxidizers and do possess the metabolic networks to degrade urea, two nitrogen wastes secreted by fishes. Thus, nitrogen-rich gill is a prominent site for microbial symbiont proliferation where the biogeochemical cycle of N_2 proceeds in a highly effective manner. The oxidation of ammonia to nitrite and nitrate is further processed to gaseous N_2 via reactions mediated by enzymes like nitrite, nitrate and nitrous oxide reductases. This biochemical

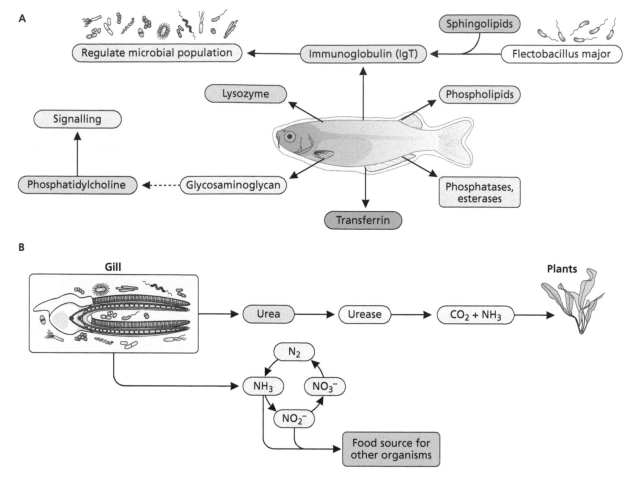

Figure 6.7. Biochemical activity within the cutaneous mucus and the gill. The molecules lodged in the mucus within the skin and gill are responsible for numerous functions. Transferrin, a glycoprotein with strong affinity for iron deprive pathogens of this essential ingredient; lysozyme can readily degrade membrane rendering microbes ineffective. Sphingolipids mediate the synthesis of the immunoglobulin IgT, a modulator of microbial population. Phosphatidylcholine is a signalling metabolite mediating numerous behavioural changes. The microbes housed in the gills have such enzymes as urease, ammonia oxidizers and nitrate reductases that contribute nitrogen nutrition and cycle.

transformation protects the fish from ammonia toxicity and replenishes the global N_2 content of the atmosphere. *Ferrovum* spp., a microbe known to harness the energy generated from the redox reaction of Fe also aids in this process and may play a role in combatting infections by sulphur-reducing bacteria like *Desulfobacter* spp. responsible for enteritis in fish (**Figure 6.7**).

6.6.3 Gut Microbes and Fish Nutritional Lifestyle

The gut houses the majority of microbes in fishes. Its concentration ranges from 10^7 to 10^{11} bacteria per gram of intestinal content with variation observed amongst the stomach, pylorus, midgut and hindgut. Nutrient retention time, pH fluctuation and O_2 gradient are factors that modulate the quantity and nature of microbes colonizing this part of the fish anatomy. During development, cellular proliferation and epithelial renewal are promoted by β-catenin, a protein stabilized with the assistance of resident microbes like *Aeromonas veronii*. This biomolecule plays a pivotal role in cellular growth and adhesion. The intestinal microbes are known to participate in digestion as they express a series of hydrolytic enzymes. In fact, the nutritional habits of fishes and the morphology of the intestinal tract are shaped by these microbial partners.

Chitinase cleaves chitin, a polymer of *N*-acetyl glucosamine found in crustaceans; cellulase hydrolyses cellulose a constituent of algae into glucose; dehalogenase contributes to the metabolism of halogenated

7.2.2 Microbiome: a heritable trait

The prenatal and postnatal evolution is reliant on the microbial communities the egg possesses. It is becoming increasingly evident that the egg acquires gut microbes during maturation. The female can transmit microbes prenatally and thus can modulate the development process. Hence, vertical transmission occurring during ovogenesis is an important determinant of the bird microbiome. In fact, the first faeces the hatchlings produce before leaving the eggshell are known to contain microbes that are associated with the gut of the female parent. The similarity in the microbial communities in these hatchings and the birds is indicative of a parental transfer during the biogenesis of the embryo. The maternal microbes from the oviduct may interact with all the egg compartments as the intestinal microbes may utilize the caecal route to reach this part of the reproductive organ. Hence, bacteria from the cloaca and the intestine can be readily transferred to the eggs. In rock pigeons and chickens such a parental transfer has been reported. For instance, in rock pigeon, the neonates and the females share approximately 20% of *Actinobacteria* and γ-*Proteobacteria*. Hence, microbiome in birds as in other vertebrates is a heritable trait that is transmitted from one generation to another.

7.2.3 Hatchlings and their microbes

As the hatchings emerge from the eggshells, they come in contact with the microbes within the nest, the female bird, the ecological habitat, the other offspring and the food they consume. The microbes they acquire play a critical role in their development. The gut microbes influence the growth, the physiology and the general fitness of the adult bird. Depending on the species of birds, the initial growth that occurs in days is characterized by very marked morphological changes. For instance, juvenile ostriches increase their body mass five-fold in the first three months. This stage of the bird lifecycle is typified by a continuous rearrangement of the microbial communities the juveniles possess. However, the microbiome attains a degree of stability during adulthood, an observation mirrored in the other visible organs. As microbial diversity, specificity and functionality vary with growth, it is evident that microbes help guide the overall ontogeny in birds. In fact, organisms devoid of or prevented from acquiring microbes during the development process suffer from diminished sizes of such organs as the lungs, heart and liver. They tend to have a poorly functioning immune system, abnormal gut morphology and altered metabolism. Owing to the rapid morphological changes accompanying juveniles, bacterial communities are relatively dynamic and more transient. The gut of young ostriches is dominated by *Akkermansia muciniphila*, a microbe known to metabolize glycoproteins and lipids important constituents of the yolk that they are exposed to early in their lives. In some birds, the residual yolk in the abdominal cavity constitutes 20% of the weight of the neonates and contributes to the recruitment of the select microbial communities. Hence, during the early stage of development, microbes mediating the metabolism of yolk are abundant and diminish as the source of food becomes disparate.

7.2.4 Less diverse microbes in hatchlings promote growth

Microbial communities in hatchlings usually tend to be less diverse and less abundant compared to those in juveniles and adults. The limited diversity is postulated to confer increased growth potential to the hatchlings. In species like the hoatzin, penguin, arctic shorebirds and great tits, taxonomic composition during weeks 1–2 is marked by a higher abundance of *Firmicutes* in comparison to *Proteobacteria*. This microbial profile may facilitate growth since *Firmicutes* are known for their ability to metabolize organic acids and mediate the deposition of fats, two features that help accelerate weight gain and contribute to ontogenetic functional shift. In fact, *Lactobacillus* spp. has been recognized to have a negative influence on growth due to its ability to modulate intestinal bile metabolism. A higher growth rate is linked to a limited microbial diversity. An increase in diverse microbial communities at an early age is accompanied by stunted growth, a condition that can be reversed with the treatment of antibiotics. As the hatchlings start to grow, a shift in diet helps shape their microbial residents. The colonization and extinction of microbial communities occur in succession until a more stable microbiome is established as adulthood progresses. The rearing environment and how the developing hatchling acquires its nutrients are important in selecting the nature of microbes that populates the gut. Mothers and hatchlings share numerous common microbes and the gut microbial communities are also transferred during regurgitation. In altricial birds, regurgitation may enable vertical transmission of the mother's microbes to the hatchlings. The nest that may house other siblings and parents coupled with the interactions with the ecological milieu can add to the microbial diversity of the growing juveniles. As precocial young birds capable of moving around after hatching leave the nest and engage in foraging, the parental transmission may be limited. Faeces in the nest are an important source of microbes that may contribute to the genesis of the hatchlings' microbiome. Following the successive colonization and extinction of microbial communities orchestrated by the genotype of the bird (species), the age of the hatchlings, the dietary switch, the rearing environment (feeding and nest), a relatively taxonomically diverse microbial assemblage becomes an integral part of the juvenile that then proceeds to the adult phase.

7.2.5 Adults and their microbial partners

The microbiome in adult bird located mainly on the skin, feathers and the intestine is moulded on the basis of a variety of factors that act as filters to select the proper microbial assemblage. Immune system, sex, genotype, habitat, climate, diet, mobility and migratory behaviour are some of the determinants responsible for the maturation of a relatively stable microbial population that subsequently fulfils numerous functions in partnership with the other organs in adult birds. As migratory birds encounter a wide range of habitats, their microbial communities tend to be more diverse than the resident birds that spend most of their lives in a relatively specific region. Furthermore, after a long flight the butyrate concentration in the blood increases, an event that decreases food intake. This situation is a modulator of the microbial gut colonization. The foraging habit of the birds is also an important factor that helps shape the microbial community, a phenomenon commonly observed in most multicellular organisms. The morphological changes associated with the gastrointestinal tract of migratory birds that usually undergoes a 30% reduction in space attract disparate microbial communities. This transition in the gut also results in different dietary requirements, a condition necessitating colonization by a unique set of specialized microbes.

Avian gut is usually dominated by the phyla *Firmicutes* and *Proteobacteria* with lower abundance in *Bacteroidetes*, *Actinobacteria*, *Tenericutes* and *Fusobacteria*. For instance, *Proteobacteria* constitutes approximately 25% of the microbial population compared to 1% in humans. In urban Canada geese, *Firmicutes* predominate while in cormorants, *Fusobacteria* is the most abundant microbes. The black-headed gulls have a high abundance of the genus *Catellicoccus*, a member of the *Firmicutes* phylum. *Firmicutes* are Gram-positive bacteria usually facultative and obligate anaerobes consisting of genera like *Bacillus*, *Clostridia* and *Mollicutes*. They are known to improve metabolic efficiency, promote nutrient uptake and produce short-chain fatty acid (SCFA). Supplementation of chicks' diets with *Bacillus subtilis*, a probiotic, results in enhancement of growth rate. Heavy and flightless birds like emus and ostriches tend to have

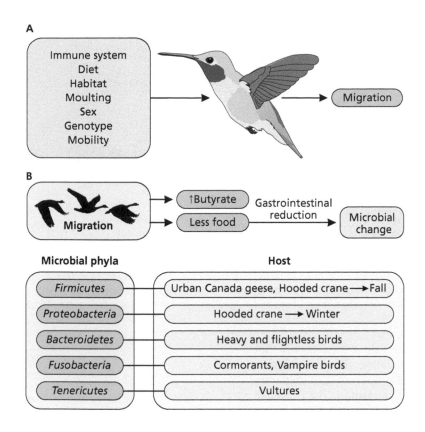

A

Immune system
Diet
Habitat
Moulting
Sex
Genotype
Mobility

Migration

B

Migration

↑Butyrate

Less food

Gastrointestinal reduction

Microbial change

Microbial phyla | Host

Firmicutes — Urban Canada geese, Hooded crane → Fall

Proteobacteria — Hooded crane → Winter

Bacteroidetes — Heavy and flightless birds

Fusobacteria — Cormorants, Vampire birds

Tenericutes — Vultures

Figure 7.4. Factors affecting microbial residents in adult birds. (a) A variety of factors including immune system, migration, habitat, moulting behaviour, genetic factors and sex are key determinants of microbes residing within and on birds. (b) In migratory birds, an increase in butyrate signals less food intake, an event resulting in major physiological changes in the gut. This is followed by the reprogramming of the gut microbe. The genetic make-up of birds is an important selector of the microbial community. For instance, Canada geese has a higher abundance of *Firmicutes*. In the hooded crane, fall is characterized by an increase in *Firmicutes* while in winter *Proteobacteria* abound.

higher amounts of *Bacteroidetes* and a microbial population that is less diverse than volant birds. The microbes belonging to this phylum are Gram-negative, proliferate in a wide range of O_2 gradient and range from being strict aerobes to obligate anaerobes. They belong to such genera as *Bacteroides*, *Prevotella*, *Porphyromonas* and *Flavobacteria*. The ability to tolerate a sharp O_2 variability and their cellulose-degrading characteristic are attributes these birds feeding on mostly a plant-based diet depend on. Thus, herbivory favours an abundance of *Bacteroidetes*. The blood-feeding vampire bird has a gastrointestinal tract that houses a high amount of *Cetobacterium somerae* of the phylum *Fusobacteria*. This microbe is known for its vitamin B_{12}-synthesizing ability, an important cofactor in various biochemical reactions. Birds like vultures reliant on carrions and meat products tend to have highly acidic stomach and gut rich in *Firmicutes* and *Fusobacteria*. They can also be recalcitrant to the tissue-degrading *Clostridia*. The nutritional habits of birds are dictated by the microbial communities they possess (**Figure 7.4**).

7.3 Climate, Sex and Moulting: Microbiome Selectors

Climate is a significant selective force in the maturation of the microbiome in adult birds. In the hooded crane, the gut is characterized by an abundance of *Firmicutes* during fall while with the arrival of winter switches the microbial communities to *Proteobacteria* that tends to predominate. The winter months also are marked by a relative increase in *Cyanobacteria* coupled with a noticeable rise in *Tenericutes* in the fall. The nature of the host species also plays an important role in the selection of the avian microbiota. For instance, siblings of the great tit in the same nest have similar microbial communities compared to conspecific young birds in other nests. Young magpies and the parasitic young great spotted cuckoos in the same nest are colonized by different microbes despite residing in the same nest. Genetic profile of the birds is at play in a manner analogous to humans.

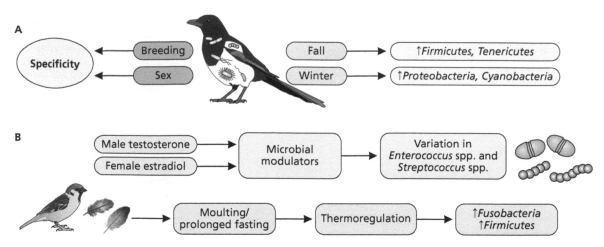

Figure 7.5. Climate, sex and moulting: microbiome selectors. (a) Climate and sex are important in the assembly of a unique set of bacteria. Breeding season also results in a shift of microbes. As fall and winter necessitate a marked change in physiological processes, different microbes are recruited. (b) Testosterone and oestradiol prompt changes in the abundance of *Enterococcus* spp. and *Streptococcus* spp. Moulting, fasting and thermoregulation help recruit disparate microbes. These phenomena signal an increase in *Fusobacteria* and *Firmicutes*.

Breeding season and the sex of the birds are selective elements that determine the specificity of the microbial population these vertebrates enter into partnership with. As males and females operate with disparate reproductive physiology, dissimilar microbes are recruited. The presence of testosterone in males and oestradiol in females may contribute to this disparity. In male and female bobwhites there is significant variation in the microbes belonging to the genera *Enterococcus* and *Streptococcus*. The levels of hormones that result in increased or decreased copulation promote an increase or decrease in microbial transmission. Sexual behaviour favouring multiple partners is also a determinant in the microbial dynamics. Hence, breeding season is associated with a rise in diversity of microbes in males.

Moulting, the period of prolonged fasting is a common phenomenon in seabirds. During this phase of their lives, these birds lose all the plumage and resynthesize new feathers. Thermoregulation and feather synthesis require energy despite the absence of food intake. The seabirds with limited plumage are predominantly confined on land and have to rely on their fat storage to survive. Maintaining body temperature when the plumage is being reconstituted necessitates a major adjustment in cellular processes. There is a significant change in the microbiome that aids these seabirds in the moulting behaviour. There is a sharp increase in *Fusobacteria* that is reflected in the faeces followed by *Firmicutes*. The lack of nutrient intake in addition to the diminished energy-generating pathways results in a major shift in microbial communities. Microbes with metabolic networks favouring tissue degradation, glycan synthesis and butyrate production are selected. The latter is responsible for energy formation and in promoting cellular development in numerous organisms. A microbial cellular component like other body parts of the seabirds undergoes marked adjustment in order to adapt to the regeneration of their plumage (**Figure 7.5**).

7.4 Gut Microbiome and Their Functions

The gastrointestinal tract houses the most microbes in birds and other vertebrates. Its morphology, varied pH and O_2 gradient provide a home to a wide assortment of microbes. This anatomical habitat, the foraging characteristic, sex, climate, age and host fitness determine the nature of the microbiome a bird possesses. The microbes in return also shape the morphological features associated with the host. Hence, all these factors come into play in the genesis of the holobiont, i.e. the host (bird) and its invisible partners. This selective pressure ensures the colonization by symbionts and commensals and diminishes the opportunity for pathogens to invade the organism. The lack of mechanical digestion in birds is compensated by the presence of the crop and the gizzard where food can be processed. The oral cavity is followed by the crop, the stomach (proventriculus, gizzard), the small intestine, the caeca, the large intestine and the cloaca. This morphological arrangement is punctuated with varying pH and a decreasing O_2 gradient with the colon being most anaerobic and the stomach presenting the most acidic condition. The cloaca is exposed to microbes from the digestive tract and the urogenital system. The acidic proventriculus and gizzard provide a proliferating environment to acidophiles. *Lactobacillus* species is an important group of microbes in this region. The small intestine harbours microbes involved in food processing and nutrient absorption. The caeca associated essentially with herbivorous birds are elongated protrusions between the boundary of the small and large intestines. This part of the digestive tract is replete with *Bacteroidetes* known for their ability to cleave glycosidic bonds associated with plant fibres. The absorption of electrolytes, H_2O and the fermentation occurs in the large intestine, a component of the digestive tract dominated by *Firmicutes*. The cloaca tends to have an abundance of *Proteobacteria*. These microbes also form a protective barrier by adhering to the epithelial walls and out-compete pathogens, thus guarding against infections. The caecal bacteria can also supplement nitrogen nutrition by metabolizing uric acid into NH_3.

7.5 Microbial Enzymes, Extraction of Nutrients and Detoxification

The digestive tract contains the most microbes in birds. These microbial partners are involved in a myriad of functions. They help in digestion as these microbes possess enzymes that are not part of the birds' genetic profile. They facilitate the metabolism of dietary polymers. In the crop, the well-developed foregut of herbivores, the pre-digestion mediated by microbes, aids in the optimization of nutrients extracted from diets. In fact, the crop of the hoatzin is endowed with methanogenic bacteria and archaea that resemble the microbes found in ruminants. As humming birds rely predominantly on a carbohydrate-rich diet, they depend on nitrogen nutrients derived from their microbial partners. Birds are often exposed to toxins in the environment especially found in plants they consume. The plant defence compounds like phenol, saponins and tannins have to be detoxified if the birds are to survive. The gut microbes are equipped with hydroxylases, sulfatases and other detoxifying enzymes dedicated to the neutralization of these toxins. Microbes residing in the intestinal tract help fortify the immune response of the host. An offshoot of the intestine near the cloaca is referred to as the bursa and is responsible for the production of β-lymphocytes. This anatomical feature is colonized by bacteria from the beginning of the development process following hatching and helps programme the immune machinery. This continues until their sexual maturity when the bone marrow is charged with the synthesis of the immune cells. This switch in function is followed by the atrophy of the bursa. Microbes belonging to the phylum *Firmicutes* play a critical role in this phenomenon as they stimulate the immune response and interact with pathogens. The varied functions of the gut microbes are essential for the proliferation and survival of the birds. This partnership between the visible and invisible components of this vertebrate ensures its biological success (**Figure 7.6**).

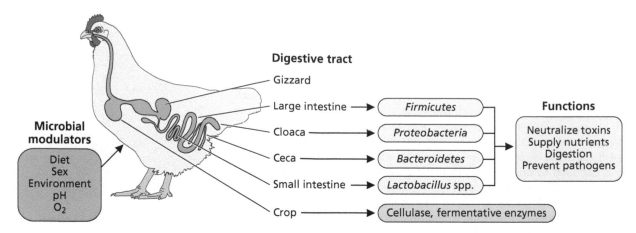

Figure 7.6. Microbes of the digestive system and their functions. The digestive tract houses the most bacteria. Owing to the disparate components of this organ and the varied environment it provides, the alimentary canal is home to a wide range of microbial communities. Furthermore, the nutritional habit of birds is a strong modulating factor of the microbes able to colonize the gut. While the crop possesses bacteria with a range of hydrolytic enzymes, anaerobic microbes are more prominent within the large intestine. The caeca have an abundance of *Bacteroidetes*, the cloaca is populated by *Firmicutes*. These microbes are involved in digestion, supply nutrients, defend against pathogens and eliminate toxins.

BOX 7.1 VAMPIRE FINCH AND VULTURES: UNIQUE DIETS AND MICROBES

Microbes in Blood-Consuming Vampire Finch

Owing to the disparate habitats they live in, birds have evolved to feed on a wide range of foods. They can consume vegetation, insects, fruits, carrion and blood. While waterfowls rely on a diet constituted of vegetation, vultures are adapted to derive their energy from carrion. Although the morphological organization of the gastrointestinal tract is reflective of these dietary habits, the effectiveness of the digestive process is reliant on the presence of the select microbial community that not only mediates the hydrolysis of complex biomolecules but also provides essential nutrients like vitamins and amino acids. Challenging diets such as blood or carrion necessitate the presence of unique enzymes. The gut microbes with a range of hydrolytic and synthetic capabilities are able to fulfill the nutritional requirements of the hosts by releasing nutrients from these unusual food sources.

The vampire finch (*Geospiza septentrionalis*), a native of the Galapagos Island, has to resort to a narrow limited diet during the dry season when food is scarce. The bird feeds essentially on blood and eggs of blue-footed boobies, a diet that may be supplemented with regurgitated fish consumed by marine animals. Blood contains a select group of nutrients like lipids and proteins with an abundance of haemoglobin, a biomolecule known to release iron and haem derivatives upon degradation. It is deficient in vitamins. This extremely specialized source of food requires a microbial community not only with the ability to release simple nutrients that can be readily absorbed but also to tolerate the toxicity associated with elevated amount of iron and viruses. *Firmicutes*, *Proteobacteria* and *Actinobacteria* are the most prominent phyla associated with this vampire bird. Bacteria belonging to the phylum *Bacteroidetes* known for their polysaccharide-degrading properties are notably rare in stark contrast to other finches that are herbivorous. *Fusobacterium* spp., *Campylobacter* spp. and *Mucispirillum* spp. are important members of the vampire finch's gut as is *Clostridium* spp. common in carnivores like vultures and alligators. *C. somerae*, a microbe with the genes to synthesize a range of vitamin B, is also present. These vitamins are critical in the metabolism

of folate, ubiquinone and methionine, biomolecules essential for cellular proliferation. Adaptation to any diet is essentially guided by the microbes the host harbours. The shaping of the microbial community within birds is in turn dependent on numerous other factors. The shift in the microbial population occurs on an ongoing basis in tandem with the availability of different food sources. This biological arrangement enables the host to adjust to changes in the environment without undergoing permanent morphological re-organization.

Vultures Depend on Their Microbes to Feed on a Carrion Diet

Vultures are another group of birds that rely on a very specific and demanding dietary niche to acquire their nutrients. They feed on carrions and are referred to as nature's cleaning crew. Without their usually unappreciated scavenging service, the decaying carcases would become a major source of contamination by pathogens known for such deadly diseases as anthrax and brucellosis. To partake in such a dietary habit that is fraught with danger, vultures depend on their microbial partners both to intake the food and promote its digestion so that proper nutrients can be extracted. The decomposing carcass is replete with pathogenic organisms and toxic constituents like aromatic hydrocarbon, botulinum toxin and putrescine they generate. The facial skin of the bird is in direct contact with the food source as the bird rips the carcass before ingesting. The risk in this exercise can result in significant contamination. To prevent such an occurrence this body part is associated with a microbial community with the necessary genetic information aimed at fending pathogens and their toxins. In fact, vultures possess more microbes on their skin than in their gut. These microbes not only produce antimicrobials, antifungals and insecticides but also dissuade colonization by deadly bacteria like *Yersinia pestis*, an agent of numerous plagues. They also possess metabolic pathways to degrade putrescine and phenolic compounds. While *Hylemonella gracilis* prevents infection by *Yersinia pestis*, *Carnobacterium* spp. and *Lactobacillus sakei* impede attacks by *Listeria monocytogenes*, a promoter of listeriosis. Flies that are part of the living landscape associated with carrions are eliminated with the aid of *Pseudomonas sakei*. *Arthrobacter phenanthrenivorans* is armed with enzymes mediating the metabolism of phenanthrene, other polyaromatic hydrocarbons and a variety of xenobiotics released by the decomposing carcass. The bacteria residing on the facial skin produce antibiotics such as tetracycline and macrolides to which they are resistant. This strategy allows the microbial partners to counter the pathogenic microbes without falling prey to the same antibiotics. Viruses known to target specific bacteria are also recruited on the facial skin to eliminate infective microbes like *Clostridium botulinum*. *Bacillus cereus* and *Streptococcus* spp. are some of the microbes responsible to providing vultures with a safe carrion meal. In the black vulture (*Coragyps atratus*), the phyla *Proteobacteria*, *Bacteroidetes* and *Firmicutes* are most abundant.

Once the food has been ingested, it needs to be processed within the digestive tract. Here too, the microbial community aids the different components of the alimentary canal in accomplishing this task. The highly acidic environment of the stomach contributes to eliminate any pathogen that may have escaped the vigilance of the facial skin microbes. Furthermore, phages specific to pathogenic microbes like *Escherichia fergusonii* and *Salmonella* spp. are also abundant in the gut in an effort to thwart any infection following the consumption of this risky diet. The microbes belonging to the phyla *Firmicutes*, *Proteobacteria* and *Fusobacteria* constitute the majority of the microbial residents in this part of the vulture's body. The presence of such genera

as *Herbaspirillum* and *Gordonia* helps in the provisioning of vitamins. They possess enzymes involved in the biosynthesis of cobalamin, riboflavin, vitamin B_5, vitamin B_6, folate and essential amino acids. The gut harbours an abundance of *Clostridia* including *Clostridium saccharolyticum*, a microbe with the ability to ferment carbohydrates into ethanol, acetic acid and H_2. *Clostridium butyricum* and *Fusobacterium varium*, producers of butanoic acid, are also members of the microbial assemblage. This SCFA is utilized as a source of energy and the modulation of the immune response. The metabolism of butanoic acid is further promoted by formyl C-acetyl transferase, an enzyme expressed by these microbes. The formation of the antimicrobial butyrate glycerides aids in the controlling pathogenic strain of *Salmonella*. Biofilm-producing bacteria also populate the gut as they promote the colonization of symbiotic microbes and impede the proliferation of pathogens. Hence, the multi-pronged approach of preventing the entry of dietary pathogens in the gastrointestinal tract, detoxifying the toxic emission from the carcase, aiding in the extraction of absorbable nutrients and supplying essential vitamins enable the resident microbes to assist vultures in fulfilling their crucial ecological role. Without the assistance of microbes, these birds will not survive or would have to evolve anatomical and physiological features to adapt to these diets. It is evident that most if not all multicellular organisms prefer to recruit their microbial partners to aid them in these tasks, a strategy affording numerous evolutionary advantages (**Box Figure 1**).

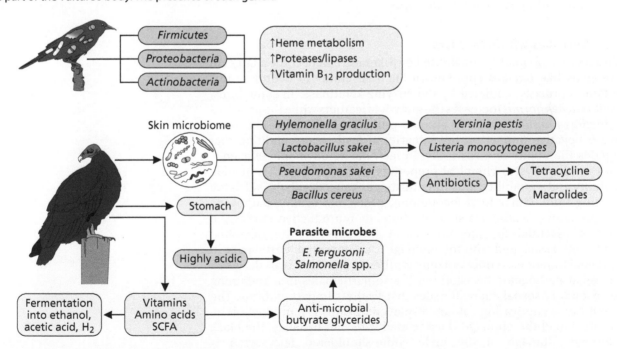

Box Figure 1. Microbial profile in a blood sucker and a carrion consumer. These unique microbial communities enable the vampire finch to consume blood and the vulture to feast on carrion.

7.6 Microbes Associated with Skin and Feathers: Contribution to Avian Life

The avian skin and the associated feathers are the primary barriers that separate the hosts from the external environment. The skin usually has a thin epidermis decorated with an abundance of feathers of varied shapes, sizes and colours. Although it is devoid of any sebaceous gland, the skin does have specialized cells known as sebokeratinocytes that are involved in the synthesis of organic compounds serving a variety of functions. The uropygial gland usually located in the dorsal section of these vertebrates also fulfils various roles including the oiling of the external anatomy. Most of the skin is covered with plumage except for the brood skin where the feather content is sparse. The microbial community of the skin appears to be host-specific and relatively homogeneous. Odour production and fending off pathogen infection are two important functions these skin-residing microbes contribute to. Allopreening, grooming of skin or feathers of a bird by another, neck pecking or other courtship behaviours do not tend to have a significant impact on the microbial population. This microbial stability may be conferred by genetic factors as odour profile in birds like songbirds is specific to each individual in a manner analogous to human scent. Hence, the signalling and recognition cues derived from odours generated with the assistance of the microbes provide important information essential for the well-being of the birds. These features aid in choosing mating partners, seeking food or evading predators. In zebra finches, the microbes such as *Acinetobacter* and *Methylobacterium* are the main contributors of these olfactory prompts.

7.6.1 Microbes within feathers

Feathers that are mainly constituted by β-keratin can become a target of pathogens like *Bacillus* spp. known for their ability to hydrolyse this protein, a process mediated by the enzyme keratinase. In some birds, *Bacillus licheniformis* increases the growth of feathers while the presence of *Bacillus pumilus* results in stunted plumage. Pathogenic infections have a negative impact on the integrity of feathers, a condition that impedes the ability of birds to fly. In fact, the aerodynamic efficiency relies on the physico-chemical properties of the feathers that include flexibility and wettability, attributes further enhanced with the aid of the preening fluid. Other than locomotion, protection against pathogens and predators, feathers also contribute to reproductive success, a function essential for the survival of the species. The microbial community associated with the plumage is an important participant in these vital functions. It does not only synthesize antimicrobials designed to combat pathogenic invasion but it also participates in maintaining the secondary sexual characteristics that feathers impart on birds. The wide array of captivating colours displayed by birds is a product of the microbial load the plumage is associated with. For instance, the black iridescent plumage of the male white-shouldered fairy wren is determined by the high diversity and low microbial load within the feathers compared to females or the brown-coloured counterparts. The preening activity supported by the uropygial gland not only acts as a selective force determining the microbial partners but also contributes to the hydrophobicity and the light-reflecting properties of the feathers. Hence, this unique integumentary arrangement constituted by a relatively thin epidermis covered with feathers is essential for flight, thermoregulation, social interaction, reproductive success and pathogenic defence. All of these attributes of the plumage are dependent on the participation of the microbial community it houses (**Figure 7.7**).

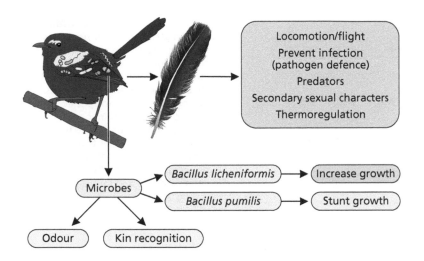

Figure 7.7. The microbial contribution to the functions of skin and feathers. The avian skin is decorated with feathers that participate in locomotion, sexual selection, thermoregulation and defence against pathogens. All these vital functions are possible due to the microbes residing within the feathers. They help in the synthesis of chemicals with characteristic odours essential for communication and kin recognition. Furthermore, these oily compounds increase the resistance of the feathers to water, a key feature for mobility. Microbes within the skin can also promote or stunt growth.

7.7 The Uropygial Gland and the Roles of Microbes Within

The uropygial gland is an avian exocrine gland that secretes a fluid containing fatty acids, alcohols, peptide antibiotics and other yet to be characterized compounds. This oily liquid is coloured and has an odour. The pungency and the colour vary with species, sex and season. For instance, during nesting season this secretion is dark in European hoopoe brooding females compared to their male counterparts known for their white uropygial fluid. The bacterial symbionts tend to be specific. In turkey, *Corynebacterium uropygiale* is the dominant microbe, while in the wood hoopoe *Enterococcus phoeniculicola* constitutes an important component of the microbial community. *Enterococcus faecalis* occurs in abundance in the gland of the European hoopoe. During the nesting season, this exocrine organ is larger and houses more microbes. The microbial population is controlled by the host and the need for the antimicrobial and anti-predator properties during nesting results in a major increase in microbial density, an evolutionary adaptation the nesting females have been bestowed with. This strategy helps provide sufficient antimicrobials like bacteriocins and impede pathogenic invasion, especially in a hole nest where the risk of infections is high. Females are known to voluntarily impregnate eggshells with this secretion, an act that ensures the success of the hatching process by preventing pathogen infection. Furthermore, the colourful eggs tend to scare away predators.

Microbes such as *Vagococcus* spp. and *Parascardovia* spp. located in the microbial community within the gland produce acetic acid, lactic acid and butanoic acid, organic acids known for their antiseptic properties. Their distinctive odours repel potential predators. Butanoic acid, the most abundant SCFA in the dark secretion is involved in a variety of physiological processes, attributes that may further contribute to the successful development of embryos. In European hoopoe, the phylum *Firmicutes* constitutes the majority of the microbes with a high abundance of *Clostridia* that is vertically transmitted from the mother's gland. The nesting environment is also an important selector of microbes residing within this exocrine system. It is clear that the dynamic microbial population housed in this body part is controlled by the host and its increased density during the nesting season provides powerful chemicals aimed at rendering the hatching process successful.

The odour emanating from the secretion is an important source of chemical information that birds rely on to choose a mate, recognize conspecifics and confuse predators. The volatile organic compounds that

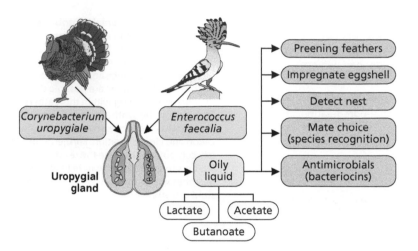

Figure 7.8. Roles of microbes within the uropygial gland and their functions. The uropygial gland houses a rich diversity of microbes known to synthesize a range of organic acids, alcohols, antibiotics and other yet to be characterized infochemicals. The nature of microbes is bird-specific, and this gland becomes larger during nesting season in order to accommodate more bacteria. The oily liquid contributes to nest detection, preening feathers and mate choice. Female birds impregnate eggshells with this liquid to deter predators and thwart microbial infection.

vary according to the different seasons are recognized by olfactory receptors. These then guide the hosts accordingly. The trace-amine-associated receptors localized in the olfactory epithelium may be responsible for detecting these stimuli. In fact, removal of the nerve connecting to these receptors and sealing the nostril results in limited social and mating behaviour. The sexual ornaments like feathers, combs and spurs also contribute to mate selection and mating exercise. These secondary features essential for the successful transmission of genetic information are under hormonal and microbial control. The preening routine may fortify these features that lead to a successful outcome. Males tend to mate preferentially with females having uropygial gland compared to those that are devoid of this exocrine system. Thus, besides releasing chemosensory cues, the uropygial gland also participates in enhancing the secondary sexual characters designed to facilitate the transmission of genetic material, central to the survival of the avian species (**Figure 7.8**).

7.8 Microbiome and Reproductive Success

The reproductive success in birds like in most vertebrates is dependent on the right choice of mate, an event that can ensure the proper transmission of genetic information. Vocal prompts, chemosensory cues and visual information all contribute to this phenomenon. The latter is manifested in form of sexual secondary features characterized by a colourful display of feathers and other external ornaments. These attributes are extravagantly manifested in peacocks known for their ability to exhibit very elaborate and eye-catching plumage. The microbial community residing within the feathers play a pivotal role in both the maintenance and the diversity of colours in these birds. Beta-keratin, the main protein component of feathers is rich in cysteine and is organized in a manner that provides a rigid structure resistant to hydrolysis. The protein matrix is infused with melanin and other pigments like carotenoids and porphyrins that are responsible for a wide array of colours ranging from green to blue. These pigments are mostly derived from diets the birds consume. The incorporation of melanin in the molecular structure of feathers not only protects against wear and tear but also adds to the visual feature associated with birds. Increased melanization confers resistance to microbial damage. This chromophore is mostly composed of eumelanin imparting a black hue and pheomelanin responsible for a brownish tone. The interactions of the melanin moieties with carotenoids result in different colourful plumages. The nanostructures constituting the feathers and their ability to absorb and reflect light also play a critical role in visual properties of this outermost avian anatomy.

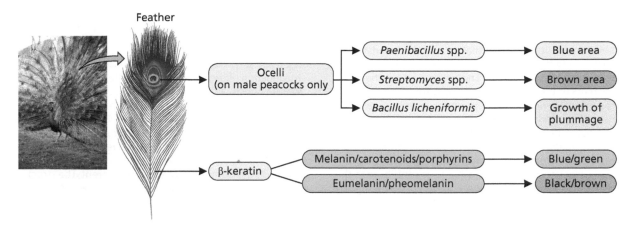

Figure 7.9. Peacock, microbes, colour display and reproductive success. The colourful feathers associated with male peacocks are secondary sexual features that are essential in attracting females and ensuring successful transmission of genetic information. Microbes residing within the external ornamental appendages contribute to the diverse hues and also aid in the healthy development of the feathers. The ocelli found only in male peacocks harbour select bacteria dedicated to specific colours. Beta-keratin a cysteine-rich protein that constitutes feathers is enmeshed with melanin and carotenoid derivatives resulting in the display of disparate colourations.

Male peacocks usually bigger in size are polygamous and can mate with numerous females that are smaller in size. Males have metallic blue feathers on top with iridescent greenish blue long trains that are decorated with circular structures known as ocelli. Following the breeding season, they moult their features. The diversity of microbes within these external ornaments is essential if the richness of colours is to be maintained. This is an important feature that females evaluate in order to choose their mating partners. The health of these exaggerated secondary characteristics is an indicator of microbial load and is reflective of the susceptibility or resistance of the host to pathogenic infection. This discerning feature ensures the production of chicks that are usually male biased. The presence of *Paenibacillus* spp. in the blue area and *Streptomyces* spp. in the brown components are indicative of healthy ocellar patches. As the male peacocks spend most of the daylight hours walking, the trains are exposed to vegetation on the ground. This activity is risky as it exposes the feathers to parasites and pathogens. Hence, it is important that the feathers are populated with the proper microbial community to thwart any infection. The abundance of *Bacillus licheniformis* is positively corroborated with growth of the plumage. The interaction of the male peacocks with the proper microbial community is essential in maintaining these secondary sexual characteristics that contribute to the success of the species. Without the contribution of the microbiome associated with these colourful appendages designed to attract females, these birds will be ineffective in transmitting their genes (**Figure 7.9**).

7.9 Chemosensory Signals: A Microbial Contribution

Odour is an important communication tool that many organisms including birds utilize to accomplish a variety of biological tasks. Location of mating partners, recognition of conspecifics, deterring predators, seeking food, relocation of burrows or nests and socialization are all dependent on the precise chemical cues generated primarily by the microbes residing on the skin, feathers and the uropygial gland. The reception and deciphering of these signals coded in the volatile organic compounds are vital if birds are to function properly. Signal fidelity is crucial as alteration of chemical messages results in inaccurate communication that can have devastating effects. The elimination of microbes with antibiotics results in change of odours and incorrect transmission of information. Like in humans, odour associated

BOX 7.2 FOSTER PARENTS AND MICROBES: CUCKOOS' BROOD PARASITISM

Avian brood parasites are birds that lay eggs in the nest of host birds where the eggs are hatched and the fledglings are raised by the foster parents. This practice allows the parasite birds to transmit their progeny without much parental input. However, such reproductive strategy can be a risky proposition as the host birds can recognize the parasitic eggs in which case the eggs are ejected or the nests are deserted leading to the demise of the parasitic species. To ensure that the eggs are hatched and properly reared, microbes associated with the parasitic eggs contribute to the diminished hatchability of the host eggs as parasite fledglings are less well fed in a mixed brood. For instance, some cuckoos that parasitize magpies' nests, lay eggs with a higher amount of *Bacillus* spp., while in the host eggs, *Pseudomonas* spp. are more prominent. Additionally, the cuckoo chicks that usually hatch earlier evict all nest content so that they become the sole beneficiary of the parental care from the foster parents. This strategy enables the survival of species post-fledgling as the juveniles are afforded the crucial parental care needed to transit to adulthood. The bacterial community within the eggs tends to be an important participant in this successful reproductive adaptation (**Box Figure 2**).

Even though the cuckoos' eggs are raised, hatched and reared by the foster magpie parents, the cuckoos have a mixed microbial profile resembling the biological and the foster parents. The hatchlings and the juveniles acquire the microbes from the environment within the nest and during the feeding process mediated by the foster magpie parents. The female cuckoo may also be responsible to vertically transmit microbes during the genesis and the laying of eggs. The cloaca via which eggs are deposited is replete with microbes and may be a possible source of the parasite cuckoo hatchlings' microbiome. In fact, eggs are known to be trans-generational carriers of microbes. As the female cuckoos have no or negligible contact with their eggs once they are laid in the magpie nests, the delivery of the microbes to their hatchlings is egg-mediated. Furthermore, these microbes may be acquired

during the development of the embryo. The cuckoo has a relatively large cloaca, a pair of appendages protruding from the junction of the small and large intestine. This anatomical feature houses a range of diverse bacteria involved in a variety of functions including nitrogen cycling, carbohydrate fermentation and immune responses. The assembly of the microbes promoted during the development of the cloaca may target the selection of microbial residents uniquely suited for the functions this organ executes and hence they are disparate from those of the foster parent. The structural feature of the cloaca of the magpie is different from that of cuckoo, a situation necessitating dissimilar microbial communities. The success of the obligate parasite brood is dependent on microbes that favour the hatching of the parasite eggs compared to the host eggs, an occurrence ensuring the parasite offspring is given the utmost attention by the host birds to reach adulthood.

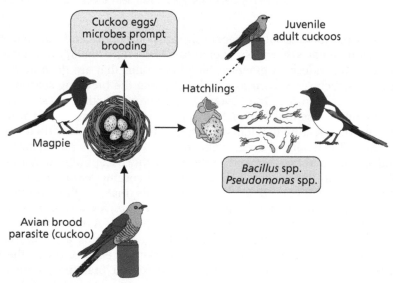

Box Figure 2. Brood parasitism: a microbial connection. The microbes that female cuckoos impart to their eggs play a crucial role in the success of these eggs to be hatched and reared by the magpie hosts.

with birds is individual specific and is dependent on the sex of the organism. Thus, each bird has an odour profile that is reflective of its genotype. These factors consequently shape the nature of the microbes residing within the birds as they are the main producers of these odoriferous compounds. For instance, symbionts residing in the uropygial glands are bird-specific and tend to be more abundant in females, especially during the brooding season. *Corynebacterium* spp., *Pseudomonas* spp. and *Burkholderia* spp. are some of the bacteria participating in the production of compounds like aliphatic alcohols and acids that have characteristic odours.

In the common blue petrel (*Halobaena caerulea*), 1-hexadecanol and 1-dodecanol help generate the specific odour profile due to the presence of microbes like *Stenotrophomonas* spp. and *Staphylococcus* spp. The specificity of these individualized odours results from the microbial selection and is utilized to assess the similarity or dissimilarity of the major histocompatibility complex (MHC). This probing of the MHC is important as numerous animals tend to preferentially mate with individuals with dissimilar MHC in order to maximize the range of antigen-binding

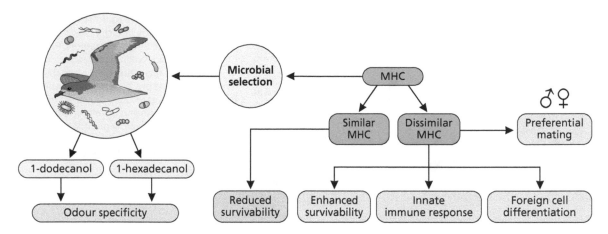

Figure 7.10. Microbe initiated chemosensory communication selection of partners and immune response. Microbial selection is key to the generation of specific odour characteristics of each bird. Responses elicited by such chemicals as 1-hexadecanol and 1-dodecanol help discern the similarity and dissimilarity of the major histocompatibility complex (MHC), a deciding factor in the choice of mate. This ensures an improved immune system and a better survivability of the offspring.

characteristics in their offspring. Such a discriminating property controlled by odours ensures the enhanced survivability of the species. It is also responsible for the innate immune response and helps differentiate between self and non-self (foreign cells). The genetic information codes for surface proteins that identify foreign bodies and present them for elimination. The microbial symbionts contributing to the characteristic odour of an individual bird are screened by the MHC to form part of the bird's cellular organization. The microbiome constituted by the microbial cells is an integral component of the bird and provides an individualized odour signature essential in chemosensory communication. The fidelity of the message and its decoding process allow the organism to accomplish a variety of tasks pertaining to reproduction and survivability (**Figure 7.10**).

7.10 Probiotics and the Poultry Industry

Poultry meat and eggs are an important food source for humans and the sustainable production of these goods is critical for global food security. Thus, it is important that the poultry industry advocates for effective strategies aimed at promoting growth of these consumable birds without comprising food quality. Promotion of gut health favouring high feed conversion ratio to high growth rate and prevention of dysbiosis are essential. The former increases sustainability while the latter limits diseases, factors resulting in improved food quality and profitability. Supplementing poultry feed with probiotics not only stimulates growth, but it also enhances egg production coupled with increased immune protection. The strengthening of the avian immune defence thwarts bacterial infection, a situation that limits the use of antibiotics. Widespread utilization of antibiotics is associated with an increase of antibiotic-resistant microbes that can further complicate the operation of the poultry industry.

Probiotics such as *Bacillus* spp. and *Lactobacillus* spp. as feed additives increase villus function and trigger improved digestion and absorption. This allows for optimal intake of nutrients from feed and fuels a better growth rate coupled with enhanced egg production. In fact, these probiotics increase the length and surface area of the villi, morphological features that ameliorate the absorption of nutrients. They are also responsible for the production of amylase, proteases, lipases and keratinases, enzymes known for their ability to maximize the extraction of nutrients from diverse feeds. The probiotics are known to improve mucin dynamics, a feature essential in maintaining a balanced gut

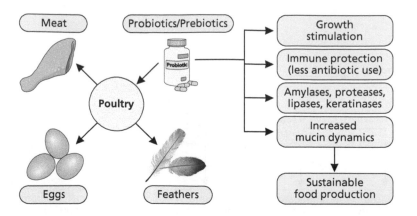

Figure 7.11. Probiotics and food security. The poultry industry is an important source of meat, eggs and feathers. Probiotics and prebiotics utilized as feed additives are essential in maintaining a sustainable industry and provide products devoid of chemicals. These additives result in improved gut health critical for growth and development. These features promote healthy poultry and mitigate the use of antibiotics that is associated with the spread of antibiotic-resistant bacteria.

microbiota and combatting pathogens. The SCFA and metabolites with hormonal properties produced by the probiotics also aid in better food quality derived from the poultry industry. The introduction of such prebiotics as oligosaccharides stimulates the exclusion of pathogens and the proliferation of favourable microbes resulting in healthy birds. Thus, probiotics and prebiotics known to modulate gut microbiome trigger improved digestive, absorption, immune and hormonal functions, factors responsible for a sustainable poultry industry, a key contributor to global food security (**Figure 7.11**).

7.11 Birds, Microbes and Global Pandemic

Birds can be an important vector of microbial dispersal. The avian transmission of bacteria and viruses is a transnational problem that poses a serious challenge to human health globally. The diversity of birds and their ability to travel long distances are two features that make these flying vertebrates a potent threat to global health. The migratory activity, perching tendency and feeding from toxic environments like polluted areas and sewer systems are effective pathways via which birds are known to disperse deadly microbes from one place to another. Diseases such as avian flu and salmonellosis are transmitted when humans come in contact with infected birds and edible products are contaminated with birds' droppings. In the case of West Nile virus spread, mosquito bites containing the virus obtained from birds are responsible for human transmission. Thus, birds are a natural repository of zoonotic pathogens that need to be monitored if global health is to be maintained.

Furthermore, the emergence of antibiotic-resistant microbes around the globe can be readily dispersed by birds as their migratory routes are not restricted by national boundaries. Hence, birds are ideal vectors that can carry these resistant bacteria from one place to another. Numerous microbes including *Klebsiella* spp. and *Salmonella* spp. harbouring genes to counter such antibiotics as sulphonamides, aminoglycosides, tetracyclines and quinolones are becoming increasingly widespread worldwide via avian transmission, a situation that can be catastrophic. The proliferation of drug-resistant bacteria is primarily due to the rapid increase in the prescription of these medications. These bacteria are then concentrated in areas with increased human activity, in wastewater treatment plants, landfill sites and livestock farms. These are rich feeding grounds for birds where the initial interaction with the antibiotic-resistant microbes is initiated and are subsequently distributed to far and wide. In fact, resistant microbes known to be common in one country have been reported in faraway regions. Although birds need their own microbes to survive and proliferate, they can be an important distributor of infectious microbes and antibiotic-resistant bacteria. Undoubtedly, monitoring of

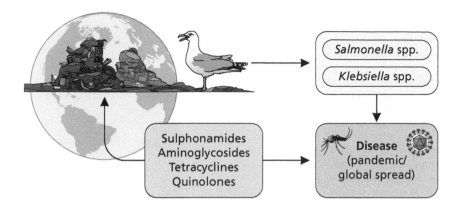

Figure 7.12. Birds, microbes and global pandemic. The ability of birds to migrate and travel long distances can pose a serious threat to global health. They feed from contaminated areas rich in microbes and in urban areas where antibiotic-resistant bacteria are prevalent. These microbes can then be dispersed far away from their initial sites. Numerous diseases are transmitted by avian vectors.

microbial population within birds will be critical if the spread of diseases and emergence of super bugs are to be curtailed. To prevent a global pandemic, it is essential that monitoring of avian microbes becomes an integral component of public health authorities worldwide (**Figure 7.12**).

7.12 Conclusions

The health, adaptability and survivability of birds are intricately tied to the microbial communities they harbour. If this fine balance is perturbed, dysbiosis occurs, a situation that can result in major negative ramifications on planet Earth. Besides the ecological roles that birds fulfil, they are also an important component of the human diet. In fact, meat and eggs supplied by the poultry industry are widely consumed globally. Hence, to maintain a sustainable food supply and to ensure food security, it is critical that the poultry enterprise remains healthy. To promote growth and stimulate egg production, antibiotics are often utilized. This practice is being albeit at a slow pace replaced by the introduction of prebiotics and probiotics in poultry farm feed. This policy results in improved yield in meat with better nutritional quality coupled with the enhanced production of eggs and a diminution in antibiotic-resistant bacteria. A molecular understanding of the different roles microbes play in the life of the avian hosts is bound to further enhance the acceptance and utilization of these microbial additives in the poultry industry, an event that will boost the global food security with minimal negative impact.

The well-being of birds that to a large extent is dictated by the microbial consortia they possess is not only essential for the nutritional needs of humans but is also critical for human health and prevention of diseases. Birds are potent vectors that can transmit infectious and antibiotic-resistant microbes as they travel long distances without any national border restriction. Their perching and feeding habits in combination with their ability to migrate are ideal vehicles birds utilize to spread infectious microbes from one region to another leading to devastating consequences. Hence, their interactions with microbial population resulting in the assembly of the selected microbes and elimination of pathogens will provide key molecular details that will pave the way to thwart any pandemic triggered by avian intervention. Microbes are integral to birds' physiology and guide the development of their various anatomical features. Although the microbial assembly within organisms necessitates the participation of disparate selecting entities including the host, the environment and the microbiome, the collection of all microbes constitutes the invisible organ. It performs functions just like the visible organs and undergoes similar constraints imposed by a multicellular lifestyle. Its contribution to the holoorganism is essential for the normal functioning of the avian host.

SUGGESTED READINGS

Al-Murayati, H., (2017) Diversity of the bacterial community and secondary sexual characters in the peacock. University Paris-Sud. doi: https://tel.archives-ouvertes.fr/tel-01578914/document..

Bonnedahl, J., & Järhult, J. D. (2014). Antibiotic resistance in wild birds. *Upsala Journal of Medical Sciences, 119*(2), 113–116. doi:10.3109/03009734.2014.905663

Chen, C. et al., (2020). Maternal gut microbes shape the early-life assembly of gut microbiota in passerine chicks via nests. *Microbiome, 8*(1), 129. doi:10.1186/s40168-020-00896-9

Dewar, M. L. et al., (2014). Influence of fasting during moult on the faecal microbiota of penguins. *PLoS ONE, 9*(6), e99996. doi:10.1371/journal.pone.0099996

El-Hack, M. E. et al., (2020). Probiotics in poultry feed: A comprehensive review. *Journal of Animal Physiology and Animal Nutrition, 104*(6), 1835–1850. doi:10.1111/jpn.13454

Funkhouser, L. J., & Bordenstein, S. R. (2013). Mom knows best: The universality of maternal microbial transmission. *PLoS Biology, 11*(8), e1001631. doi:10.1371/journal.pbio.1001631

Garcia-Mazcorro, J. F. et al., (2021). Composition and potential function of fecal bacterial microbiota from six bird species. *Birds, 2*(1), 42–59. doi:10.3390/birds2010003

Geltsch, N. et al., (2018). Common cuckoos (*Cuculus canorus*) affect the bacterial diversity of the eggshells of their great reed warbler (*Acrocephalus arundinaceus*) hosts. *Plos One, 13*(1), 1–13. doi:10.1371/journal.pone.0191364

Grond, K., et al., (2018). The avian gut microbiota: Community, physiology and function in wild birds. *Journal of Avian Biology, 49*(11), 1–19. doi:10.1111/jav.01788

Hauffe, H. C., & Barelli, C. (2019). Conserve the germs: The gut microbiota and adaptive potential. *Conservation Genetics, 20*(1), 19–27. doi:10.1007/s10592-019-01150-y

Hird, S. M. (2017). Evolutionary biology needs wild microbiomes. *Frontiers in Microbiology, 8*(725), 1–10. doi:10.3389/fmicb.2017.00725

Javůrková, V. G. et al., (2019). Plumage iridescence is associated with distinct feather microbiota in a tropical passerine. *Scientific Reports, 9*(1), 1–10. doi:10.1038/s41598-019-49220-y

Javůrková, V. G. et al., (2019). Unveiled feather microcosm: Feather microbiota of passerine birds is closely associated with host species identity and bacteriocin-producing bacteria. *The ISME Journal, 13*(9), 2363–2376. doi:10.1038/s41396-019-0438-4

Leclaire, S. et al., (2017). Odour-based discrimination of similarity at the major histocompatibility complex in birds. *Proceedings of the Royal Society B: Biological Sciences, 284*(1846), 20162466. doi:10.1098/rspb.2016.2466

Maraci, Ö, et al., (2018). Olfactory communication via microbiota: What is known in birds? *Genes, 9*(8), 387. doi:10.3390/genes9080387

Martín-Vivaldi, M. et al., (2017). Acquisition of uropygial gland microbiome by hoopoe nestlings. *Microbial Ecology, 76*(1), 285–297. doi:10.1007/s00248-017-1125-5

Mendoza, M. L., et al., (2018). Protective role of the vulture facial skin and gut microbiomes aid adaptation to scavenging. *Acta Veterinaria Scandinavica, 60*(1), 1–19. doi:10.1186/s13028-018-0415-3

Michel, A. J. et al., (2018). The gut of the finch: Uniqueness of the gut microbiome of the Galápagos vampire finch. *Microbiome, 6*(1), 167. doi:10.1186/s40168-018-0555-8

Pearce, D. S. et al., (2017). Morphological and genetic factors shape the microbiome of a seabird species (Oceanodroma leucorhoa) more than environmental and social factors. *Microbiome, 5*(1), 1–17. doi:10.1186/s40168-017-0365-4

Peralta-Sánchez, J. M. et al., (2019). Egg Production in Poultry Farming Is Improved by Probiotic Bacteria. *Frontiers in Microbiology, 10*(1042), 1–13. doi:10.3389/fmicb.2019.01042

Rahman, M. M. et al., (2020). Coronaviruses in wild birds – A potential and suitable vector for global distribution. *Veterinary Medicine and Science, 7*(1), 264–272. doi:10.1002/vms3.360

Rocklöv, J., & Dubrow, R. (2020). Climate change: An enduring challenge for vector-borne disease prevention and control. *Nature Immunology, 21*(5), 479–483. doi:10.1038/s41590-020-0648-y

Roggenbuck, M. et al., (2014). The microbiome of new world vultures. *Nature Communications, 5*(1), 1–8. doi:10.1038/ncomms6498

Rowe, M. et al., (2021). The reproductive microbiome: An emerging driver of sexual selection, sexual conflict, mating systems, and reproductive isolation. *Trends in Ecology & Evolution, 36*(1), 98. doi:10.1016/j.tree.2020.10.015

Ruiz-Rodríguez, M., et al., (2018). Gut microbiota of great spotted cuckoo nestlings is a mixture of those of their foster magpie siblings and of cuckoo adults. *Genes, 9*(8), 381. doi:10.3390/genes9080381

Sergeant, M. J. et al., (2014). Extensive microbial and functional diversity within the chicken cecal microbiome. *PLoS ONE, 9*(3), e91941. doi:10.1371/journal.pone.0091941

Veelen, H. P., et al., (2018). Microbiome assembly of avian eggshells and their potential as transgenerational carriers of maternal microbiota. *The ISME Journal, 12*(5), 1375–1388. doi:10.1038/s41396-018-0067-3

Waite, D. W., & Taylor, M. W. (2015). Exploring the avian gut microbiota: Current trends and future directions. *Frontiers in Microbiology, 6*(673), 1–12. doi:10.3389/fmicb.2015.00673

Whittaker, D. J. et al., (2016). Social environment has a primary influence on the microbial and odor profiles of a chemically signaling songbird. *Frontiers in Ecology and Evolution, 4*, 1–15. doi:10.3389/fevo.2016.00090

Whittaker, D. J. et al., (2019). Experimental evidence that symbiotic bacteria produce chemical cues in a songbird. *Journal of Experimental Biology. 222*(20), jeb.202978. doi:10.1242/jeb.202978

Yang, Y., et al., (2016). Characterising the interspecific variations and convergence of gut microbiota in *Anseriformes* herbivores at wintering areas. *Scientific Reports, 6*(1), 1–11. doi:10.1038/srep32655

MAMMALIAN MICROBIOMES: THE ROLES THEY PERFORM

Contents			Keywords

Keywords

- Mammals
- Microbiome
- Behaviours
- Nutrition
- Development
- Immune
- Climate Change

8.1 Introduction

8.1.1 General characteristics of mammals

Mammals are vertebrates with mammary glands and usually have a hairy skin. They are warm blooded and possess a respiratory system designed to transport O_2 to all body parts with the assistance of a four-chamber heart. They are mostly viviparous, i.e. they give birth to live young. In placental mammals, the embryo develops within, while in marsupials the development of the embryo occurs in a pouch located outside the body. They have a nervous system controlled by the brain and have a larynx for emission of sounds. The skin can have a range of colorations implicated in a variety of functions including thermoregulation, camouflage, sexual

DOI: 10.1201/9781003166481-8

selection and communication. Some sloth may appear green as they house algae with whom these mammals live in a symbiotic relationship.

Mammals vary in mass from 2 g in the case of the bumblebee bat to more than 15 tonnes associated with the blue whale. The size of the former is 40 mm, while the latter can measure up to 30 m. Their lifespan also shows a large variation ranging from 2 years in the case of a shrew to approximately 200 years in the case of the bowhead whales. They are adapted to live in diverse habitats and can be arboreal, aerial, aquatic, terrestrial and subterranean. Their nutritional habits enable mammals to play an important role in the ecosystem as they are carnivorous, herbivorous, omnivorous and insectivorous. They are known as nature's ecological engineers. The herbivores have a major imprint on the forest landscape and assist in the genesis of grassland, savannahs and modulating forest densities. These natural environments shape the establishment of different life forms. Mammals also contribute to the dispersal of seeds and help control pests. These activities are beneficial to the agriculture industry. Some of these vertebrates can be utilized as life stock and as a source of dairy products. A range of medical benefits including blood and organ supply can also be derived from mammals. Hence, mammals are an important component of our ecosystem that needs to be fully appreciated. Although most of their visible organs have been uncovered, the workings of their microbiome are only now beginning to be understood. These multicellular organisms depend on their microbial partners to live. They constitute an important cellular component of all mammalian systems and participate in a range of physiological functions. An average adult human has 2–3 kg of microbes, while cows possess up to 40 kg of these invisible cells (**Figure 8.1**).

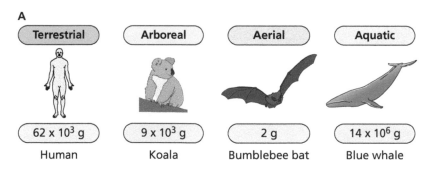

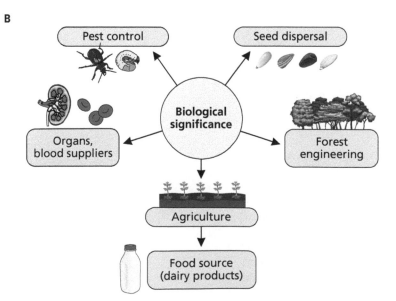

Figure 8.1. Mammals: general features and ecological importance. (a) Mammals vary in shape, size and the habitats they live in. A flying bumble bat weighs around 2 g, while an ocean-dwelling blue whale has a mass of several tonnes. They are viviparous and have mammary glands. (b) These vertebrates contribute to the well-being of the ecosystem they reside in. They are involved in seed dispersal and in insect control. They are important ecological engineers and can also be consumed. Mammals are suppliers of organs and blood to the medical system.

8.2 Genesis of Microbiome in Mammals

Like in most multicellular organisms, microbes are part of mammals from the very beginning. They are associated with the development of the embryo. The transmission of microbiota from the mother to its progeny is evident in such diverse organisms as insects where the female insect sprays eggs with microbes and in birds where the microbial transfer occurs during fertilization or the brooding phase. In mammals, the maternal microbes acquired by new borns are restricted to the epithelial surfaces and the invaginations in the digestive tract. The immune system that recognizes self from non-self is primed by this microbial infusion to impede the colonization of most internal organs. The microbiome essentially occupies the external space (skin, the gastrointestinal (GI) tract, oral cavity, lungs and sexual organs) and is primarily shaped by the environment (biotic and abiotic), the diet and the maternal microbial profile. This arrangement enables the microbial community to adjust to any perturbation and to respond accordingly. Hence, the microbial assemblage can be readily dismantled and reconstituted with diverse microbial partners depending on the environmental cues. This invisible organ is highly malleable. A baby feeding primarily on the maternal milk has a different set of GI microbes than those consuming solid foods following the appearance of teeth a few months later.

8.2.1 Microbial community during fertilization

The first contact of microbes takes place during fertilization. Even though the uterus is well guarded by the immune system, the bacterial cells associated with the uterine cervix may enter with the sperm or during the initial implantation of the embryo. This process must be intricately modulated as any intake of fortuitous microbes can be risky and reduces fecundity. This may result in miscarriages. The maternal transfer of microbes from the vagina and the gut during gestation may be a well-conserved route mediating this process. In placental mammals, the vagina dedicated to reproduction, the anus serving as an opening for faeces and the urethra charged with elimination of urine are distinct openings with disparate functions. The birth canal is adjacent to the rectum, a location that may facilitate the transmission of vaginal and perineal faecal microbes from the mother to the infant. The rupture of the chorioamniotic membrane provides another conduit via which this microbial exchange can take place. During birth, the skin is covered with microbes and the mouth is also a microbial passage. Hence, microbes from the mother and mothers before her are passed on to infants.

8.2.2 Postnatal microbiome development

The postnatal development is further punctuated by the inoculation of microbial population during breast-feeding and the milk infants consume. The latter contains lactose, glucans and metabolites like urea and oxalate that favour the proliferation of a select group of microbes such as *Bifidobacterium* spp. The human oligosaccharides are essential prebiotics shaping the development of babies. Studies with non-human mammals devoid of microbes have revealed that these subjects suffer from various abnormalities. During labour, lactic acid bacteria known to utilize lactose are transferred. These microbes aid in priming the nervous system, metabolic processes and hormones as well as the morphological development of the neonates. The maturing of the sensory and motor capabilities is also guided by the microbial community. As vitamin K cannot readily diffuse through the placental barrier, microbes may be an important source of this critical ingredient essential for the proper functioning of the circulatory system. This moiety involved in the coagulation of blood is not synthesized by humans and has to be supplied by microbes or from food intake. Vitamin B_{12}, pivotal in cellular

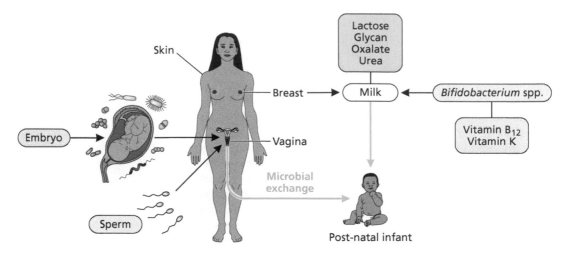

Figure 8.2. Microbiome genesis from conception to postnatal phase. Microbes are an integral component of humans from conception to death. Fertilization of the sperm and the ovule and the subsequent implantation of the embryo are accompanied by microbial residents. Microbial exchange between the mother and the baby occurs via the vagina, skin and breast. The maternal milk is also a source of microbes the infant is exposed to. Additionally, the constituents within the milk like oxalate and glycans help select for the proper microbes. These microbes including the *Bifidobacterium* spp. are pivotal in supplying vitamins, moieties not produced by humans.

metabolism, is also another molecule derived from food or produced by bacteria. Hence, these vital moieties that are not readily made available from the intake of milk must be obtained primarily through microbial population transmitted by the mother. It is not unlikely oligosaccharides in the milk fuels the proliferation of these bacteria within the infant gut (**Figure 8.2**).

8.2.3 Microbiome during the formative years and adulthood

The place of birth whether it is a hospital, a family home or a farm is an important contributor to the microbial community a baby is initially inoculated with. The use of medications or antibiotics coupled with the nature of the family structure is a selective microbial filter. The presence of grandparents and other siblings can affect the constituents of the microbiome. Female siblings known to possess a higher amount of *Bifidobacterium fragilis* and *Lactobacillus* spp. than male counterparts may shape the microbial constituents accordingly. Skin to skin contact, dietary supplementation with probiotics and prebiotics and presence of pets in the home all have an influence on the nature of microbes the baby has. The intake of solid foods following the development of the initial teeth results in a major shift of the microbiota in the hindgut. This change is guided by the nature of foods taken as nutrient rich in fibres tend to favour a different microbial community compared to a fat-rich diet. At this stage, the chemical cues provided by the microbes enable the host to develop accordingly. For instance, herbivorous mammal will have a longer hindgut than a carnivore or an omnivore. As the age-dependent release of hormones occurs, microbial colonization shifts. During puberty, the growth of hair and the sebaceous glands create a novel environment within the skin. The increase in lipids and NaCl promote a different microbial community compared to the pre-puberty stage. In females, the onset of menstrual cycle is associated with a major change in the chemical environment of the vagina, a situation conducive to the proliferation of a select group of microbes like *Staphylococcus aureus* and *Escherichia coli*. It important to note that this organ is well guarded by a microbial population consisting of *Lactobacillus* spp. and *Prevotella* spp. responsible for production of antimicrobial compounds aimed at any opportunistic invasion by pathogens.

Hormonal variation, lifestyle and social interactions that mammals experience during adulthood result in a significant change in microbial

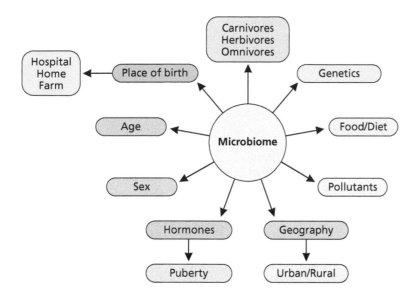

Figure 8.3. **Factors affecting the assembly of microbial communities.** Like all visible organs, the assembly of the invisible microbiome is governed by a variety of parameters. Age, place of birth, geographical location, diet and occupation of parents are important contributors to the nature of the microbiome. Nature of foods and microbes dictate the morphology of the digestive tract. This organ system is longest in herbivores and shortest in carnivores. Sex, puberty and genetics also determine the constituents of the bacteria within the microbiome. Pollution, rural settings and urban environments can all guide the genesis of the microbiome.

partners. Climate change coupled with exposure to environmental pollutants also has an influence on the composition of the constituents of the microbial assemblage. Rural and urban dwellers have distinct microbial profiles reflective of the geographical locations where these regions are situated. These factors also affect most if not all of the visible organs on an ongoing basis. The ageing process is also mirrored in a shift in the components of the microbiome. There tends to be a decrease in the diversity of microbial population. This phase of life is characterized by an increase in *Clostridium* spp. with a concomitant decrease in *Bifidobacterium* spp. In fact, in humans, the downward shift in diversity starts manifesting itself at the age of 40 years. Healthy conditions are associated with an abundance of the genera *Christensenella*, *Akkermansia* and *Bifidobacteria*, a microbial profile observed in centenarians. The life-promoting attributes of these microbes have been recognized in diverse population groups within a wide geographical area. The vertical transmission mediated by females through the vaginal interaction, the maternal milk and the birthing process lay the initial foundation for the microbiome to be an integral component of the normal functioning body. The invisible organ then is further fine-tuned by a variety of environmental factors and physiological cues from the host. This cellular cross-talk allows for the genesis of a holoorganism where all the constituents are cognizant of their biological responsibility to contribute to an effective functioning system. Only foreign microbial invasion, inherent genetic defects, nutritional imbalance and injuries can disrupt the integrity of the complex multicellular entity (**Figure 8.3**).

8.3 Energy Harvesting and Ruminants–Microbial Partnership

The significance of the microbial community in the life of mammals is nowhere more evident than in the digestive tract. In fact, the morphology of this organ is known to have evolved with its microbial residents. It is the part of the mammalian body that houses the most microbes. Its diverse pH, fluctuating O_2 gradient, chemical environment and anatomical features provide an ideal niche to be colonized by a wide variety of microbial consortia. They thrive within the GI canal and perform vital functions, essential for the survival of the host. Their absence results in abnormal processes and eventually death. The assembly of the microbial community is dictated by diets which in turn help guide the structural adjustment of the alimentary canal, an organ whose development is to a large extent shaped by the resident microbes. In nearly all mammals, the harvesting of energy is mediated by microbes irrespective of the diets consumed. For instance, how can pandas survive without bacteria aiding in the digestion of bamboo plants or whale can feed on crustaceans in the absence of microbes with the ability to cleave chitin, a polymer of N-acetyl glucosamine.

8.4 Ruminants Cannot Exist without Their Microbes

The crucial role gut microbes play is vividly exposed in ruminants. These organisms are herbivores restricted to a plant-based nutrition for which they possess few enzymes coded by their own genome to degrade the complex polymers like cellulose, hemicellulose and lignin. These moieties are degraded into monomers by microbial enzymes housed in the stomach. The herbivorous dietary habit has compelled ruminants to evolve a unique GI system composed of a foregut and a hindgut. The former comprises the rumen, the reticulum, the omasum and the abomasum while the latter is made up of the small intestine and the large intestine. The volume of the rumen which is approximately 185 l houses most of the microbes. The wall of the rumen (mucosa) is an important absorption site with finger-like projections referred to as papillae dedicated to this task. The omasum is smaller and has a leaf-like structure involved in the filtration of water while the abomasum (the true stomach) is about 27 L and performs similar functions executed by the stomach in monogastric mammals like humans. The ruminal compartment has little or no O_2, a temperature of 39°C, a pH range of 5.7–7.3 and a diverse chemical composition constituted of plant-based feeds at different stages of digestion. This unique environment with multiple ecological niches is populated by a disparate microbial community tasked with unlocking the nutritional value in the plant polymers that ruminants nourish on. These polymers consisting of cellulose, hemicellulose, xylans, pectins, lignin, tannins and other complex carbohydrates are processed by the microbial residents into volatile fatty acids (VFAs), vitamins, monosaccharides, amino acids and a vast range of metabolites the host depends on.

The microbial community is made up of a heritable core group of microbes consisting of bacteria, archaea, fungi, protozoa and viruses. Bacteria constitute the bulk of the microbial assemblages that are relatively well preserved across breed, geographical locations and diet. The phyla *Bacteroidetes*, *Firmicutes* and *Proteobacteria* are the dominant microbes. The microbes are vertically transmitted by the females. At birth, the ruminant has an undeveloped reticulo-rumen resembling a monogastric anatomical feature. The initial colonization begins with the microbes associated with the dam's vagina, skin, the colostrum and the environment. The consumption of the mother's digesta is a common route how the new born is directly inoculated with the maternal microbes. The introduction of solid feed results in the further maturation of the microbiome, a process that occurs in tandem with the development of the rumen. The digestion of complex carbohydrates prompts the growth of the rumen wall, while VFAs like propionate and butyrate stimulate the development of the papillae, the invaginations associated with the rumen. These partners, i.e. the microbes and the host, rely on the chemical signals emitted by each other to eventually mature into an effective functioning system aimed at digesting complex carbohydrates and utilizing the nutrients generated. This symbiotic relationship enables the host to complete its lifecycle that will be severely hampered or impossible in an environment devoid of microbes.

8.5 Microbes, Important Allies of Ruminants

Ruminococcus spp., *Prevotella* spp., *Fibrobacter* spp. and *Clostridium* spp. are some of the microbes that colonize the rumen as they tend to be anaerobes and/ or facultative anaerobes. Methane-metabolizing archaea are also an important constituent of the microbial population. The diversity of these core microbial species may vary and is dependent on the season, diet or geographic regions. During winter when only low-quality feed is available, the genus *Clostridia* is abundant. In the spring, the genus *Prevotella* becomes dominant microbes. Similar microbial shift is evident if the ruminant is in the wild or in an enclosed or in a captive setting. The microbes are associated with the ruminal mucosa, papillae and the feed particulates. Some of these microbial communities form biofilms where they congregate in order to accomplish the task of breaking down the polymers into monomers. The biofilm essentially composed of exopolysaccharides provides an ideal environment for the microbes to generate all the nutrients that the ruminant requires to grow and produce milk. Other microbes tend to float freely in the ruminal fluid. This arrangement allows a diverse assemblage of microbial cells to play a pivotal role in ensuring that the ruminant can live on a herbivorous diet and has a body rich in protein content. The majority of the energy need of the host is extracted from the VFAs the microbes generate. Monosaccharides like glucose, xylose or fructose liberated by glycosidases are converted to pyruvate that is subsequently processed via the tricarboxylic acid and acrylate pathways to elaborate the VFAs.

Cellulosomes, complex enzyme systems anchored to dockerin proteins, are especially elaborated by some of the microbes to dismantle the lignin cellulose derived from plant biomass. Xylanases and pectinases are some of the other enzymes that participate in the hydrolysis of the complexes recalcitrant to cellulase activity. The simple sugars are then metabolized to produce a variety of metabolites essential for the host. To optimize the production of value-added moieties from the plant biomass, large particles on top of the rumen are chewed again during rumination. This decreases the particle size and increases the surface area, features that are more conducive to enzyme-mediated hydrolysis. The concomitant production of saliva also contributes to the modulation of the acidity within the rumen, an important factor for the optimal functioning of the microbial community. The biosynthesis of VFAs is coupled to the liberation of such gases as CO_2, CH_4 and H_2, a situation that can affect the energy budget of the host. The abundance of microbes like *Succinivibrio* spp. may dictate levels of CH_4 being emitted. This gas is a potent modulator of climate change. The microbial community is also involved in provisioning the host with vitamins and nitrogen metabolites. In fact, most of the proteins and amino acids needed for muscle build-up and milk formation are of microbial origin. Further digestion continues on in the abomasum and the intestines. The acidity of the abomasum ensures that the ruminal bacteria do not populate the small and large intestine. The symbiotic microbes housed in the rumen are the main reason why ruminants have adopted an herbivorous lifestyle. They depend entirely on these tiny organisms for their energy need and a variety of essential metabolites. Hence, without microbes, ruminants will be non-existent. This partnership has enabled ruminants to convert photosynthetic plants into proteins and milk. This example clearly argues for the meta-organismic nature of all multicellular entities where the host genomic information couples with the microbial genetic traits to accomplish key physiological tasks. Hence, ruminants like all mammals are made up of own cells and that of all the microbes they harbour (**Figures 8.4 and 8.5**).

8.6 Mammalian Diets, Digestive Tracts and Microbial Communities

The diets that mammals have evolved to utilize for their energy needs are to a large extent dependent on the nature of the microbial communities they harbour. The microbiomes and host digestive tracts work in tandem from the developmental stage to the ageing process in order to maximize energy

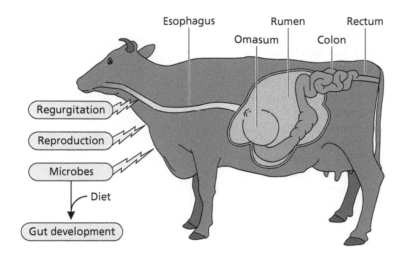

Esophagus   Rumen   Rectum
Omasum   Colon

Regurgitation
Reproduction
Microbes
Diet
Gut development

Figure 8.4. Ruminant digestive tract and microbes. Ruminants like all mammals are dependent on their microbes to extract nutrients from the herbivorous lifestyle they have adopted. In fact, the digestive tract has evolved with microbes and consists of a well-defined rumen whose development is guided by microbes. The omasum and the abomasum serve in further maximizing this digestive system. Microbial transmission during reproduction, diet and nourishing of the newborn by regurgitation help shape the assembly of the microbiome.

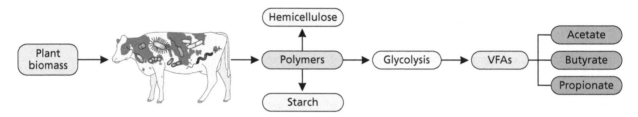

Hemicellulose

Plant biomass → Polymers → Glycolysis → VFAs → Acetate / Butyrate / Propionate

Starch

Figure 8.5. Microbial transformation of plant biomass into ruminant metabolites. The plant biomass ruminants consume is rich in cellulose, hemicellulose and lignin. Microbes within the rumen possess such enzymes as cellulase, amylase and hemicellulase that can liberate glucose and xylose. These metabolites are then converted by glycolysis, tricarboxylic acid cycle and fatty acid metabolism into volatile fatty acids (VFAs) that are then processed into energy by the host. Acetate, propionate and butyrate are the main VFAs generated.

production from the foods consumed. Host phylogeny also plays a role as related species tend to exhibit similar digestive attributes. For instance, the giant panda known to rely on bamboo plants possesses a GI system resembling carnivores than herbivores as this mammal has evolved from carnivorous ancestors. Although phylogenetic and dietary factors are pivotal in the genesis of microbial constituents within the host, selective determinants guided by age, sex, size and environment also shape the constituents of the microbiome. Bats, the only flying mammals tend to share numerous similarities with the avian microbiome. Bactrian camels with their characteristic two humps reside in arid and semi-arid regions and have to contend with poor grazing conditions where the climate is prone to sharp temperature changes. Their diets include shrubs, bushes and halophytes that are made up of lignin–cellulose complexes. These herbivores have a three-chamber stomach devoid of the omasum found in ruminants. The dominant microbial phyla consist of *Firmicutes*, *Verrucomicrobia* and *Bacteroidetes*. The foregut is populated with *Firmicutes* and *Bacteroidetes* while the ileum and the large intestine house a microbial population rich in *Firmicutes* and *Verrucomicrobia*. The presence of species like *Clostridium bifermentans*, *Bacteriodes fragilis* and *Akkermansia* spp. enables the release of maximal amount of nutrients from this habitat with scanty vegetation. Although the single-humped dromedary camels have a similar stomach like the Bactrian camels, they possess microbial communities having phyla of comparable abundance to ruminants. In these camels, *Bacteroidetes*, *Firmicutes* and *Proteobacteria* are the most prominent microbes. In donkeys, the monogastric herbivores, *Firmicutes* are most widespread with a marked presence of *Lactobacillus* spp. in the stomach most probably due to the acid-tolerant attribute of these bacteria. Nearly 70% of the energy is obtained from the organic acids produced from microbial fermentation in the large intestine, a situation in sharp contrast to ruminants. Small herbivores including the flying squirrel which lives on tree-top and feed on seeds, fruits and flowers has a simple stomach but possesses an enlarged

BOX 8.1 RUMINANTS, MICROBES AND CLIMATE CHANGE

Ruminants play a pivotal role in global food security as they are a major source of protein-rich nutrients. However, they also contribute significantly to the climate change that planet Earth is being subjected to. These mammals are potent emitters of greenhouse gases (GHGs) like CH_4, CO_2 and N_2O. In fact, CH_4 which is 30 times more detrimental to the environment than CO_2 as a GHG, is produced in the rumen during methanogenesis. This metabolic process enables the host to fulfill its energy need as the accumulation of H_2 and CO_2 have a negative impact on these mammals. The release of CH_4 promotes the energy-making machinery to proceed despite resulting in feed inefficiency and dietary loss. The microbial consortia comprising archaea such as *Methanobrevibacter* spp. are the main generator of this gas. The formation of CO_2 and H_2 during the metabolism of feed consisting of plant polysaccharides is converted into CH_4 and concomitantly supplies acetate and propionate that the ruminant readily utilizes to extract ATP and propel its growth. Hence, there needs to be a balance between the H_2 producers and utilizers if the release of CH_4 is to be controlled and the efficiency of feed enhanced. Fibrolytic bacteria like *Ruminococcus* spp. and *Eubacterium* spp. are H_2 producers while *Bacteriodes* spp. are H_2 consumers. *Fibrobacter* spp. are not known to produce H_2. The microbiome of the rumen can thus be reconfigured to modulate the concentration of H_2, a moiety that drives the synthesis of CH_4. *Lactobacillus* spp. known for their lactate-producing capability, *Selenomonas* spp. with their propionate-forming ability and formate-consuming bacteria are utilized as probiotics to mitigate CH_4 output of ruminants by modulating H_2 homeostasis.

Feed additives that inhibit enzymes and precursors involved in the biosynthesis of CH_4 are also an important mitigation strategy. While 3-nitrooxy propanol interferes with the enzyme methyl coenzyme M reductase, NO_3 and SO_4 act as a sink for H_2. Young plants, less-digestible fibres, grains, legumes, high-starch nutrient, essential oils and C_4 grass are some of nutrients supplied in order to manipulate the amount of CH_4 released by ruminants. Additionally, feeds devoid of nutrients generating methanol, choline and methylamine are favoured. Exogenous enzymes such as cellulases, hemicellulases, antibiotics and tannins are also utilized in this greenhouse gas mitigation effort. Even vaccines against the methane-producing microbes are being contemplated. Hence, maintaining a fine balance between the microbial consortium, the dietary supplements and the availability of the reducing power of H_2 is pivotal to diminishing the global warming trend and attaining food security worldwide (**Box Figure 1**).

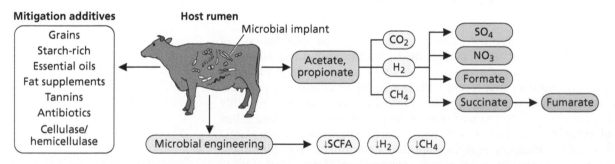

Box Figure 1. Mitigation strategies to curb CH_4 release in ruminants where microbes are important partners.

caecum to enable the maximal processing of nutrients and elimination of toxins. The dominant *Firmicutes* phylum is constituted of bacteria with enzymes to detoxify plant toxins and produce short-chain fatty acids (SCFAs). Sloths are arboreal mammals with one of the slowest metabolic rates. They are primarily engaged in folivory, a nutritional behaviour that exposes these mammals to elevated amounts of cellulose. They have a rumen-like foregut with a very lethargic digestive process aimed at promoting fermentative reactions with the concomitant synthesis of organic acids.

Myrmecophagy, a dietary habit common in mammals like pangolins and armadillos necessitates a completely different morphological and microbial organization. These organisms primarily consume social ant-like termites and ants and are endowed with an elongated muzzle with an extensible tongue. As limited fragmentation and mastication are required, these animals have undergone a reduction or loss of teeth. *Firmicutes*, *Bacteroidetes*, *Proteobacteria* and *Tenericutes* are the dominant phyla with an enrichment of such microbes as *Prevotella* spp., *Klebsiella* spp. and *Streptococcus* spp. This microbial assemblage allows for the degradation of the chitin and protein, two components abundant in insects. Baleen whales are sea-dwelling mammals and are filter feeders living essentially on a diet of fishes and crustaceans. These proteins and chitin replete foods are digested with the assistance of a microbial community akin to terrestrial

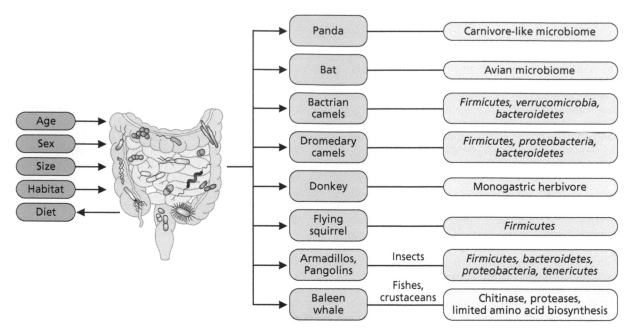

Figure 8.6. Determinants of gut microbiome in mammals. The gut microbiome of mammals is shaped by numerous factors including diet, phylogeny, size, habitat and sex. The bat, a flying mammal possesses a microbiome resembling that of birds while pandas known to consume bamboo leaves have a microbiome akin to carnivores. Bactrian camel's microbiome is rich in *Verrucomicrobia* and the dromedary camel has an abundance of *Proteobacteria*. Insect eaters like the pangolin has digestive tract with *Tenericutes* and crustacean-consuming baleen whales have microbes with the ability to degrade chitin and proteins.

carnivores with biochemical pathways limited in the biosynthesis of glutamate but enriched in the catabolism of this acidic amino acid. However, the intake of chitin, a polymer of *N*-acetyl glucosamine selects for some microbes common to herbivores with the ability to cleave this complex carbohydrate and ferment the products into nutrients that can be readily absorbed by the host. To promote these processes, this marine mammal has a blind-ended caecum between the ileum and the colon. Diet, phylogenetic features and metabolic rate are the important determinants contributing to the microbial community residing in the mammalian GI tract (**Figure 8.6**).

8.7 The Skin Microbes

The skin is the primary physical barrier that protects the interior biological features from the external environment, helps maintain the body temperature, keeps the inside moist, thwarts microbial invasion and prevents nutrients to escape. Non-human mammals tend to have a denser fur with the outermost layer referred to as the epidermis interacting with the sebaceous and sweat glands. The former releases oily exudates. The apocrine gland associated with the hair and the eccrine gland linked to the dermal pore constitute the sweat glands and are involved in the secretion of a variety of chemicals including NaCl, HCO_3^-, urea, steroids and fatty acids. The skin is the largest organ in virtually all mammals and is characterized by invaginations, folds, appendages and follicles. This diverse landscape coupled with a continually changing chemical environment provides a unique habitat for microbes to reside. The selection of the microbial partners is dictated by maternal transmission, geography, diet and interactions in the open (wild) or in captivity. Here again, an intricate balance has to be struck between the host and the microbes in order to exclude pathogens in the microbial assemblage. In humans, the phylum *Actinobacteria* dominates, while microbes belonging to the phylum *Chloroflexi* are most abundant in animals. Racoons and cattle have higher amounts of *Micrococcus* spp. and *Staphylococcus* spp., while animals residing in high altitudes have an elevated abundance of *Arthrobacter* spp.

BOX 8.2 GUT MICROBES, METABOLITES AND DISEASES

The GI tract is home to majority of microbes in humans and is known to produce a range of metabolites from dietary substrates. These microbes not only synthesize bioactive metabolites, but they can also transform the products generated during cellular metabolism. Additionally, they can supply precursors to synthesize modulators like hormones and neurotransmitters that have a profound influence on health and diseases. Obesity, dyslipidaemia, hypertension and ageing are some of medical conditions emanating from these microbial activities. SCFAs, tryptophan derivatives, bile acids, spermidine and branched-chain amino acids are some of the metabolites that can evoke significant physiological complications. Bile acids are obtained during normal hepatic metabolism of cholesterol. Primary bile acids such as chenodeoxycholic acid and cholic acid linked to glycine and taurine moieties play an important role in the digestion and absorption of lipids and vitamins. These bile acids are deconjugated and reabsorbed while some are excreted. However, the accumulation of conjugated bile acids can result in dysbiosis. For instance, taurocholic acid fosters the bloom of *Bilophila wadsworthia*, a sulphite-reducing bacterium. This situation that culminates in increased intestinal permeability and inflammation is attributed to the impairment in the activity of the microbial enzyme bile salt hydrolase. The modulation of this enzyme dictates the homeostasis of bile acids and its conjugates, metabolites known for a variety of diseases. For instance, increased level of bile acids in serum is an indicator of obesity. Dietary tryptophan can be transformed into serotonin, indole derivatives and kynurenine metabolites responsible for potent physiological responses. Spermidine, a metabolite that promotes autophagy plays an important role in longevity and is also modulated by gut microbes. The microbial activity in the GI tract can aid or disrupt the homeostasis of bioactive metabolites with severe consequence on the health of the host (**Box Figure 2**).

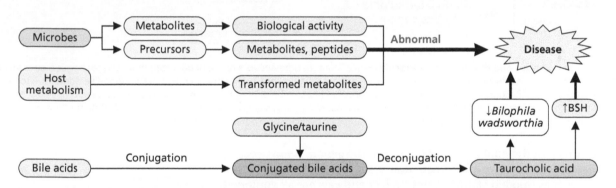

Box Figure 2. Metabolites derived with microbial assistance result in dysbiosis. BSH: bile salt hydrolase.

and *Paenibacillus* spp. The humpback whale's skin is dominated by the genera *Psychrobacter* and *Tenacibaculum*.

Although the dermal microbes help combat pathogens by producing antimicrobials and participate in a variety of other functions, their contribution to the odour-releasing ability of mammals is crucial for the survival of the hosts. The chemical constituents associated with these odours signal mate recognition, dissuade predators, establish hierarchy and communicate other behavioural cues. This chemical profile is species and individual specific. In humans, the presence of the *Staphylococcus horminis* generates a musk-like urinous odour due to the metabolism of steroids into androstenol and androstenone derivatives. The microbial production on the odorant trimethylamine is important in murine reproductive processes. Hyaenas are carnivorous mammals and some species like the spotted hyaenas live in a structured hierarchy termed 'clan' with 30–40 members. Despite possessing vocal, tactile and visual signalling modalities, they have evolved intricate scent profiles that elicit a variety of behavioural responses. The subcaudal scent pouch deposits an odorous secretion referred to as 'paste' that is recognized by clan members. It has a chemical profile consisting of VFAs, esters, hydrocarbons, alcohols, aldehydes and organic acids which vary with sex, age and reproductive stage. For instance, pregnant females have a scent made up of compounds such as pentanoic acid, hexanoic acid and heptanoic acid while male's odour is attributed to acetic acid, propanoic acid and butanoic acid. These chemical signatures are the products of the microbial communities residing

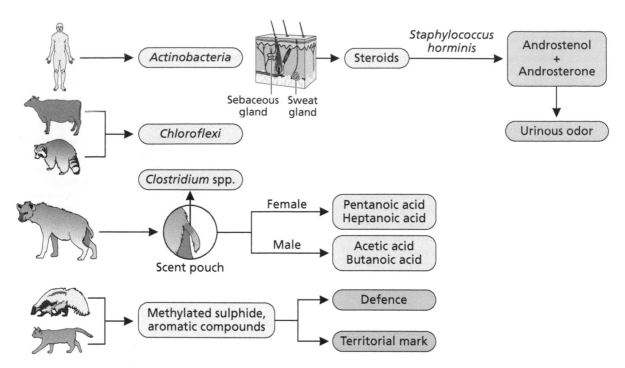

Figure 8.7. Skin microbes and the tasks they execute. Skin of mammals possesses a range of microbes that secrete chemical cues involved in a variety of functions. *Staphylococcus horminis* is a human dermal resident known to be the architect of the urine-like odour due to its interaction with steroids. Racoons and cattle have an abundance of bacteria belonging to the *Chloroflexi* phylum. Hyaenas have a scent gland replete with microbes responsible for volatile fatty acids that distinguish males from females. The methylated sulphides and aromatic compounds derived from the microbial residents in skunks and cats participate in defence and territorial integrity.

in sacs localized in the anal region of the hosts. The genera *Anaerococcus*, *Propionibacterium* and *Clostridium* are dominant in the spotted hyaenas. Mammals such as hooded skunk, honey badger and cats are all equipped with anal sacs replete with bacteria that enable them to generate the appropriate odour that evokes the desired response. The methylated sulphides and modified aromatic compounds have characteristic pungent smell that can be readily eliminated upon treatment of the scent sacs with antibiotics. The microbial residents of the skin are the main producers of odours emanating from all mammals including humans. The precise information laden in these volatile signals are deciphered by the olfactory-sensory apparatus that subsequently triggers behavioural changes in the hosts. Although the exact composition of the microbes synthesizing these information-laden chemicals is currently being catalogued, the precise molecular pathways how they evoke a comportment shift in the hosts have yet to be fully elucidated (**Figure 8.7**).

8.8 Microbes Fortify the Body's Immune Response

The immune system guards the host from pathogenic invasion and helps discern self from foreign entities (non-self). The resident microbes that are harboured within the host not only occupy space or produce antimicrobial compounds to prevent colonization by opportunistic microorganisms but they also prime the immune cells for their surveillance activities. The adaptive immune system is primarily spearheaded by the T lymphocytes, synthesized in the thymus gland and B lymphocytes, elaborated by the bone marrow. These cells are known to recognize foreign molecules (antigens) via their corresponding receptors. These defence cells are presented the antigens by proteins encoded by the major histocompatibility complex (MHC) genes which to some extent are shaped by the resident microbes. The antigen-presenting cells associated with the intestine help coordinate the establishment of the microbial community and restrict their

BOX 8.3 MICROBES AND SOCIAL BEHAVIOURS IN MAMMALS

Microbes play an important role in controlling social behaviours in numerous organisms including mammals. Their ability to generate chemical cues that are subsequently decoded by the hosts and living systems in an ecosystem is crucial how the constituents living in an ecological niche interact. Group residency, individual identity, recognition of conspecifics, hierarchical organization, aggressive behaviour, territorial boundary, cooperative foraging or hunting, mating success and parental care are all dictated by the extracorporeal messengers usually elaborated by the microbes residing within the organisms or in dedicated sacs usually located in the caudal region in a variety of mammals. These chemical signals have organoleptic properties and act on the sensory organs. Volatile organic compounds and amine derivatives are the main information-rich moieties that act extra corporeally while peptides like oxytocin and arginine vasopressin deliver their message within the organism. Oxytocin elicits bonding behaviour and helps foster maternal relationship with the infant. On the other hand, arginine vasopressin in male hamsters provokes aggressive behaviour. *Lactobacillus reuteri* evokes oxytocin-mediated positive responses while *Lactobacillus rhamnosus* orchestrates emotional changes.

Trimethylamine (TMA) is a potent social signalling molecule that is produced from choline via microbial interaction. This quaternary trimethyl amino moiety is found in urine, semen, faeces, blood and cerebrospinal fluid. It can readily escape the confines of the body into the surrounding environment when present in concentrations as low as 0.5 ppb. This highly odoriferous amine with a distinctive fishy smell interacts via the trace-amine-associated receptors localized within the olfactory apparatus and the central nervous system. The Western diet with its high content in fats and proteins possess high levels of carnitine and choline, two precursors of TMA. The flavin monooxygenase localized in the liver transforms this amine into its oxide derivative (TMAO), a moiety that is emerging as a key indicator of cardiovascular diseases. A mutation in this enzyme results in elevated amounts of TMA, a metabolic disorder referred to as the fish odour syndrome or trimethylaminuria and leading to rejection and social isolation. However, in mice, this chemical is utilized as a sex attractant. Female mice tend to have a very marked increase of TMA in the urine aimed at the male counterparts resulting in the vocalization of courtship calls in the ultrasonic range. P-tyramine, a microbially derived product from tryptophan prominent in urine of carnivorous animals elicits an averse behaviour in mice. Thus, mammals enlist the assistance of their microbial partners to generate chemical signals that trigger a plethora of social behavioural changes. Modulation of the microbiota, an integral component of the holobiont, has profound social repercussion on the host (**Box Figure 3**).

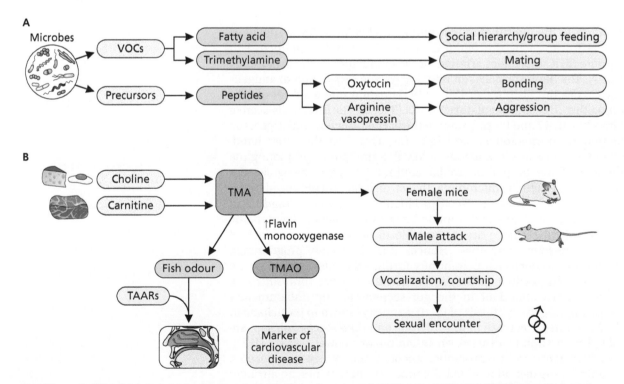

Box Figure 3. Microbial chemicals signal social behaviours in mammals. TMA: trimethylamine; TMAO: trimethylamine oxide.

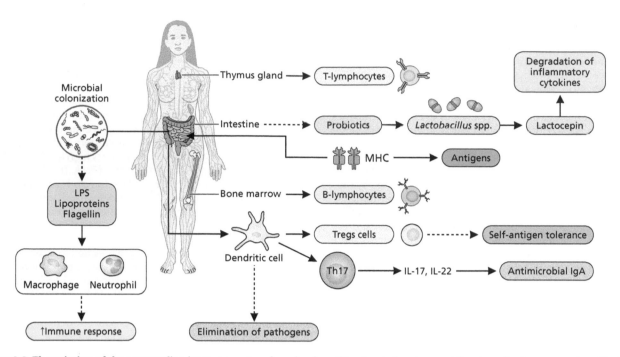

Figure 8.8. The priming of the mammalian immune system by microbes. Microbes help prime and fine-tune the immune system. The T and B lymphocytes and their further differentiation are guided by in part by microbial input in an effort to recognize the resident microbes and eliminate pathogens. Lactocepin derived from *Lactobacillus* spp. contribute to degradation of inflammatory cytokines while LPS (lipopolysaccharides), flagellin and lipoproteins generated from resident bacteria activate the immune response. Microbial stimulation of the dendritic cells promotes Tregs, interleukin (IL) production and pathogen elimination.

colonization to the intestinal mucus and epithelial layer. Any pathogenic evasion is thwarted by the dendritic cells. These cells equipped with appropriate MHC molecules arrange for the elimination of the invaders by the T cells. The dendritic cells not only induce the production of antigen-specific T cells but also trigger their differentiation into a subset of cells known as the Th17 and regulatory T cells (Tregs). The Th17 cells secrete cytokines like IL-17 and IL-22, promote the barrier function and trigger the production of antimicrobials and IgA. The Tregs, on the other hand, maintain tolerance to self-antigens, a key determinant in promoting the assemblage of the proper microbial community. Tregs recognize the antigens derived from the microbiota. In fact, an integrase produced by the genus *Bacteriodes* stimulates T-cell production. *Bacteriodes fragilis* and *Clostridium* spp. are known to promote Tregs differentiation. Gnotobiotic (devoid of microbes) mice are susceptible to infections due to lack of priming by resident microbes. Hence, this deliberate programmed acquisition of the microbial community fortifies the effectiveness of the immune system against pathogens. *Faecalibacterium prausnitzii*, a microbial species abundant in the gut secretes an anti-inflammatory protein prompting the synthesis of IL-10, a moiety known to participate in the Th17 response. Decrease in this microbe and *Bifidobacterium* spp. are associated with Crohn's disease, an inflammatory bowel disease (IBD). *Lactobacillus paracasei*, a probiotic, encodes the protease lactocepin responsible for the degradation of inflammatory chemokines, an attribute that is utilized to combat IBDs. Microbially derived molecular patterns such as lipopolysaccharides (LPS), lipoproteins and flagellin are known to promote the production of macrophages and neutrophils, important participants of the innate immune response (**Figure 8.8**).

8.9 Microbial Metabolites and Cellular Immunity

Metabolites generated by the microbial residents play an important role in modulating the immune response. These metabolites can be of

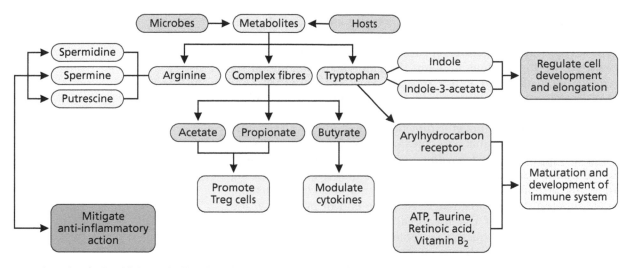

Figure 8.9. The role of microbial metabolites in cellular immunity. Metabolites generated by resident microbes and host-derived metabolites transformed by the microbial partners are potent modulators of the immune system. Arginine derivatives can mitigate anti-inflammatory activity while metabolites obtained as a consequence of tryptophan metabolism can promote cell elongation and immune system maturation. Taurine, ATP and vitamin B_2 also contribute in the modulation of the immune response.

microbial origin or derived from the host but modified by the microbes or synthesized from nutrients as a result of microbial activity. The microbially mediated metabolism of the aromatic amino acid tryptophan results in the formation of such metabolites as indole, indole-3-acetate and indole-3-propionic acid. These ligands act via the aryl hydrocarbon receptor system to modulate the immune activity and impede the colonization by pathogens like *Candida albicans*. Bacterial sphingolipids and capsule polysaccharides also help regulate the host defence mechanisms. The complex carbohydrate-derived SCFA like acetate and propionate produced by *Bacteroidetes* and butyrate synthesized by *Firmicutes* also contribute to calibrate how the immune system is deployed. While acetate and propionate promote the Tregs, butyrate limits the release of pro-inflammatory cytokines stimulated by LPS. The metabolites putrescine, spermidine and spermine resulting from arginine metabolism contribute to mitigate anti-inflammatory action and to the maintenance of intestinal mucosa where resident microbes are housed. Bile acids and products of bile acid metabolism like taurine are also instigators of numerous immune-related activities. ATP secreted by microbes interacts with the purine receptor isoform (P2X7) to regulate ion channels localized in immune cells. The role of microbes in nurturing and calibrating the immune system occurs during the initial development of the host. The maternal metabolites and microbiota enhance the immune cells during foetal life. Following birth, the development of intestinal lymphoid tissues and the maturation of myeloid cells are triggered by microbial metabolites like retinoic acid and vitamin B_2. The resident microbes and the mammalian host work in tandem to render the immune response effective and help in the recognition of self from non-self. A disruption of this cross-talk is an important precursor of numerous diseases. Hence, microbes and the metabolites they generate not only shape the development of the mammalian immune system but they also help fine-tune the precise surveillance strategies whereby in-house microbial population is nurtured while pathogens and antigen-emitting entities are destroyed (**Figure 8.9**).

8.10 Microbes in the Reproductive Process

The mammalian sexual organs house a range of microbial residents. The female reproductive organ possesses more microbes than the male counterpart. The microbiome of the semen contains anaerobic microbes like *Prevotella* spp. and facultative anaerobes such as *Lactobacillus* spp., *Pseudomonas* spp. and *Finegoldia* spp. The uterus and the endometrium are colonized by the vaginal and gut-residing microbes. The phyla *Bacteroidetes*, *Proteobacteria* and *Firmicutes* are most abundant. The vaginal microbiome is dominated by *Lactobacillus* spp. including *Lactobacillus crispatus*, *Lactobacillus iners* and *Lactobacillus jensenii* and is involved in preventing pathogenic infection, maintaining an acidic pH and generating ROS. In mammals like cows and ewes, *Aggregatibacter* spp., *Streptobacillus* spp. and *Lactobacillus* spp. are more prevalent. After fertilization and during pregnancy, the microbes may help in nurturing and protecting the embryo and foetus. An abundance of *Lactobacillus* spp. is usually common during pregnancy. Near the end of pregnancy, the microbiome is reflective of the microbial population pre-conception. Women with diverse vaginal microbial profiles during pregnancy are at risk for pre-term births. Even though the maternal gut microbes during the first trimester resembles the microbial population pre-conception, the microbial assemblage within this organ changes as the pregnancy proceeds. There is an overall increase in *Proteobacteria* and *Actinobacteria*. This shift is reflected in the change of metabolism during the gestation period.

In fact, women who are overweight while pregnant are characterized with an increase in *Bacteriodes* spp. and *Staphylococcus* spp. The placenta harbours microbes belonging to the phyla *Firmicutes*, *Tenericutes*, *Proteobacteria*, *Bacteroidetes* and *Fusobacteria*, a microbial constituency akin to that found in the oral cavity and the endometrium. *Streptococcus* spp. and *Lactobacillus* spp. are associated with foetal membranes, umbilical cord and amniotic fluid. Radiolabelled microbes like *Enterococcus fecium* fed to pregnant animals can be traced within the amniotic fluid and the meconium. During vaginal delivery, the new born is enriched with *Lactobacillus* spp., *Provetella* spp. and *Bifidobacterium* spp., a common maternal microbial signature. Although microbes within the sexual organ vary with age, sex hormones and diet, an increase in *Actinomyces* spp. and *Enterococcus* spp. has been shown to be linked to infertility. An elevated amount of *Corynebacterium* spp. is associated with endometriosis. In general, infertile women possess a higher abundance of *Ureaplasma* spp. and *Gardnella* spp. In vitro fertilization procedures tend to be less successful with a microbial population rich in *Gardnerella vagilis* and *Atopobium vaginae* coupled with a concomitant decrease in *Lactobacillus* spp. Probiotics including *Lactobacillus rhamnosus* may help shift the balance towards a positive outcome as they impede the proliferation of opportunistic bacteria. Although the role of the microbes residing within the sexual organs is just beginning to be unravelled, it is becoming increasingly clear that they contribute and help determine the outcome of the reproductive process. Despite the prominent role microbes residing in vagina are known to play in the conception and development of the foetus, the importance of the penile microbiome has yet to be fully appreciated (**Figure 8.10**).

8.11 Microbes: The Invisible Endocrine Modulators

Owing to their ability to synthesize biomolecules that can serve as signalling moieties, precursors to neurotransmitters and hormones, microbial communities associated with mammals are known to influence a wide variety of behavioural and social changes. It is relatively well established that numerous parasites use these strategies to modulate the social behaviours of their hosts. The single-celled *Toxoplasma gondii* alters the fear response in rodents. Upon infection, rodents become friendly to their feline predators, a behaviour favouring the continuation of the lifecycle of the parasite in a different host. Tape worm-infected fishes leave their

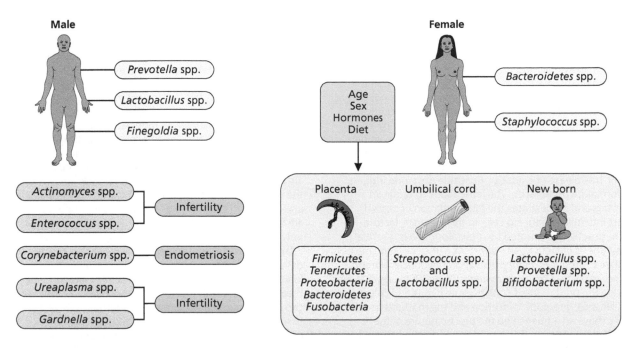

Figure 8.10. Microbes and the reproductive process. Microbes play a pivotal role in the reproductive outcome. While the presence of *Lactobacillus* spp. results in a successful pregnancy, an abundance of *Enterococcus* spp. and *Gardnerella* spp. can lead to infertility. The female sexual organ has more bacteria than the male counterpart. These microbes contribute to the prevention of vaginal infection. The placenta and the umbilical cord are also known to possess microbes. Maternal transmission of microbes is a key component promoting the development of the newborn.

cold-water environment for warm water, a physical conduct designed to promote the proliferation of the parasite in a milieu conducive to reproduction. Even in humans, viral infection is known to upregulate the expression of cholesterol 25-hydroxylase, a key mediator of Alzheimer's disease. Thus, it is not surprising that the microbial community residing within mammalian systems and other higher organisms help shape mood, cognition and social behaviour. Germ-free mice are known to suffer from neurological and learning deficiencies. Microbes play an important role in neurogenesis and the maturation of the central nervous system (CNS). The gut–brain axis is linked by the vagus nerve, a component of the parasympathetic nervous system that facilitates communication between the CNS and gut microbes.

Lactobacillus rhamnosus (JB-1) and *Bifidobacterium longum* decrease anxiety-like behaviour. Microbes are known to produce neurotransmitters like 5-hydroxytryptamine (serotonin), *N*-methyl D-aspartate, glutamate, GABA, dopamine and 4-ethyl sulphate. In fact, the latter contributes to anxiety-like behaviour. *Clostridium* spp., *Ruminococcus* spp. and *Lactobacillus* spp. have the ability to decarboxylate tryptophan into tryptamine, a metabolite modulating the level of serotonin. The latter can then influence the mood and appetite of the host. An increase in *Bacteroides* is characterized with depression, a situation resulting from an increase in dopamine. The SCFA resulting from microbial activity influences the homeostasis of the brain-derived neurotrophic factor (BDNF), a moiety associated with numerous neurological functions. The intake of prebiotics like fucosylated oligosaccharides and galacto-oligosaccharides results in an increase in BDNF. This emerging connection between microbes and the functioning of the brain will pave the way for better understanding of a variety of neurological disorders and microbial therapy may help alleviate these medical conditions.

The metabolites produced by the gut microbes also have an impact on the production of numerous hormones synthesized by the enteroendocrine cells located in the mucosal lining of the small intestine. These specialized cells

BOX 8.4 BREAST AND MILK MICROBIOME

Mammary glands are characteristic anatomical features specific to mammals and they serve numerous functions including the production of milk. This lactose-rich nutrient is the main nourishment of the newborn and does not only orchestrate the development of the infant but is a critical vehicle to transfer maternal microbes to the progeny. The breast tissue is composed of moist intricate intra-mammary ecosystem conducive for microbial proliferation especially during lactation period. During late pregnancy and following labour, the release of oxytocin, a peptide hormone triggers the formation of prolactin. The latter binds to prolactin receptors and orchestrates the differentiation of normal breast into milk-producing cells. In this instance, lactose synthase activity promoted by α-lactalbumine is essential in the synthesis of lactose, an important ATP-generating moiety in infants. Although *Lactobacillus* spp., *Bifidobacterium* spp., *Lactococcus* spp., *Corynebacterium* spp. and *Propionibacterium* spp. are the predominant constituents of milk and colostrum, diet, lifestyle, geographic location and the mother's health tend to have an influence on the precise nature of the microbial community. The mammary ducts, the nipples, the skin and the suckling by the infant also play a prominent role in the establishment of the microbiota. A Mediterranean diet favours *Lactobacillus* spp., while a Western nutritional habit is more prone to colonization by *Ruminococcus* spp. in the milk. The nature and concentration of oligosaccharides in human milk (HMOs) that may be sialylated or fucosylated depend on the resident microbes. The presence of *Bifidobacterium breve* promotes HMOs with sialic acid group, while *A. muciniphila*

nurtures the synthesis of oligosaccharides modified with fucose. These HMOs help shape the infant's gut microbiome as they are targeted at select microbes. It is important to note that infants possess lactase, an enzyme dedicated to cleave the disaccharide lactose, but they are unable to digest HMOs.

The nature of the milk changes with the development of the infant. Its chemical and microbial constituents are tailored to prompt a healthy outcome in the growing infant. The immune system, the GI tract and the neuronal network are all guided by evolving factors within milk. For instance, the antimicrobial activity within milk afforded by bacteriocins and lactoferrin diminishes as the infant's immune responses are fortified with age. Thus, the mammary glands do not only provide custom-made milk for proper growth of the infant but also transmit maternal microbes that prompt the development of numerous morphological and physiological features. During various disease conditions affecting the breast, there is a shift in microbial population, a situation that can help diagnose any mammary abnormality. Breast tumour tissue is enriched with *Methylobacterium radiotolerans* compared to healthy breast where *Sphingomonas* spp. are abundant. The microbiota housed within the breast and breast's milk are assembled in an organized fashion in an effort to ensure the success of the newborn and consequently the survival of the parental genetic information. The microbial community reconfigures in tandem with the age of the infant in an effort to guide the developmental process and any dysbiosis is a potent indicator of a diseased situation (**Box Figure 4**).

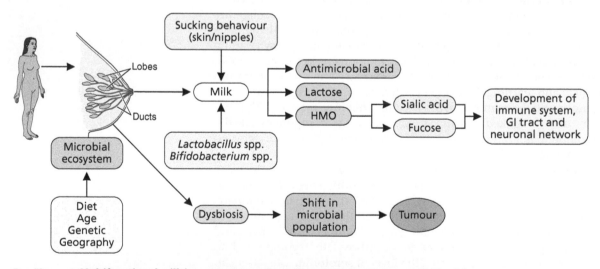

Box Figure 4. Multifunctional milk is a potent contributor to infant development and health.

produce glucagon-like protein 1 (GLP-1), peptide tyrosine (PYY) and serotonin. The intake of food, satiety and appetite are controlled by these hormones. The enterochromaffin cells dispersed throughout the GI tract sense nutrients and secrete serotonin that controls the mobility of the GI, a physiological process partly responsible for abnormal eating behaviours and consequently obesity. Modulating the activity of tryptophan hydroxylase I (TPHI) has a direct impact on the concentration of serotonin. An increase in body mass index of pregnant women is associated with an increase in *Bifidobacterium* spp. and *Staphylococcus* spp. This condition is attributed to an increase in TPHI activity. GLP-1 slows intestinal transit in response to nutrients like glucose and augments the release of insulin with a concomitant reduction of glucagon, an event regulating glucose homeostasis and the energy budget of the host. PYY, secreted by the L cells of the lower small intestine and the colon that suppresses appetite by stimulating the neuropeptide Y neurons is affected by SCFA. In fact, butyrate increases the level of PYY in the blood. The gut microbes possess enzymes that are involved in the metabolism of bile acids. The bile salt hydroxylases can modulate the secondary bile acids, hydrophobic moieties known to be reabsorbed by passive diffusion, thus impeding their elimination. For instance, *Bacteroidetes* interact selectively with taurine-conjugated bile acids. Modulation of bile acid pool has a major impact on the host metabolism. Probiotics like *Lactobacillus rhamnosus* promote weight loss due to their ability to produce conjugated linoleic acid and stimulate the expression of uncoupling proteins. Hence, the gut microbes generate numerous metabolites and their precursors that modulate various biological processes impacting behavioural and social responses of mammals (**Figure 8.11**).

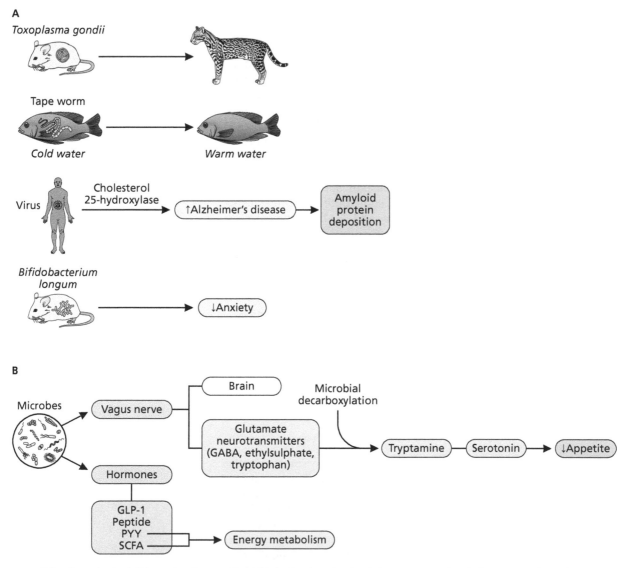

Figure 8.11. Microbes: the invisible endocrine system. (a) Infection by microbes is known to regulate behaviours and social interactions in their host. The presence of *Toxoplasma gondii* in rodents makes the hosts friendlier to cats, a behaviour designed to promote the survival of the microbe. Similar behavioural changed is observed in tape worm-infected fish. In this instance, the fish travels to a warm environment, a feature conducive to proliferation of the parasite. Viral infection in human activates an enzyme that promotes Alzheimer disease. Mice possessing *Bifidobacterium longum* have lower anxiety. (b) Microbial product generated in the digestive tract communicates with the brain via the vagus nerve. Glutamate, GABA (γ-amino butyrate) and tryptophan can all affect cerebral activity. The microbial decarboxylation of tryptophan results in tryptamine that influences appetite. Microbes also produce hormones that have a direct impact on energy metabolism. GLP-1: glucagon-like protein; PYY: peptide tyrosine tyrosine, SCFAs: short-chain fatty acids.

8.12 Microbiome and the Ageing Process

Ageing is a natural process resulting in physiological deficiencies that lead to various pathologies and eventually death. It is a complex interplay of numerous factors including the host genome, the resident microbes, lifestyle, diet and the environment. Genomic instability, telomere shortening, epigenetic alteration, loss of protein homeostasis, uncontrolled formation of free radicals and DNA damage are some of the biochemical pathways mediating the decline of bodily functions that result in ageing and eventually death. The microbial community is an important participant in this event as microbes, integral components of the host have been shown to directly affect ageing. For instance, young fishes (*Nothobranchius furzeri*) have an abundance of *Bacteroidetes* while the old ones are dominated by *Proteobacteria*. The transplantation of *Bacteroidetes* prolongs the lifespan of older fish. In humans as in most organisms, ageing is associated with a decrease in microbial diversity. There tends to be a decrease in *Bifidobacterium* spp. and an increase in *Clostridium* spp. and *Lactobacillus* spp. Centenarians have a more diverse microbial population with an enrichment in *Christensenella* spp., *Akkermansia* spp. and *Bifidobacterium* spp.

Furthermore, ageing is characterized by a decrease in the differentiation of stem cells, in immune response and in protein degradation pathways. SCFA and tryptophan metabolites are known to modulate numerous processes associated with ageing. Microbial-generated ROS are implicated in cellular proliferation, ubiquitin-mediated proteasome pathway and immune signalling. While sulphate-reducing bacteria are a source of H_2S, a promoter of longevity, some gut microbes transform nutrients into products like ellagitannins found in pomegranate that are known to improve mitochondrial function and extend lifespan. Spermidine induces autophagy, prevents protein aggregation and contributes to proteostasis, a situation common during ageing. Folate and NAD participate in epigenetic mechanisms by modifying histones and promoting sirtuin activity. These reactions are linked to the prolongation of life. Caloric restriction improves longevity by impeding LPS synthesis and increasing insulin sensitivity, events known to be influenced by gut microbes. Aromatic amino acids like phenylalanine and tryptophan and their derivatives are known to influence the ageing process. In fact, a drastic reduction in tryptophan production is observed after the age of 34–54 years. Enzymes involved in tryptophan metabolism that are elevated in infants are virtually absent in elderly population. Indole metabolites derived from tryptophan with the aid of tryptophanase are known to increase longevity as they are involved in T-cell differentiation and neurological functioning (**Figure 8.12**).

8.13 Hibernation and Microbial Contribution

Seasonal variation in food availability is the main driver why some animals hibernate. Hibernation that entails fasting or limited intake of food coupled with marked reduction in energy utilization is usually preceded by intensive feeding when food is plentiful. This adaptation triggered by lack of nourishment is designed to maintain the internal organs so that the organism can survive the periodic scarcity of food. Although this process is characterized by punctual anatomical changes, there is a significant reconfiguration of the microbiome. During the feeding season, microbes promoting the storage of energy predominantly as lipids are favoured while microbes aiding in the metabolism of waste products are preferred during the fasting period. Host-derived nutrients like mucin and oligosaccharides play a pivotal role in assembling the desired microbial communities. For instance, microbes enabling the host to synthesize lipids from plant polysaccharides are dominant in the spring and summer while during the

A

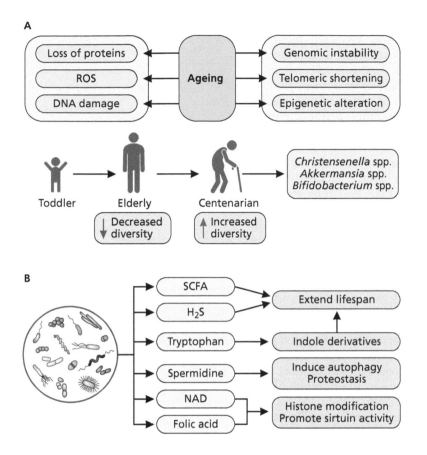

Figure 8.12. Microbiome in the ageing process. (a) Ageing is characterized by numerous biochemical events such as loss of proteins, increase in ROS, genomic instability and shortening of the telomeres. Many of these processes are directly or indirectly mediated by bacteria involved in the production or transformation of ageing metabolites. The diversity of the microbiome decreases as age progresses. Centenarians have a more diverse microbial population, a situation responsible for their longevity. (b) The ability of microbes to regulate metabolites like NAD, tryptophan, spermidine and SCFA has a major impact on ageing. While NAD influences the activity of sirtuins, spermidine controls protein homeostasis. Tryptophan derivatives known to be generated by microbial activity help extend longevity.

hibernating winter months, microbes with the ability to catabolize and recycle the host's nutrients are preferred. During the feeding season, an abundance of *Firmicutes* is frequently observed. *Bacteroidetes* is the dominant phylum in most hibernating mammals including bears and squirrels when food is scarce. Calorie-restricted diet promotes the colonization of the gut by *Bacteroides fragilis* and impedes the proliferation of *Streptococcus* spp., a situation prevalent in hibernating mammals. The hibernation-induced changes in the immune system and the epithelial barrier function also help determine the proper microbial community. The increased levels of cholesterol, bile acids and triglyceride indicate a net preference of fatty acid as a source of energy. These moieties do not only liberate more energy following metabolism but they are also anhydrous, thus more readily amenable for storage. In fact, 1 g of fatty acid liberates at least twice the amount of ATP than 1 g of carbohydrate. The elevated amount of succinate indicates the involvement of this dicarboxylic acid in gluconeogenesis during hibernation as it can be readily converted into oxaloacetate, a key precursor in the synthesis of glucose.

8.14 Microbes and Energy Utilization

The changes in the GI tract characterized by the reduction in the length of the villi and the atrophy of the caecum are accompanied with a suppression of metabolic activities. Although mammals are homeothermic with a temperature ranging from 35°C to 37°C during active season, a lower variable body temperature is evident during the fasting months. This helps limit the utilization of energy and prolong consumption of the stored calories. The increased expression of glucose transporter (SGLT-1) and Na$^+$/K$^+$ ATPase pump contribute to the enhanced uptake of glucose necessary to maintain functioning organs, especially in conditions where the oxygen tension is low as in the case of hibernating organisms. The presence of microbes with the ability to salvage urea nitrogen is important during hibernation since the intake of any nutrient is virtually restricted. While the host urea cycle enzymes are sharply reduced, microbes possessing urease are promoted. The NH_3 released from urea is readily channelled towards amino acid and protein synthesis. *Akkermansia muciniphila*, a microbe belonging to the *Verrucomicrobiota* phylum is abundant in hibernating mammals. This microbe can readily consume host-derived mucin as a source of energy and helps promote non-shivering thermogenesis due to its ability to modulate the conversion of white adipose tissue into the beige counterpart. The beige adipose tissue like the brown adipose tissue possesses mitochondria and has the ability to generate heat. In fact, the shift to beige adipose tissue coincides with the increased expression of the uncoupling protein 1 (UCP-1), a biomolecule participating in thermogenesis. The UCP mediates the conversion of the proton gradient formed during oxidative phosphorylation into heat and not into ATP, an ingredient not in high demand in immobile hibernating hosts with low metabolic rates.

8.15 Mucins, the Microbial Selectors

Mucins are glycosylated proteins that are rich in such amino acids as cysteine, threonine and serine. The intestinal epithelium is coated with these thick, viscous and sticky glycoproteins. These O-linked oligosaccharides macromolecules not only serve as a protective barrier, they also attract select groups of microbes that contribute to the integrity of the intestine. Only microbes with the capacity to degrade these moieties into monosaccharides and amino acids are able to reside within these biofilm-like structures. Furthermore, the cysteine residues generated as a consequence of mucin degradation liberates sulphate. This environment nurtures sulphate-reducing bacteria like *Desulfovibrio* spp. that are prominent in hibernating mammals. The different metabolites like H_2S known to orchestrate numerous biological processes also help shape the microbial population conducive to the hibernating needs of the hosts. In this instance, the microbial assemblage is characterized by an increase in the phyla *Bacteroidetes* and *Verrucomicrobiota* during hibernating months in such mammals as hamsters, squirrel and bears. This reconfiguration of the microbial constituent coincides with a decrease in SCFAs. Butyrate and propionate that are elevated in summer months undergo a sharp decline in the fasting season and most probably results in the modification of the GI tract that is subsequently primed for reduced digestive function. The microbial community is an integral component of hibernating mammals and the change in its constituents allow the host not only to maintain its body temperature but also to prevent any significant damage to its visible organs, conditions essential for their survivability during the fasting season. This reprogramming of the intestinal microbes coupled with the morphological features associated with the alimentary tract is central to the hibernating success of these mammals. In this instance, both the visible and the invisible organs work in tandem to elicit a successful outcome for the holoorganism (**Figure 8.13**).

8.16 Temperature Fluctuation and Microbial Community

Abiotic stress is a major modulator of the microbial constituents in any multicellular organisms. Temperature change, climate variation, CO_2 gradient, salt concentration and pH are some of the parameters determining the nature of the microbial constituents within a holobiont. Extreme temperatures or drastic change in pH are known to disrupt microbial communities, an event having important implications in terms of energy homeostasis, host's fitness, digestive performance and other essential

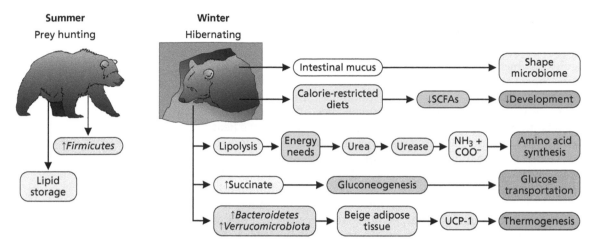

Figure 8.13. Reprogramming of microbial consortia in hibernating mammals. Mammals hibernate in order to survive a scarcity of foods. During summer months, the hibernating mammals store foods as lipids due to their high calorie content. Microbes are pivotal in this adaptation as they produce metabolites that promote thermogenesis by stimulating the expression of uncoupling protein-1 (UCP-1) and gluconeogenesis essential during diminished O_2 intake. The microbes also help recycle urea into ammonia and subsequently into amino acids. During summer months there is an abundance of *Firmicutes*, while *Bacteroidetes* and *Verrucomicrobiota* predominate in the hibernating season.

BOX 8.5 ENERGY EXPENDITURE, THERMOGENESIS AND MICROBES

Energy expenditure, energy harvest, thermogenesis and metabolites derived or transformed by the microbiome are intricately linked. The cross-talk between these metabolites and the cellular signalling networks can shift the balance from a lean host to an obese one and from a cold-adapted mammal to a hibernating one. The ability of the gut microbial population to extract energy-producing precursors like SCFAs and lactate from high-fibre diets modulates the synthesis of ATP and triglycerides within the host. This, in turn, dictates whether cellular metabolism follows an anabolic route of storing energy or a catabolic pathway aimed at utilizing energy. The fine-tuning of these metabolites modulated by metabolic networks operating within the microbial communities is pivotal to the physiological situations the holobiont is exposed to. The release of increased SCFAs liberated from the dietary fibres favours fat storage while the diminished levels of SCFAs promote ATP consumption. The presence of *Xylanobacter* spp. is known to liberate energy-rich components from non-digestible fibres as they possess cellulases and hemicellulases. Hence, the microbiota has a major influence on food intake.

The transformation of primary bile acids into secondary bile acids, a reaction mediated by the gut microbes results in a significant change in the white adipose, brown adipose and beige adipose tissues. These lipid-replete organs serve as energy storage and heat generator. The white adipose tissue is essentially a triglyceride depot while the latter two lipid-rich tissues possess mitochondria involved in β-oxidation of fatty acids and thermogenesis. The bioactive bile acids interact with their cognate receptors leading to an increase in cAMP and the eventual activation of thyroxine (T_3), a hormone known to trigger thermogenesis. When exposed to cold, some mammals undergo a marked shift in their microbial population leading to a decrease in fat content and an increase in calorie-rich moieties in faeces. This reshaping of the gut microbes

increases the level of circulating bile compounds in the blood with the concomitant stimulation of thermogenesis. The presence of *A. muciniphila* has been linked to the browning of the white adipose tissue, punctuated by a decrease in lipids. Metabolites derived or modified by microbes are integral components of the signalling networks in the holobiont and are modulated by thermal exposure. Such an interaction helps maintain the energy expenditure and body temperature in mammals (**Box Figure 5**).

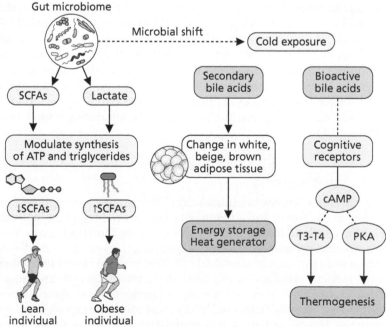

Box Figure 5. **Metabolites generated by microbes essential in energy metabolism and temperature regulation.**

functions. For instance, cold-induced change in bile acids results in a shift in microbial consortium. Corals from warm water harbour more *Mycobacterium* spp., a microbe that can increase the risk of infection. Some organisms have evolved various strategies to adapt to cold temperatures. Social animals engage in different cooperative behaviours to maximize fitness when exposed to harsh environments. They gather food together, use vocal signals to warn their conspecifics of predators and engage in social thermoregulation during winter. The latter reduces heat loss and energy expenditure. Huddling is a practice that has been adopted by numerous mammals including voles, the non-hibernating rodents. They maintain a lower resting metabolic rate and invoke molecular networks to promote non-shivering thermogenesis. They possess predominantly brown adipose tissue during cold-winter months. These fat bodies are replete with mitochondria that are charged not to produce ATP but utilize the resulting membrane potential following the transfer of electrons to O_2 into heat. This process is only possible because these mitochondria localized in the adipose tissue are known to express high amounts of uncoupling proteins. In this instance, H_2O and heat are generated. Additionally, such an adaptation triggers a microbial shift that further helps decrease the energy need of the host.

The Brandt's voles (*Lasiopodomys brandtii*) are small non-hibernating herbivores known to huddle in groups of 20 to survive winter. Physical contact increases skin thickness, allows less heat loss from the surface and helps maintain a higher surface temperature. This behaviour promotes less consumption of energy than when they are separated and promotes the lowering of the core body temperature, features contributing to save on energy output. The temperature on the surface is higher than the core body temperature during this grouping activity. In the winter, these mammals possess microbial communities with higher phylogenetic diversity. There is a major reconfiguration in microbes housed in the caecum. There is a marked increase in *A. muciniphila*, a microorganism known to metabolize mucins and create an environment for the nurturing of a microbial population dedicated to promote the low metabolic rate within the host. The optimization of the energy budget of the host is facilitated by the utilization of lipids, the main energy reserve. In fact, a decrease in adiposity is observed as lipolysis, followed by β-oxidation of fatty acid is enhanced. This process requires an adequate supply of O_2 and mitochondria with increased UCP expression, coupled with a diminution in ATP synthesizing activity. Microbes with the ability to supply a range of metabolites can propel the mitochondrion towards a heat-generating organelle. This shift in microbial community lays the molecular foundation to thermogenesis, a process critical in maintaining body temperature in these non-hibernating warm-blooded mammals during winter months (**Figure 8.14**).

8.17 Conclusions

Mammals are complex organisms derived from the evolutionary process that has given rise to small flying bats, ocean-dwelling giant blue whales weighing nearly 15 tonnes and the average 70-kg thinking human. Despite their size and geographical location spanning the sea, air or land, they all house a wide plethora of microbial cells that in some cases outnumber the hosts' 'regular' cellular components. These microbial cells are intimately linked to the well-being of the host organisms and their demise or unorganized assembly triggered by different conditions result in unhealthy outcomes. Hence, to have a proper understanding how these mammals function, it is important that the roles of microbial consortia residing in various organs be delineated. Such molecular information will provide a better appreciation as to how these microbes are at the centre of the

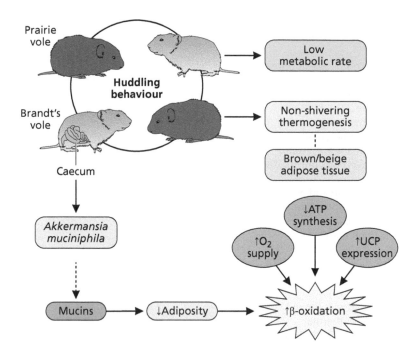

Figure 8.14. Thermoregulation in non-hibernating mammals: microbial influence. The non-hibernating rodents like the Brandt's voles adopt a huddling behaviour in order to maintain their body temperature during winter. This behavioural change is coupled to a significant shift in the microbial community aimed at promoting heat-generating metabolic pathways. The enhanced expression of UCP with the concomitant increase in β-oxidation favour thermogenesis. The elevated presence of *Akkermansia muciniphila* within the caecum stimulates mucin degradation and a decrease in adiposity. Heat generation, decrease metabolic rate and huddling contribute to the survival of these rodents in winter.

evolutionary changes living systems on this planet have and are undergoing. This insight will inform how mammals and humans in particular behave molecularly when they are challenged by adverse biotic and abiotic stress. The former is an ongoing concern, especially under the current situation where a global viral pandemic has debilitated activity worldwide and have killed millions of people. The imbalance created by the viral entry in the human body triggers a disproportionate response of the immune system that is at the root cause of most of the symptoms elicited following infection by SARS-CoV-2, a situation eventually leading to the death of the host. The deciphering of the immunological networks selecting the 'good' microbes and neutralizing the parasitic ones will be a step in the right direction. The communication between the visible and invisible organs and how they interact will also provide clues on how this information may be hijacked by viral or other parasitic microbes. This will further the understanding of supra-organisms like mammals and demonstrate the crucial role microbes play in fostering the harmonious functioning of all the organs within. Even though the various organs are specialized to perform a dedicated task, microbes with their flexible molecular capacity can readily respond to changes and enable the host to adapt, an attribute essential for survival. Thus, the multi-tasking microbes are integral components of all mammals as their roles have evolved in tandem with the hosts' anatomy and physiology and most likely will guide further modification dictated by various biological and physical events.

SUGGESTED READINGS

Agus, A. et al., (2020). Gut microbiota-derived metabolites as central regulators in metabolic disorders. *Gut, 70*(6), 1174–1182. doi:10.1136/gutjnl-2020-323071

Bana, B., & Cabreiro, F. (2019). The microbiome and aging. *Annual Review of Genetics, 53*(1), 239–261. doi:10.1146/annurev-genet-112618-043650

Cani, P. D., & Vos, W. M. (2017). Next-generation beneficial microbes: The case of Akkermansia muciniphila. *Frontiers in Microbiology, 8,* eCollection 2017. doi:10.3389/fmicb.2017.01765

Cani, P. D. et al., (2019). Microbial regulation of organismal energy homeostasis. *Nature Metabolism, 1*(1), 34–46. doi:10.1038/s42255-018-0017-4

Carey, H. V., & Assadi-Porter, F. M. (2017). The hibernator microbiome: Host-bacterial interactions in an extreme nutritional symbiosis. *Annual Review of Nutrition, 37*(1), 477–500. doi:10.1146/annurev-nutr-071816-064740.

Chevalier, C. et al., (2015). Gut microbiota orchestrates energy homeostasis during cold. *Cell, 163*(6), 1360–1374. doi:10.1016/j.cell.2015.11.004

Clarke, G. et al., (2014). Minireview: Gut microbiota: The neglected endocrine organ. *Molecular Endocrinology, 28*(8), 1221–1238. doi:10.1210/me.2014-1108

Delsuc, F. et al., (2013). Convergence of gut microbiomes in myrmecophagous mammals. *Molecular Ecology, 23*(6), 1301–1317. doi:10.1111/mec.12501

Fernández, L et al., (2020). The microbiota of the human mammary ecosystem. *Frontiers in Cellular and Infection Microbiology, 10*, eCollection2020. doi:10.3389/fcimb.2020.586667

Fernández, L., & Rodríguez, J. (2020). Human milk microbiota: Origin and potential uses. *Nestlé Nutrition Institute Workshop Series Milk, Mucosal Immunity, and the Microbiome: Impact on the Neonate*, 1–11. doi:10.1159/000505031

He, J. et al., (2018). Characterizing the bacterial microbiota in different gastrointestinal tract segments of the bactrian camel. *Scientific Reports, 8*(1). doi:10.1038/s41598-017-18298-7

Hoye, B. J., & Fenton, A. (2018). Animal host-microbe interactions. *Journal of Animal Ecology, 87*(2), 315–319. doi:10.1111/1365-2656.12788

Hylander, B. L., & Repasky, E. A. (2019). Temperature as a modulator of the gut microbiome: What are the implications and opportunities for thermal medicine? *International Journal of Hyperthermia, 36*(Sup1), 83–89. doi:10.1080/02656736.2019.1647356

Kean, E. F. et al., (2017). Odour dialects among wild mammals. *Scientific Reports, 7*(1), 1–6. doi:10.1038/s41598-017-12706-8

Kordy, K. et al., (2020). Contributions to human breast milk microbiome and enteromammary transfer of *Bifidobacterium breve*. *Plos One, 15*(1), e219633. doi:10.1371/journal.pone.0219633

Lutz, H. L. et al., (2019). Ecology and host identity outweigh evolutionary history in shaping the bat microbiome. *MSystems, 4*(6), 1–16. doi:10.1128/msystems.00511-19

Martin, A. et al., (2019). The influence of the gut microbiome on host metabolism through the regulation of gut hormone release. *Frontiers in Physiology, 10*(428), eCollection 2019. doi:10.3389/fphys.2019.00428

Min, B. R. et al., (2020). Dietary mitigation of enteric methane emissions from ruminants: A review of plant tannin mitigation options. *Animal Nutrition, 6*(3), 231–246. doi:10.1016/j.aninu.2020.05.002

Mitchell, S. C., & Smith, R. L. (2016). Trimethylamine—The extracorporeal envoy. *Chemical Senses, 41*(4), 275–279. doi:10.1093/chemse/bjw001

O'Hara, E. et al., (2020). The role of the gut microbiome in cattle production and health: Driver or passenger? *Annual Review of Animal Biosciences. 8*, 199–220. doi: 10.1146/annurev-animal-021419-083952

Ost, K. S., & Round, J. L. (2018). Communication between the microbiota and mammalian immunity. *Annual Review of Microbiology, 72*(1), 399–422. doi:10.1146/annurev-micro-090817-062307

Ross, A. A. et al., (2019). The skin microbiome of vertebrates. *Microbiome, 7*(1), 1–15. doi:10.1186/s40168-019-0694-6

Ruiz-Ruiz, S. et al., (2019). Functional microbiome deficits associated with ageing: chronological age threshold. *Aging Cell, 19*(1), e13063. doi:10.1111/acel.13063

Samsudin, A. A. et al., (2012). Cellulolytic bacteria in the foregut of the dromedary camel (Camelus dromedarius). *Applied and Environmental Microbiology, 78*(24), 8836–8839. doi:10.1128/aem.02420-12

Sarkar, A. et al., (2020). The role of the microbiome in the neurobiology of social behaviour. *Biological Reviews, 95*(5), 1131–1166. doi:10.1111/brv.12603

Sharon, G. et al., (2014). Specialized metabolites from the microbiome in health and disease. *Cell Metabolism, 20*(5), 719–730. doi:10.1016/j.cmet.2014.10.016

Singhal, R., & Shah, Y. M. (2020). Oxygen battle in the gut: Hypoxia and hypoxia-inducible factors in metabolic and inflammatory responses in the intestine. *Journal of Biological Chemistry, 295*(30), 10493–10505. doi:10.1074/jbc.rev120.011188

Suzuki, T. A. (2017). Links between natural variation in the microbiome and host fitness in wild mammals. *Integrative and Comparative Biology, 57*(4), 756–769. doi:10.1093/icb/icx104

Swartz, J. D. et al., (2014). Characterization of the vaginal microbiota of ewes and cows reveals a unique microbiota with low levels of lactobacilli and near-neutral pH. *Frontiers in Veterinary Science, 1*, eCollection 2019. doi:10.3389/fvets.2014.00019

Tapio, I. et al., (2017). The ruminal microbiome associated with methane emissions from ruminant livestock. *Journal of Animal Science and Biotechnology, 8*(1), 1–11. doi:10.1186/s40104-017-0141-0

Theis, K. R. et al., (2013). Symbiotic bacteria appear to mediate hyena social odors. *Proceedings of the National Academy of Sciences, 110*(49), 19832–19837. doi:10.1073/pnas.1306477110

Tomaiuolo, R. et al., (2020). Microbiota and human reproduction: The case of male infertility. *High-Throughput, 9*(2), 10. doi:10.3390/ht9020010

Wu, W. et al., (2021). Microbiota regulate social behaviour via stress response neurons in the brain. *Nature. 595*(7867) 409–414. doi:10.1038/s41586-021--03669–y

Zhang, X. et al., (2018). Huddling remodels gut microbiota to reduce energy requirements in a small mammal species during cold exposure. *Microbiome, 6*(1), 1–14. doi:10.1186/s40168-018-0473-9

PLANTS AND THEIR MICROBIAL COMMUNITIES

9

Contents

Keywords

- Plants
- Microbes
- Nutrients
- Germination
- Abiotic Stress
- Defence
- Flowering Signals
- Carnivores

9.1 Introduction

9.1.1 Plants: general characteristics

Plants are multicellular organisms known to synthesize their own nutrients. They are composed of eukaryotic cells and possess a specialized organelle termed chloroplast. The latter derived through endosymbiosis of cyanobacteria contains the light-absorbing chlorophyll, an important component of the photosynthetic machinery. Plants belong to the kingdom of *Plantae* and can be further classified into vascular and non-vascular groups. Vascular plants have a root and shoot system with the xylem and phloem. These two components connect with all the other parts of plants like stem, leaves, flowers, fruits and root. They help transport nutrients,

DOI: 10.1201/9781003166481-9

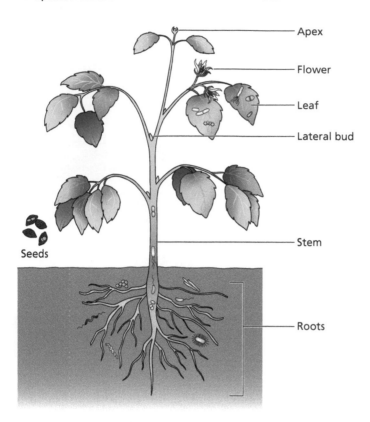

Apex

Flower

Leaf

Lateral bud

Seeds

Stem

Roots

Figure 9.1. Structural features of a typical plant. Most plants consist of roots, stem and leaves. They produce seeds via flowers or cones. The leaves tend to have a characteristic green colour due to the presence of the light-harvesting chlorophyll. Roots are involved in the absorption of H_2O, minerals and nutrients, while the stem facilitates their transport throughout the organism. Microbes are essential components of this holobiont and are localized within and on the plants.

minerals and H_2O. The non-vascular plants usually do not have roots and transport nutrients via osmosis. Seedless plants reproduce with the aid of spores. The seed-producing plants can be further divided into gymnosperms and angiosperms. The former generates their seeds in cones while the latter derived their seeds from flowers. These multicellular organisms with the ability to produce their own nutrients comprise a wide variety of specialized cells that are assisted by microbial cells in performing an array of essential functions. Plants need their microbes to survive. Like all multicellular organisms, they depend on their microbial partners to create diverse living entities that contribute to the ecological and economic well-being of the world (**Figure 9.1**).

9.1.2 Ecological and economic importance

Plants are essential for the survival of numerous organisms on this planet. Their ability to capture solar energy and convert it into nutrients is the basis of life on which all animals depend. For instance, herbivores that are a food source for numerous other organisms cannot live without their main source of energy, plants. The photosynthetic process involved in the fixation of CO_2 and H_2O into carbohydrates in the presence of sunlight is associated with the concomitant liberation of O_2. This plant-driven activity is crucial for the lifestyle of all aerobic organisms. The forest systems and savannahs they generate are unique habitats that a variety of animals, birds and other species are reliant on in order to propagate. Without these diverse ecosystems shaped by plants, the rich biodiversity of planet earth will be non-existent as the impact on the global climate will be disastrous. Plants also help maintain the level of atmospheric CO_2 and act as a sink of air pollutants. They contribute to the quality of soil by generating humus and depositing nutrients that are important determinants of soil fertility. Roots possess a plethora of microbes that not only mediate the detoxification of environmental toxins but also play a pivotal role in the biogeochemical cycles of such minerals as phosphorus, calcium and nitrate. Through the process, plants regulate the water content of the atmosphere, lakes and

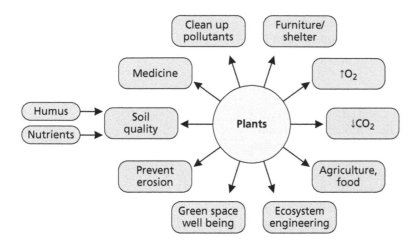

Figure 9.2. **Economic and ecological importance of plants.** Plants serve as a major source of food that help feed the world. They regulate atmospheric O_2, CO_2 and essential for the survival of most organisms. Improving soil quality and engineering the ecosystem are also important tasks plants perform. They supply numerous medications and contribute to the detoxification of pollutants.

rivers. Humidity of the soil is another contribution the activity of plants modulates and ensures the growth of organisms residing in their specific habitat. Land denuded of plants undergoes rapid erosion creating an ecosystem incompatible with the sustenance of life. Deforestation is a major cause of environmental and ecological disasters.

One of the main benefits of plants is their role in agriculture, a planned human activity that supplies food to the world population and livestock. The economic impact of agriculture worldwide is wide ranging. Human civilizations have engaged in agrarian practice since time immemorial. Staples like rice, wheat and potato coupled with fruits and vegetables have been feeding the world population since the emergence of humans. Lumber, cotton, medicinal plants and flowers can be converted into a variety of consumer products and are important economic drivers in different corners of the world. The recent contribution of plants to the biofuel industry has added another layer to this economic twist. Green space in urban setting is part of human fabric and contributes to a healthy being. Even in offices, hospitals and roof tops, the presence of plants has become a critical aspect of the architectural landscape, thus shaping human activities. Hence, the ecological and economic significance makes it imperative that the holistic nature of plants that include its visible components such as roots, stem, leaves, flowers and fruits and its invisible parts constituted of microbes be fully appreciated. These photosynthetic holoorganisms like all multicellular living systems work as coordinated units living in harmony with their ecosystem in in disparate regions of the globe (**Figure 9.2**).

9.2 Acquisition of Microbial Partners

Plants, like all multicellular organisms have evolved in close association with microbes. These microbial communities are located both on the surface and within the host. While endophytes, the internal dwellers are located inside the plants, the epiphytes colonize the extracellular components. The microbial cells are integral constituents of roots, stems, leaves, flowers and fruits. Although some microbes can be transmitted vertically usually from seeds like the colonization of maize roots by *Enterobacter asbutiae* localized in kernels, most of the microbial community is assembled from microbes in the geographical environment where the plants are located. The rhizosphere refers to the microbial organisms located in the soil and is the main recruiting site for the microbiome found within roots. The phyllosphere, the area above the ground houses microbes associated with the stem and the leaves. Likewise, the anthosphere, the flower environment, spermosphere, the area around the seeds, and carposphere, the external milieu of the fruits are the sources of microbes from where these plant components derive their microbial residents respectively. The interaction of the plants with the diverse flora and fauna also provides a recruiting opportunity for the microbes constituting the plant holobiont. There are approximately 10^{11} microbial cells per gram of root.

9.2.1 Factors shaping microbial assembly

As the soil matrix is the major supplier of microbes, soil types coupled with various chemical and physical parameters like moisture, aeration, temperature, pH, O_2, CO_2 and light intensity are important determinants in the microbial assembly. Plant genotypes, age, root exudates, availability of nutrients, phytochemicals, root morphology, microbial interaction and pathogens are filters that select for the appropriate microbes being housed on and within the host. Although there is stochastic variation, the dominant bacterial phyla tend to be *Actinobacteria, Bacteroidetes, Firmicutes, Proteobacteria* and *Acidobacteria*. For instance, grapevine roots have an abundance of *Proteobacteria, Acidobacteria* and *Actinobacteria* while maize roots harbour predominantly *Proteobacteria, Firmicutes* and *Bacteroidetes*. The phyla *Ascomycota* and *Basidiomycota* are the most abundant fungi. The floral parts, the leaves and the vegetative foliar provide a unique environment for the microbes to associate with while the developing roots and cracks within are ideal sites for microbial colonization. *Ralstonia, Burkholderia* and *Pseudomonas* are the prominent endophytes in grape berries, stems and leaves are associated with *Pseudomonas, Sphingomonas* and *Frigoribacterium*. Almond, grapefruit and pumpkin flowers possess an abundance of *Pseudomonas*. The microbial community is essentially composed of core and satellite (commensal) microbes. For instance, the core microbes of potato, *Solanum tuberosum*, are constituted by *Bradyrhizobium* spp., *Sphingobium* spp. and *Microvirga* spp., while in grapevine *Pseudomonas* spp., *Micrococcus* spp. and *Hyphomicrobium* spp. are some of the microbes selected as core irrespective of the soil type or climatic conditions (**Figure 9.3**).

9.2.2 Chemical signals shaping microbial constituents

Regardless of the nature of plants involved, their root exudates are rich in a diverse variety of chemical compounds that play a key role in selecting and recruiting the proper microbial consortium designed to propel growth and proliferation. These moieties are laden in information programmed to attract favourable microbes and impede invasion by pathogens. Nearly 20% of the fixed CO_2 and 15% of processed N_2 are secreted in the exudates. Monosaccharides like glucose, galactose and fructose and organic acids such as malate, citrate and fumarate are part of the exudate. Amino acids, phenolic derivatives, quaternary amines, fatty acids and nucleotides are also present in the radical secretion. These molecules are recognized by microbes having the appropriate receptors. Following the sensing of the signal, the microbe travels to the root, proliferate and colonize a component of the plant radical structure. The presence of simple sugars is elevated during the early developmental stage. Galactose is known to be a chemo-attractant for *Bacillus subtilis*, a link facilitated by the phosphoenol transfer system, while amino acids target *Sinorhizobium meliloti* in alfalfa (*Medicago sativa*) with the assistance of its amino acid sensor. Further, this bacterium also expresses the enzyme L-proline dehydrogenase mediating the conversion of proline into glutamate, a moiety that can readily serve as a source of carbon and nitrogen. In fact, the exudates of alfalfa are rich in proline and proline-containing compounds like stachydrine that promote bonding with this rhizobium.

Malate mobilizes *Pseudomonas putida* and organic acids in banana root exudate select for *Bacillus amyloliquefaciens*. Hydroxy-benzoate discharged by plants to solubilize Fe and P also determines the microbial residents of roots. This catechol recruits numerous microbes including

A

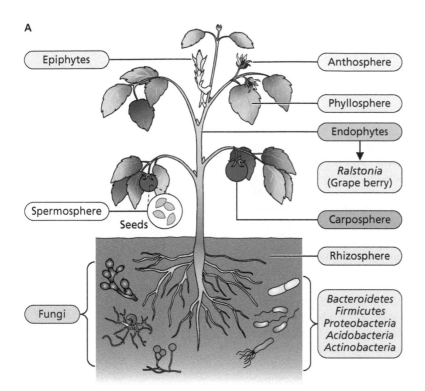

Figure 9.3. Plant microbial communities: location and selection. (a) Microbes colonize virtually all components of plants. Those localized outside are referred to as epiphytes while endophytes reside within the plants. *Ralstonia* is an important endophyte of grape berry. The rhizosphere harbours the most microbes including fungi. Carposphere is the environment within fruits occupied by bacteria while anthosphere is microbial space provided by flowers. (b) Temperature, pH, O_2, CO_2, moisture and genotype help shape the plant microbiome. Root lysate and soil composition are involved in the selection of the microbial partners. Plants do also acquire their microbial partners from the flora and fauna found in their ecosystem.

B

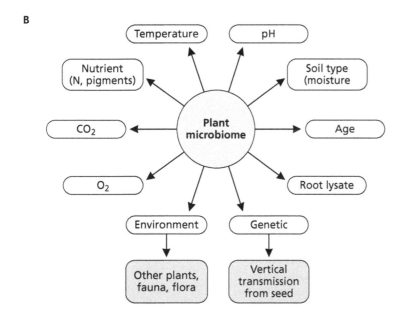

Pseudomonas spp. Ethylene (C_2H_4), a plant hormone and choline, a quaternary amine are also involved in the recruitment of microbes. In situations where C_2H_4 impedes host–microbe interaction, inhibitors are produced to counter this phytohormone. *Bradyrhizobium elkanii* secretes rhizobactin to inhibit C_2H_4 synthesis in the host purple bush-bean plant. Degradation of 1-aminocyclopropane 1-carboxylate, a C_2H_4 precursor by the enzyme 1-aminocyclopropane 1-carboxylate deaminase (ACC) into α-keto-butyrate and ammonia may also contribute to increased nodulation. The dependence of the members of the consortium on each other is another factor driving the assemblage of the microbial community. This syntrophic feature compels some strains of *Pseudomonas fluorescens* to associate with *Saccharomyces cerevisiae*. The yeast depends on thiamine produced by the bacterium while the fungal trehalose is utilized by the microbe. Numerous symbionts have reduced genetic profiles compared to their free-living counterparts as this strategy fortifies the partnership and forces the constituents of the holoorganism to be dependent on each other.

9.3 The Innate Immune System and Microbial Assembly

The innate immune system is an important guardian that regulates the entry of bacteria. Plants produce antimicrobials like phytoanticipins to dissuade any pathogenic infection and possess pattern recognition receptors that detect any bacterial intrusion that is elicited by the defence response. Reactive oxygen species (ROS) production and the release of enzymes targeting the microbes are upregulated. However, these defensive strategies are downregulated to facilitate the entry of a symbiont like *Rhizobium* spp. The host recognizes the microbial Nod factors with the lysine motif responsible for mitigating the innate immune response. For instance, *Gluconacetobacter diazotrophicus*, an endophytic N_2-fixing bacterium, colonizes the roots of rice plants (*Oryza sativa*). To accomplish this intimate interaction, the microbe requires the gene *gum*D coding for an enzyme mediating the synthesis of the exopolysaccharide (EPS) that communicates with the surface of the root. Without this precise chemical conversation, the two organisms will not be able to forge this partnership. Specific signals like salicylic acid aimed at pathogens and jasmonic acid directed at insects also contribute to this surveillance mechanism. Some pathogens like *Pseudomonas syringae* produce counter chemical information such as coronatine to blunt the plant-derived signals. Quorum-sensing molecules regulating microbial proliferation are sensed by plants and help modulate the population of the microbiota. *N*-acetyl homoserine lactones are involved in this bacterial population control while lipochitooligosaccharides contribute to the selection of fungal population. These various factors involving plant exudates, microbial sensors, innate immune system and quorum-sensing work in tandem to ensure the proper selection and the precise population of microbes are assembled into a community designed to contribute to the welfare of the host and add to its capability to perform a variety of functions and survive (**Figure 9.4**).

9.4 Rhizosphere and Its Microbes

The microbial population residing on and within the roots are assembled from the soil microbes and those transmitted vertically from seeds. The rhizosphere, the environment around the roots is populated by the microbes from the bulk soil that subsequently acts as a reservoir for the recruitment of epiphytes and endophytes. The initiation of this process starts with microbes associated with seeds. These microbes facilitate the exit of seeds from dormancy and orchestrate their germination. Cytokinins and other signalling molecules trigger the development of the root and the shoot. Coffee and rice seeds harbour *Bacillus* spp. and *Pseudomonas* spp. This microbial community shifts dramatically upon emergence from dormancy. The root tips select microbes, a process guided by the chemical signals in the exudate. The root tips and zones feature enzymes with the ability to degrade dead cells and exhibit differential profiles in chemical signals with a varying pH gradient aimed at eliciting a chemotactic response from appropriate microbes. Low pH and carboxylate-rich rhizosphere are dominated by *Burkholderia* spp. Citrate and oxalate in exudate responsible for the solubilization of PO_4 and Fe are important determinants in the filtering of the preferred microbes. The sensing mechanisms within roots and microbes are able to cross-talk and establish an intimate contact. For instance, to acquire fungi, roots secrete strigolactones that promote the attachment of chitin within the fungal cell wall resulting in cytosolic rearrangement in the host. This molecular reprogramming favours the intracellular penetration of the microbe

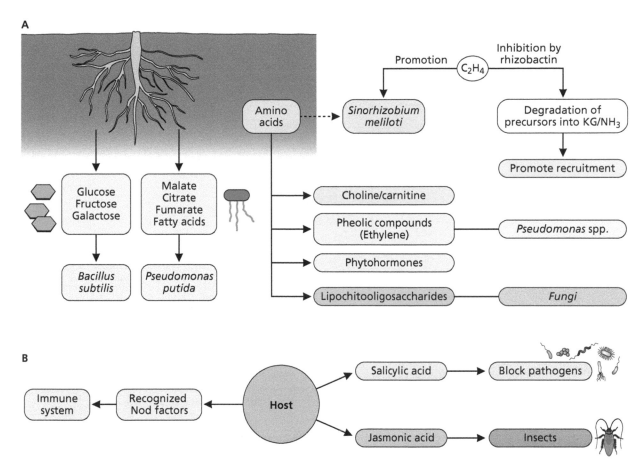

Figure 9.4. Chemical signals and microbial colonization: a fine balancing act. (a) Root exudates are rich in numerous chemicals mediating the selection of the microbial partners. Monosaccharides attract *Bacillus subtilis* and organic acids signal the assembly of *Pseudomonas putida*. Ethylene (C_2H_4), carnitine, choline and phenolic compounds are also involved in the genesis of the microbiome. The selection of fungi is facilitated by lipochitooligosaccharides. (b) The modulation of the plant immune system further fine-tunes the association between plants and microbes. The nodulation (Nod) factors prime the entry of symbiont microbes while salicylic acid blocks the uptake of pathogenic microbes. Jasmonic acid plays a role in dissuading insect invasion.

culminating in arbuscule formation. It is important to note that signals generated by arbuscular mycorrhiza fungus like *Rhizophagus irregularis* can stimulate bacterial growth and help shape the microbial community. Nitrogen-rich exudate derived from fungi is a major driver of the microbial residents in the rhizosphere. The flavonoids from the host and the Nod factors from the microbes are recognized and translated into morphological changes within the root tips. The suppression of the immune system and the reorganization of the intracellular component promote the intracellular localization of the endophytes. The endophytes are either housed in nodules or transported via the phloem and xylem to other cellular compartments (**Figure 9.5**).

9.5 Microbes Residing in the Phyllosphere

9.5.1 Shoots and leaves

The part of a terrestrial plant above the ground is referred to as the phyllosphere and it comprises the stem, the leaf, the flower, the fruit and any other aerial appendages. The microbial community occupying this area can either be epiphytes, residing on the surface or endophytes lodged within the plant tissues. The sessile nature of plants has led to the establishment of microbial population fulfilling a variety of essential functions. These microbes help extend the adaptability of plants to a range

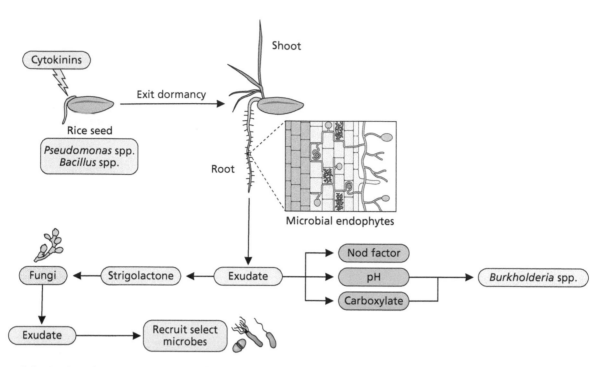

Figure 9.5. Colonization of roots. Microbes found in seeds promote the exit of their hosts from dormancy and the development of roots and shoots are triggered by cytokinins. The root exudate attracts fungi and bacteria. Fungi also secrete chemicals programmed to select microbes while the modulation of the plant immune system helps in the translocation of the endophytes to the appropriate tissues.

of biotic and abiotic stress. The recruitment of the microbial constituents takes place both in the soil and the air. Following the selection of the microbial partners by roots where the immune system and radical exudate are key determinants, the microbes travel through the xylem and phloem to be incorporated into the phyllosphere. This bacterial movement has been visualized with the assistance of *Burkholderia* spp. tagged with the green-fluorescent protein. This microbe populates the common grape vine *Vitis vinfera*. It possesses a range of cell wall-degrading enzymes including endoglucanases that enable the microbe to lodge within the aerial components of the plant. *Burkholderia* spp. is also an important N_2 fixer and is found in such angiosperms as *Pavetta* spp. and *Ardisia* spp., where it colonizes vegetative shoot tips and new leaves. The structures like stomata, lenticels and hydathodes that allow plants to exchange gas and contribute to the transpiration process are ideal entry points for microbes recruited from the air. The guard cells that control the opening of the stomatal pores are responsive to light, humidity and CO_2 levels, factors that all act as filters in the assemblage of the microbial community. The glucose, fructose and sucrose contents on leaf veins and surface appendages act as potent selectors of the microbial community being assembled on and within the host. The aerial environment composed of the atmosphere, rain and wind is an important source of bioaerosol-laden microbes ready to be dispersed. In fact, these bioaerosols are known to disseminate microbes from the Sahara Desert to high-altitude regions of Europe during dust storm. The large foliar area with its moist milieu is ideal to attract microbes. The N_2-fixing endophyte *Herbaspirillum seropedicae* enters through the stomata and resides in the mesophyll of leaves in pineapple plant (*Ananas comosus*). In sugar cane, nitrogen fixation mediated by diazotrophs occurs in the xylem, while *Bacillus amyloliquefaciens* harboured in shoot meristems of vanilla orchid (*Vanilla phaeantha*) helps combat pathogenic invasion. The interactions with herbivores, insects and other organisms are also factors that help shape the microbial residents of the phyllosphere.

9.5.2 Flowers and fruits

Flowers and fruits are the other components of the plants harbouring various microbial constituents. Flowers provide an ephemeral nutrient-rich environment with a longevity lasting approximately two weeks in case of apple trees. The bloom and the senescence periods are associated with different concentrations of monosaccharides, lipids, amino acids and other chemo-attractants. Blooming exposes the male (stamen) and the female (pistil) reproductive organs where microbial population can vary. Wind, rain, temperature and pollinators can all help determine the nature of the microbial community. The microbial signature of the pollinators and the flowers they visit are also factors determining the microbial populace. Apple flower is abundant in *Deinococcus thermus* and *Saccharibacterium* spp. The presence of *Pseudomonas* spp. and *Enterobacterium* spp. are also common in some flowers. The microbial succession occurring during these flowering episodes plays a critical role in shaping the microbial population of the fruit and seed. The microbes associated with the latter is involved in the vertical transmission of these invisible cellular components that become part of the plant hologenome. One of the early microbial consortia consists of *Methanosarcina* spp., an archaeal methanogen that survives on methanol, a by-product of the cell wall formation.

Fruits serve the very important function of transmitting the genetic information of plants stored in the seeds. The skin, the pulp and the seed are structural space where microbes reside. They are dominated by such phyla as *Proteobacteria*, *Actinobacteria* and *Firmicutes*. *Burkholderia* spp., *Pseudomonas* spp. and *Rhizobium* spp. are some of the prominent bacteria. Consumption of one apple may result in the intake of 100 million bacteria. The presence of auxins, cytokinins and gibberellins coupled with fruit ripening molecules like ethylene and abscisic acid promoting colour and odour help govern the microbial community. The latex in mango fruit in conjunction with the network of tiny ducts of the exocarp and mesocarp provides an environment conducive to bacterial proliferation and are home to numerous endophytes. These bacterial constituents contribute to the soluble sugar content, fruit mass, water concentration and palatability, factors that are critical for shelf-life and marketability of fruits. It is clear that microbes are an important component of the cellular system constituting plants. These cells are not only distributed on the host but they are intimately bonded with several tissues where they participate in the well-being of the holoorganism (**Figure 9.6**).

9.6 Functions Plant Microbes Perform

9.6.1 Suppliers of nutrients

The microbial cells execute diverse tasks that are part of the normal life of plants. They synthesize or contribute to the synthesis of such hormones as auxins and C_2H_4 that are involved in the growth and development of the host. Indole-3-acetic acid (IAA), an auxin is known to orchestrate embryogenesis and flower development. The modulation of the phytohormone, C_2H_4, by microbial enzymes contributes to senescence and fruit ripening. Microbes produce a range of nutrients and cofactors essential for the growth of the host. Riboflavin found in the exudates of rice and alfalfa promotes cellular metabolism and stimulates respiration. *Rhizobium* spp., *Bulkholderia* spp. and *Frankia* spp. are nitrogen fixers that fulfil the need of the host in amino acids and proteins. These microbes utilize mineral nutrients like vanadium, molybdenum, tungsten and iron to fix N_2 into NH_3 that is transported into the other plant tissues to be assimilated into proteins by such enzymes as glutamate dehydrogenase, glutamine synthetase and glutamate synthase. Various other microbes including *Paraburkholderia* spp. residing in wheat and soya plants

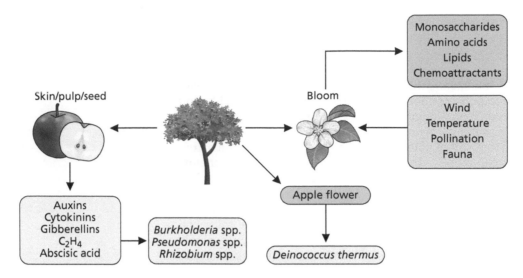

Figure 9.6. Distribution of microbes within the phyllosphere (shoot, leaf, fruit, flower). The phyllosphere acquires its microbes both from the soil and the air. The root exudate and the immune system help determine the nature of the microbes within the shoot. The blooming flowers which are associated with a variety of nutrients are pivotal in selection of microbes. The microbial partners of the fruits and seeds are dictated by phytochemicals within. Microbes are associated with almost all the components of plants. Wind and temperature contribute to the assembly of the microbiome.

solubilize phosphate, produce siderophores to facilitate the uptake of Fe and help augment photosynthetic pigments.

9.6.2 Microbial defensive arsenal

The plant growth-promoting bacteria (PGPB) are important components of the defensive strategy and participate in combatting pathogenic infections. They secrete antibiotics, proteases, hydrogen cyanide (HCN) and sulphur compounds aimed at opportunistic microbial colonization. Dimethylsulphoxide reductase and cysteine desulphurase expressed in these symbionts inhibit insects and other parasites. The antifungal thiopeptides target cell wall synthesis in the *Fusarium oxysporum*, a microbial agent involved in numerous plant diseases. The stacking of space they occupy acts as a physical barrier preventing the attachment of pathogens, and the biofilm they generate impedes the dispersal of spores. *Pseudomonas* spp. and *Bacillus* spp. are resistant to *Fusarium* pathogen. Chitinase and other hydrolytic enzymes aimed at fungal and insect outer layer are important tools bacteria utilize to guard against foreign attacks in plants. The quorum-sensing signals some disease-promoting microbes invoke to module their proliferation is neutralized with AHL lactonase. This suppresses the growth of these organisms. Insect invasion can be countered by microbial symbionts producing pyrethrins known for their toxicity against insects. Some plant symbionts like *Pseudomonas* spp. and *Bacillus* spp. are known to produce volatile organic compounds (VOCs) that can kill herbivorous insects or attract the natural predators of the insects. *Pseudomonas syringae* synthesizes undecanol and 3-hexen-1-ol in order to help protect plants from foraging predators. The microbial sentinels guard plants against any biotic assault by utilizing an array of chemical weapons ensuring that their hosts have a healthy lifestyle.

9.6.3 Combatting abiotic stress

The microbial residents enhance the ability of the host to survive not only biotic assault but also a plethora of abiotic stress such as salinity, drought, pollutants, pesticides and hydrocarbons. The genetic diversity of the resident microbes enables the host to respond rapidly to these challenges and improves its fitness. These attributes are essential especially in sessile organisms like plants, bound to the ecosystems they are located in and with limited mobility.

Salt stress poses a major burden on plants. Although halophytes have devised a variety of physiological and morphological strategies to overcome challenges posed by excess of Na^+ and Cl^-, the microbial community residing on and within the host also aid in combatting the dangers of a saline environment. Efflux pumps to siphon off these toxic ions coupled with transport systems facilitating the uptake of Ca^{2+} and K^+ are upregulated. The H_2O deficiency resulting from this stress is accompanied by the enhanced host-driven synthesis of C_2H_4, a phytohormone involved in senescence. To counter this, the microbial enzyme ACC deaminase is promoted in an effort to decrease the synthesis of ACC, a precursor of C_2H_4 and the formation of IAA, a moiety stimulating growth is activated. The production of EPS by numerous microbes facilitates the regulation of the micro-environment in the rhizosphere and prevents these toxic ions to access the root surface. Additionally, these hydroxyl and carboxylate-rich moieties help concentrate nutrients and minerals that are readily absorbed by the host. *Bacillus* spp., *Pseudomonas* spp., *Halomonas* spp. and *Kusheria* spp. are common bacteria assembled within the microbial communities residing on and within the roots of plants growing in saline environments. Accumulation of osmolytes such as proline, glutamate and other compatible solutes limits the loss of water and contributes to the maintenance of a functioning ion concentration intracellularly. While protein stabilization aids in the adaptation to fluctuating temperature, the sequestration and biotransformation of toxic metals by the root-dwelling microbes neutralize the danger posed by metal pollutants. The mineralization of hydrocarbons and polyaromatic compounds mediated by a range of microbe-encoded oxidases and hydroxylases enables plants to proliferate in habitats contaminated with organic pollutants. Hence, the presence of microbes on and within plants extends the adaptability and fitness of the hosts. This strategy promoted by the microbial population ensures the hosts can extend the range of the habitats can colonize and can increase their survivability (**Figure 9.7**).

9.7 Carnivorous Plants Need Their Microbes

Carnivorous plants are adapted to survive in habitats deficient in nutrients. Although they are photosynthetic, they acquire a range of chemical compounds from living organisms like microbes, algae, insects and other small creatures. To capture their prey, they are equipped with traps that are essentially modified leaves where the organisms are sequestered and 'digested', i.e. nutrients are extracted in a manner analogous to the hydrolytic process occurring in the stomach or digestive tract. In *Ulticularia vulgaris*, a rootless aquatic plant, the nutrient catchment is executed by numerous tiny traps (1–5 mm long) present in segmented leaves. These are liquid-filled bladder-like structures in which the lumen is isolated from the environment by a thick cell wall to prevent the escape of any organism and to maximize digestion. The tropical monkey cup (*Nepenthes* spp.) and the pitcher plant (*Sarracenia purpurea*) utilize a fluid-filled pitcher as a trap to lure insects. These pitchers are modified leaves with the capability to fix CO_2. The Venus flytrap (*Dionaea muscipula*) possesses snap traps decorated with trigger hairs that upon interaction with the prey close rapidly in order to immobilize the catch within the sticky surface. These aerial traps have an acidic pH ranging from 2 to 6 and are colonized by different microbial communities depending on the ecological niche where the plants are localized.

9.8 Aquatic Plant Carnivore: Microbial Metabolism

Uticularia vulgaris that is reliant on bacteria, fungi, algae and protozoa cultivates some of the microbial prey and utilizes them for its nutritional needs. Algae such as *Euglena* spp. are abundant and are consumed by the

A

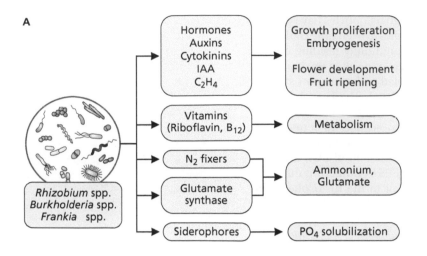

Figure 9.7. Functions microbes perform for plants. (a) Microbes supply essential nutrients like NH_3 and vitamins that fuel the growth of the host. They can promote the production of numerous phytohormones involved in development and flowering. Microbes are also a source of organic acids and siderophores fuelling the uptake of Fe and other trace metals. (IAA: indole-3-acetic acid.) (b) Plant microbiome helps combat insects and predators by producing a range of toxins. The microbial partners produce antibiotics to dissuade pathogenic invasion and enable the host to extend its survivability. They are part of the defence system of plants. (AHL: N-acyl homoserine lactone.)

B

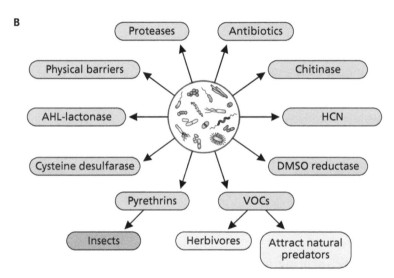

host. This aquatic carnivore has an unusually small genome and depends on its microbiome to fulfil its energy requirement. The prominent phyla are *Proteobacteria, Acidobacteria, Firmicutes, Chloroflexi* and *Cyanobacteria*. The young traps tend to be virtually sterile but as they developed a constant supply of organic nutrients derived from photosynthesis, this creates an environment for the selective assemblage of the microbial constituents responsible for digesting the prey. The proliferation of the algae in the vicinity of the trap provides an easy access to food, a situation that can be further supplemented by bacteriovory. Microbial community comprises members with enzyme systems to degrade algal, bacterial and fungal components. The prominence of the Bacteriovoracaceae and Myxococcaceae family results in the degradation of other bacteria that are subsequently utilized by the host. The presence of *Peptostreptococcus* spp. known to harbour high amounts of pyrroline 5-carboxylate reductase and threonine aldolase is involved in the metabolism of peptide and amino acids. The released NH_3 is fixed by the plant into a variety of biomolecules. The microbes with cellulase and xylanases coupled with enzymes having the proficiency to metabolize galactose are reflective of the algae-consuming attribute of the plant. This monosaccharide, an important algal constituent can be readily shunted into primary metabolism with the aid of UDP-glucose 4-epimerase that is upregulated within the trap microbiome. To promote microbial aggregation, colonization and trapping, biofilm formation is promoted by enzymes like transaldolase and UDP-glucose-6-dehydrogenase. Methanogens are also known to be part of this community in order to metabolize CH_4, a potential product from polysaccharide degradation.

9.9 Microbes within the Pitcher Plant

The pitcher plant residing in barren soil requires a source of N and P, nutrients predominantly acquired from insects. The microbial assemblage prominent in this insectivorous plant is replete with *Bacillus* spp., *Sphingobacterium* spp. and *Pseudomonas* spp. These microbes possess metabolic pathways involved in supplying these nutrients from insects. The selection of microbial partners is achieved with the help of antimicrobial and antifungal compounds designed to favour the establishment of these microbes. The pitch fluid has an acidic pH with chitinolytic, proteolytic, amylolytic and xylanolytic enzymes. The snap trap possesses *Burkholderia* spp., *Acidocella* spp. and *Bradyrhizobium* spp. with phosphatases, nucleases, peptidases and chitinases that are involved in the digestion of the insects, a slow digestive process that can take up to 10 days. Following the release of the nutrients that are subsequently shuttled to the host cellular systems, the trap opens for the next catch. As the milieu where the microbial community resides is acidic, the microbes display the presence of proton pumps, MoxR-ATPase-like chaperones proteins to guard against molecular damage and biosurfactant-producing ability to survive this environment. Methylotrophic bacteria utilizing methanol are also part of this microbial assemblage. Without the presence of these microbiota carnivorous plants would not be able to adopt this lifestyle and the microbial cells helping extract the essential nutrients in short supply are central to the survival of these holoorganisms. Even though these microbial cells are not part of the plant genome, they are fully integrated within the anatomical features and physiological processes of the host and contribute to functions vital in the normal workings of the hologenome. The genetic information of the microbial cells is harnessed in such way as to complement that of the host and hence responds to the cross-conversation emanating from these diverse cellular entities. In this instance, the microbes are more or less the digestive tract dedicated to the extraction of nutrients from the insects the organism traps or 'ingests' (**Figure 9.8**).

9.10 Microbes Associated with Desert Plants

9.10.1 *Morphological and physiological adaptation*

Deserts and arid lands provide a unique set of challenges that living organisms have to adapt to if they are to survive and proliferate. The dearth of H_2O associated with these habitats has elicited major morphological and physiological adjustments enabling plants to grow in these environments. The root system is elaborate and penetrates deep under the surface with extensive lateral roots aimed at extracting most amount of H_2O. This increased surface area coupled with the upregulation of aquaporin maximizes H_2O uptake from the ground. On the aerial component, the leaves are equipped with stomata aimed at reducing gas exchange and retaining moisture. Furthermore, they possess the biochemical machinery to perform the Crassulacean acid metabolism whereby photosynthesis occurs at night, H_2O consumption is optimized and photorespiration is restricted. Despite these intricate processes dedicated to survival in environments with limited availability of H_2O, plants have mobilized the assistance of a diverse community of microbes to overcome the multiple complications imposed by this abiotic stress. Water stress also results in increased osmotic tension, a rise in oxidative stress, a lack of essential ions and cellular injury.

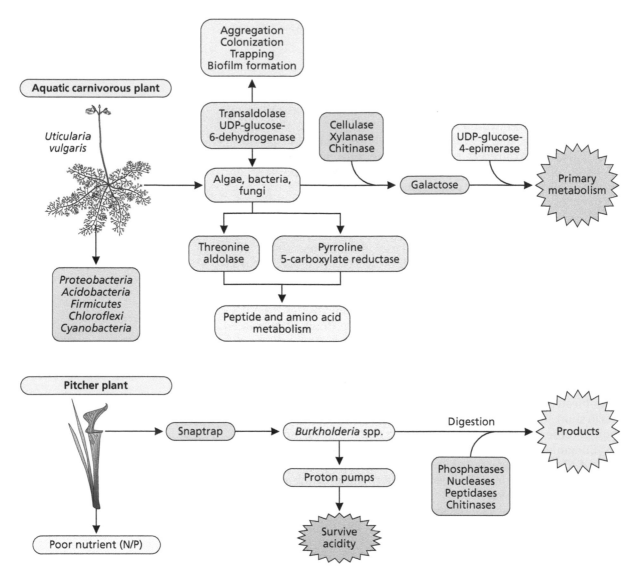

Figure 9.8. Microbial digestion aids carnivorous plants. Some aquatic carnivorous plants cultivate algae, protozoa and fungi that they consume. These organisms are a source of carbohydrates and amino acids that are released with the aid of the microbiome constituted of diverse phyla. Galactose, an important component of algae is metabolized to glucose. The pitcher plant has snap traps enriched with microbial partners dedicated to the digestion of insects and are adapted to the high acidic environment.

9.10.2 Microbial community within the rhizosheath

The rhizosheath is a characteristic of some plants like the spear grass thriving on sandy desert soils and helps combat unfavourable moisture conditions. This radical cover goes beyond the plant roots, provides mechanical support, conserves H_2O, promotes nutrient uptake and harbours microbial and macroscopic organisms. The plant exudate and mucilage enable microbes to aggregate on the sand particles from the bulk sand. They are subsequently selected to reside on the surface and other structural components of the host. *Actinobacteria*, *Proteobacteria*, *Firmicutes* and *Chloroflexi* are the more prominent phyla. *Streptococcus* spp. belonging to the *Actinobacteria* phylum is abundant in the rhizosheath while the endophytes located within the tissues are members of the *Firmicutes* phylum. There tends to be a prominence of Gram-positive microbes in desert plants that are characterized by thick cell walls. This morphological adaptation coupled with significant modification of the peptidoglycan structure is more resistant to the osmotic stress triggered by the scarcity of H_2O and to the increased level of oxidative stress associated with these arid conditions. The decrease in diversity and increase in abundance of these thick-walled microbes enable the hosts to better cope with the dearth of water prevalent in this condition.

In fact, when confronted with periodic droughts, sorghum plants tend to assemble microbial communities on the root surfaces consisting of *Streptomyces coelicolor* and *Streptomyces ambofaciens*. These microbial dynamics are dictated by plant exudates that select these organisms due to their ability to fix N_2, produce EPS and be responsive to chemotactic signals. The solubilization of P and the secretion of siderophores like desferrioxamine are also important attributes of the microbes residing in the rhizosphere. The increased expression of K^+/ATPase pump and the transport system for the uptake of alkanesulphonate are features that some of these microbes possess. While the former helps modulate the intracellular ion gradient, the latter contributes to the synthesis of the phytohormone H_2S, a reaction facilitated by the enzymes alkanesulphonate monooxygenase, sulphite reductase and cystathionine β-synthase. The enhanced ability to synthesize trehalose, glutamate, acetoin, chaperone proteins, glutathione and ROS detoxifying enzymes like catalase and superoxide dismutase is an attribute these microbes have. The pooling of glycerol-3-phosphate (G3P) a compatible solute and a constituent of cell wall is orchestrated by the decreased activity of G3P dehydrogenase and G3P acyltransferase. This strategy of enhanced compatible solute synthesis and accumulation are vital as the lack of H_2O also triggers ionic stress. The expression of polyketide synthase, cytochrome P-450 and chitinase plays a key role in preventing opportunistic organisms to colonize the roots. Thus, the selection of microbes with the ability to produce compatible solutes, antibiotics, phytohormones, EPS and mobilize nutrients enables the desert plants to survive the stress posed by the arid landscape. These microbial communities that are part of the desert plant not only help acquire H_2O via the elaborate rhizosheath, but they also produce enzymes and metabolites to combat the dangers associated with other biochemical challenges imposed by a poor H_2O ecosystem. The holobiont as a complete living entity is able to survive a dearth of H_2O, a basic ingredient of life on this planet. This would be impossible without the molecular interconnectivity of these disparate cellular entities with their own distinct genomes (**Figure 9.9**).

9.11 Medicinal Plants and Their Microbial Residents

9.11.1 Taxol and artemisinin: cancer and malaria medications

Medicinal plants are an important source of bioactive compounds that are utilized worldwide to cure or prevent diseases and to improve health. These benefits can be derived by consuming fruits and edible components such as roots, leaves and bark either raw or dried or processed. Drinks and other preparations concocted with any of these plant constituents tend to release these chemical ingredients that are readily accessible to the body. Microbes residing on and within the plants either produce these moieties with therapeutic properties or contribute to their synthesis. Taxol, an anti-cancer medication is elaborated by numerous plants including *Ginkgo biloba* and *Taxus brevifolia*. The latter harbours *Taxomyces andreanae* and the former is colonized by *Altermania* spp. These microbes are involved in the formation of this compound, a potent disruptor of cytoplasmic reorganization. Although cultures of *Aspergillus terreus*, a fungal endophyte housed within the coniferous plant *Podocarpus gracillor* have been shown to produce taxol, the yield is dependent on elicitors from the host and its other microbial partners. The flux of acetyl CoA appears to be an important determinant of the productivity of this anti-cancer medication. Artemisinin, a bioactive molecule elaborated by the herbaceous plant *Artemisia* disarm the malaria parasite *Plasmodium falciparum* by producing ROS

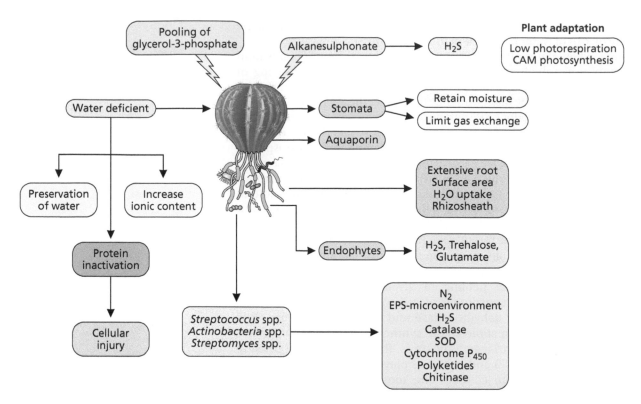

Figure 9.9. Desert plants: ultrastructural and molecular adaptation. Microbes are an integral component of plants living in arid conditions. This lack of H_2O imposes significant challenges that necessitate major physiological and biochemical adaptation. The opening of the stomata and the activation of the photosynthesis at night are key in this strategy. Microbes produce exopolysaccharides (EPS), trehalose and glutamate to maintain ionic balance. Enhanced expression of aquaporins and root system allow for an increase in the H_2O uptake.

intracellularly. The production of this metabolite is promoted by the endophyte *Pseudonocardium* spp. as this microbe modulates the expression of the host cytochrome P-450 monooxygenase, an important participant in the genesis of the compound with anti-malarial properties. Priming of chamomile seedlings with strains of *Bacillus subtilis* and *Paenibacillus polymyxa* stimulates the production of such flavonoids as apigenin and its derivatives. These compounds possess a variety of medicinal properties including anti-inflammatory, anti-bacterial and anti-oxidative attributes. The medicinal product elaborated is the result of the holobiont with its host and microbial partners working in tandem by either supplying the enzymes or the precursors.

9.11.2 Cannabidiol and tetrahydrocannabinol: the microbial link

Cannabis plants (e.g. *Cannabis sativa*) are cultivated worldwide due to the commercial importance of the chemicals, seeds and fibres they generate. The main bioactive ingredients associated with these plants are cannabidiol (CBD) and Δ-9-tetrahydrocannabinol (THC). Although the bioactive chemicals are found almost everywhere within the plants, they are most abundant in the trichomes of the female flower buds. The biosynthesis is mediated either by the polyketide or the terpene-based pathway. The cannabis plant harbours microbial assemblages in tissues, organs, seeds, roots, leaves and stem that contribute to a variety of physiological functions including the synthesis of the cannabinoids, known to dissuade herbivores from attacking the host. While such fungal endophytes as *Aspergillus* spp. and *Penicillium* spp. are present within leaves, the petioles and the seeds are laden with *Staphylococcus* spp. and *Bacillus* spp. respectively. They produce siderophores to mobilize iron, HCN to repel predators and organic acids to solubilize Ca and P. Various stimulants like UV light, C_2H_4 and

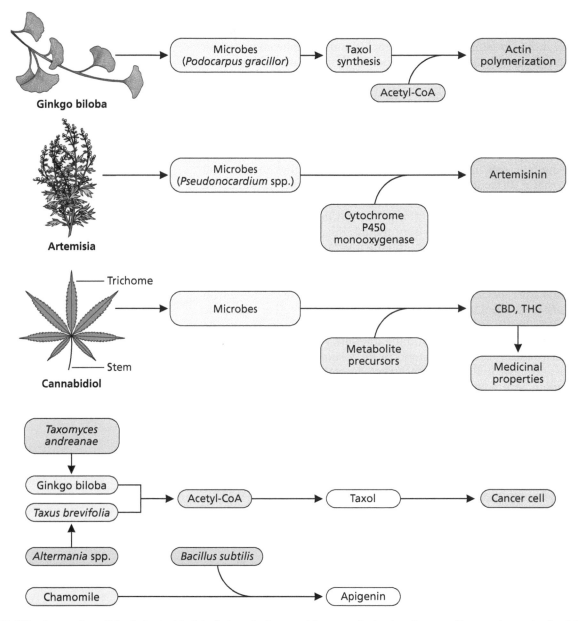

Figure 9.10. Microbes and medicinal plants. Medicinal plants harbour a wide range of microbes that contribute to the synthesis of these chemical products. *Ginkgo biloba* known for the production of taxol possesses microbes with the ability to produce acetyl CoA, a key ingredient in the synthesis of this anti-cancer compound. Artemisinin from *Artemisia* and cannabidiol from *Cannabis sativa* are synthesized with the aid of microbial input. *Bacillus subtilis* is a contributor to the synthesis of apigenin in chamomile. (CBD: cannabidiol; THC: Δ-9-tetrahydrocannabinol.)

GABA result in enhanced synthesis of THC and CBD. Although the role of the microbiome in the elaboration of these valuable chemicals has yet to be fully elucidated, it is evident that the resident microbes participate directly or indirectly in the synthesis of these compounds with important commercial value. It is not unlikely that the manipulation of the microbiota may result in higher yield of these medicinal compounds. The syntheses of these valuable chemicals are made possible by intimate interactions between the host and its microbial residents (**Figure 9.10**).

9.12 Root and Microbe Communication

The majority of microbes residing within plants are derived from the habitats the hosts are living in. The roots acquire their microbial community from the bulk soil. The microbes in the soil are deliberately selected to either make their home on the root surface or within the intracellular component. Microbes like *Rhizobium* spp., *Pseudomonas* spp. and *Azospirillum* spp. are initially attracted to the roots by specific exudates such as flavonoids, organic acids, amino acids and phenolic compounds. For instance, *Azospirillum* spp. responds positively to malate and benzoate, moieties known to guide this microbe to the root surface. The colonization of root tends to proceed generally in two stages. The primary phase is characterized by non-specific binding and adsorption of the microbes. The chemotactic attraction to the root is aided by the flagella and pili responsible for the swimming and swarming behaviours that propel the responsive microbes to the roots. Microbes unable to decode the chemical cues stay behind in the bulk soil. The weak forces constituted of physico-chemical and electrostatic attractions enable the molecules in the bacterial cell envelope and the root surfaces to interact. This initial weak and reversible attachment is consolidated into a stronger connection during the second phase whereby stable colonization is promoted. This step leads to tight and irreversible bonds between the roots and the microbes.

The biopolymers on the outer layer of the microbes consisting essentially of polysaccharides, lipopolysaccharides (LPS), cellulose fibrils and proteins communicate with specific receptors on the components of the roots with the assistance of divalent metals like Ca^{2+} and Mg^{2+}. For instance, polysaccharides bind to lectins. In this instance, the nature of monosaccharide dictates the specificity of the host and the microbe. Acidic polysaccharides are essential for the infection of pea roots by *Rhizobium leguminosarum*. The presence of arabinose in EPS and *N*-acetyl glucosamine in LPS is important for the binding of *Azospirillum* spp. to root surfaces. *Pseudomonas* spp. possesses numerous adhesion proteins that facilitate stronger attachment. The formation of biofilm helps aggregate microbes of similar traits at the expense of other organisms including pathogens. Once the chemical signals have been received, acted upon and vetted by the immune system, the host is ready to house the microbes intracellularly or on the surface. This precise communication is central to the genesis of the holobiont where the microbes are an integral component of the daily life of the visible host (**Figure 9.11**).

9.13 Microbes Modulate Flowering in Plants

Flowering is a pivotal aspect in the life of all flowering plants. Flowers ensure seed production and consequently secures the survival of the species. Furthermore, this phenomenon is closely linked to crop yield and agricultural productivity. Genetic network, metabolites, temperature, light and microbes all contribute to the timing as to when a plant is going to flower. The precise timing is critical since the reproductive success of flowers is dependent on others factors like pollinators and weather pattern like drought, frost, rain and wind. If pollinators are not around or the weather is not conducive, the formation of seeds can be compromised. A winter chill is essential to trigger flowering in perennials while a frost can destroy petals and reproductive tissues. Plants produce numerous metabolites and hormones to modulate flowering. Cyanogenic glycosides, gibberellic acid, salicylic acid, cytokinins, abscisic acid and polyamines all can affect flowering time. Cyanogenic glycosides generate HCN, a moiety known to promote exit from dormancy and to induce flowering. HCN plays an important role in manipulating the level of ROS, a mediator of flower biogenesis by inhibiting catalase activity. Inhibition of this enzyme results in an increase of ROS and leads to the expression of genes responsible for flowering. While gibberellic acid promotes flowering, abscisic acid impedes this process. Potassium nitrate is a chemical utilized to trigger flower as it supplies N and contributes in increasing ROS intracellularly.

Owing to their ability to modulate N homeostasis and generate bioactive molecules, microbes can directly or indirectly impact the time when the host is going to flower. The increase or decrease in the bioavailability of N can

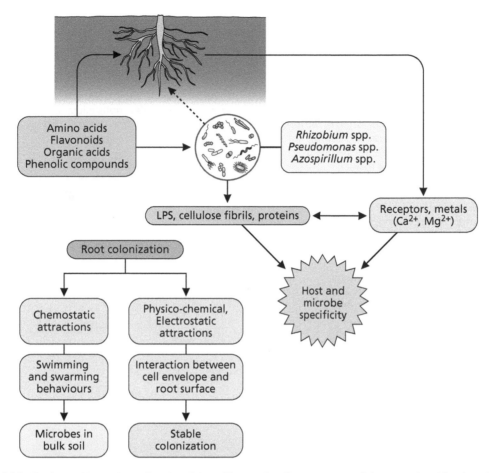

Figure 9.11. Microbial attachment to roots–molecular picture. The exudate from roots containing organic acids, phenol derivatives and flavonoids help select microbes from the bulk soil. Following this initial interaction, the precise signal is decoded by the host and the microbe resulting in the housing of the microorganism either intracellularly or extracellularly. Receptors, ligands and cofactors are involved. This intricate communication guards against pathogenic entry.

influence the synthesis of hormones involved in the initiation of flowering. Nitrogen metabolism punctuated by the formation of NH_3 and urea coupled with the denitrification of NO_3 can promote or impede the production essential nitrogen biomolecules mediating flowering. The conversion of tryptophan into IAA results in a decrease in the production of gibberellic acid, an event associated with plant growth and with the inhibition of flowering. For instance, *Arthrobacter* spp. known to secrete IAA favours plant growth at the detriment of flower formation. Glycerol is another metabolite generated by microbes that can influence flowering time. Bulbs are condensed plants living in an underground environment. Upon glycerol treatment of the Easter lily (*Lilium longiflorum*) bulb, the flowering genes are inactivated and the dormancy genes are activated. Microbes regulating the homeostasis of this polyhydric compound can either promote dormancy or trigger flowering. Microbial residents can thus be harnessed to influence flowering at a desired time and eventually to promote fruit production (**Figure 9.12**).

9.14 Light Influences Plant Microbiome

9.14.1 Visible light and microbial community

Plants depend on light to survive. This photo energy is manipulated in an effort to generate chemical energy and nutrients that the plants utilize for their daily biological activities. To facilitate this interaction with light energy mostly emanating from the sun, plants are equipped with numerous chromophores that aid in trapping this energy. The chloroplast is a specialized

BOX 9.1 PLANT–MICROBE–PATHOGEN COMMUNICATION

The communication between the plant and the resident microbial community is critical if the latter is to provide any defence against pathogens. Upon invasion by parasitic fungi like *Rhizoctonia solani* known to infect roots, signalling molecules such as jasmonic acid or derivatives of salicylic acid emitted by the host trigger the expression of microbial genes responsible for chitinase synthesis and a range of secondary metabolites toxic to fungi. The former is involved in the degradation of chitin, a major component of fungal cell wall. Whenever plants are attacked by insects or herbivores, they recruit microbial partners to assist them in fending these predators. The secretion of root exudates designed to privilege the selection and proliferation of microbial community dedicated to the synthesis of insecticides and promoting the growth of the plant is an important strategy orchestrated by defence signals such as volatile organic acids (VOCs) and C_2H_4. Methyl salicylate, a plant volatile messenger attracts natural predators of insects with the goal of killing and suppressing the insect population. These VOCs also affect the neighbouring plants and prompt them to prepare their defence mechanisms aimed at countering this imminent danger. In this instance also, a shift in the plant microbiome with the ability to generate HCN and phenazine-1-carboxylic acid chemicals known act against pathogens is observed. This communication among the residents of the holobiont enables the host to thwart insect invasion, fungal pathogens and herbivores. Such a strategy is critical in a sessile organism rooted to the ground and unable to escape. The elaborate signalling network allows the detection of the danger, the recruitment of the microbial soldiers and the elimination of the invaders. The chemical warning system is also decoded by the neighbouring plants prompting a commensurate defence response. Hence, plants and their resident microbes constitute a complex multicellular living system with an intricate communication network that ensures the survival of the holoorganism (**Box Figure 1**).

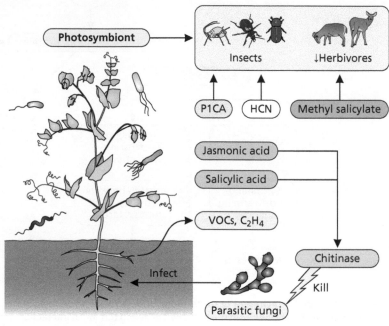

Box Figure 1. Signalling network and defence against predators: plants and their microbial guardians.

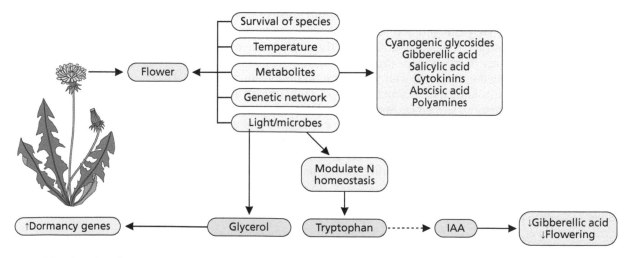

Figure 9.12. Microbes time flower formation in plants. Microbes regulate the synthesis of numerous metabolites that have a direct influence on flowering. Hydrogen cyanide (HCN), abscisic acid, polyamines, tryptophan and glycerol can modulate flowering. The conversion of tryptophan into indole-3-acetic acid results in a decrease in flowering. An increase in glycerol is associated with dormancy.

organelle consisting of such light-attracting moieties as the chlorophyll, the photosystem I (PSI), the photosystem II (PSII) and carotenoids in the light-harvesting complex (LHC). While the latter absorbs a range of visible light, chlorophyll absorbs blue and red light. These chromophores communicate with the solar energy and contribute to the fixation of CO_2 and H_2O into reduced organic molecules via a series of biochemical processes. Thus, light-induced changes can affect nutrient gradient on the leaf surfaces and the production of VOCs. The exudates secreted by the roots are also dependent on this interaction with light. In fact, 15%–40% of the photosynthetic compounds are liberated within the root system. These chemicals subsequently help shape the microbial community the plant is composed of. Activation of the plant photoreceptors by specific light wavelength results in the upregulation and downregulation of genes responsible for a plethora of physiological responses. White light stimulates conidia formation and favours the proliferation of *Aspergillus* spp., microbes such as *Pseudomonas* spp. are promoted in the presence of white and blue light. Upon exposure to ultraviolet radiation, peanut plant is predominantly colonized by *Clavibacter michiganensis* and *Curtobacterium flaccumfaciens* while an abundance of *Enterobacter cloacae* in rice plant is evident.

9.14.2 Light, microbes and metabolites

The production of bioactive metabolites like glucosinolates, phenolics and GABA coupled with the increased formation of ROS tend to trigger a change in the microbial community. Genes responsive to the phytohormones, jasmonic acid and salicylic acid that evoke the elaboration of defence proteins can also contribute to the modulation of the microbes associated with the host. For instance, red light impedes invasion of *Arabidopsis* by *Pseudomonas syringae* while tomato subjected to green light is resistant to *Pseudomonas cichori*. The activation of glutamate decarboxylase, an enzyme mediating the decarboxylation of glutamate into GABA by a specific light wavelength can promote an increased diversity of microbes in plants. Blue and green light trigger the accumulation of ascorbic acid and sulphur-containing metabolites like glucosinolates. While the former is a potent anti-oxidant, the latter can fend against pathogens. Light of varying wavelengths (300–800 nm) communicates with specific photoreceptors in plants and elicits a change in a range of molecular components including organic acids, nutrients, phytohormones, antimicrobials and immune modulators. These, in turn act as important filters shaping the microbes the host recruits. Hence, light is a very potent agent dictating the nature of microbes associated with photosynthetic organisms (**Figure 9.13**).

9.15 Strigolactones and Triterpenes: Radical Communicators

Plants are sessile organisms and are virtually condemned to the place where they set roots. This lifestyle is fraught with dangers as they are exposed to attacks by a wide range of motile organisms. To combat this situation, plants have evolved an intricate stratagem devoted to the synthesis of myriad of chemicals that enable them to thwart the aggressors and recruit friendly partners. The latter not only supply various essential nutrients plants need to proliferate but dissuade and eliminate invaders whether they are pathogenic microbes or macro-organisms like insects, birds or mammals. Strigolactones (SL) and triterpenes (TP) are members of this chemical arsenal at the disposal of plants to be deployed in an effort to secure their safety. Strigolactones are synthesized from carotene as a precursor in order to yield carlactone that is subsequently processed into a diverse range of SL with the assistance of cytochrome P-450 monooxygenases, oxidoreductases and dioxygenases. The latter utilizes α-ketoglutarate as a co-factor. These bioactive molecules with a broad

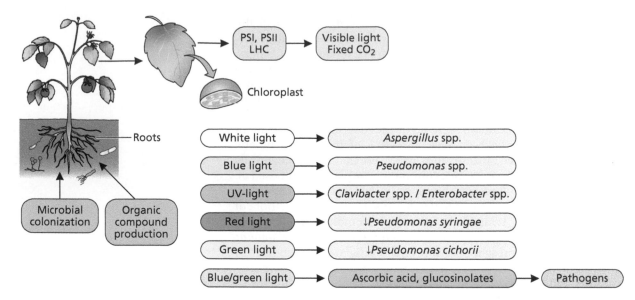

Figure 9.13. Light, a microbial selector: impact on plants. Plants possess various chromophores such as PSI, PSII and LHC that absorb light with different wavelengths. This photo energy is transformed in disparate metabolites released in the root exudates responsible for selecting diverse microbes. Light of different wavelengths can attract disparate bacteria. For instance, red light impedes colonization by *Pseudomonas syringae*, while white light promotes colonization by *Aspergillus* spp.

structural diversity are involved in numerous functions including plant development, germination of parasitic weeds, inhibition of dormancy, promotion of shoot branching, regulation of root architecture and recruitment of symbionts. Defects in SL synthesis hasten dormancy, promote excessive shoot branching and result in the establishment of ineffective microbial communities. These carotenoid-based signalling molecules elicit the secretion of chitinooligosaccharides and flavonoids that selectively enable the colonization of the rhizosphere by the appropriate bacteria and fungi. They promote spore formation, stimulate mitochondrial metabolism and facilitate the root cortical cells to accommodate the highly branched hyphae referred to as arbuscles. The branching activity of *Gigaspora* spp. (rice plant) is increased while in alfalfa (pea plant) nodulation by *Sinorhizobium* spp. is enhanced. Plants like *Arabidopsis thaliana* express the genes mediating the synthesis of TP such as thialianin and arabidin. These molecules have a role in fending off pathogen invasion and promoting the proper microbes to establish colonies within the rhizosphere. The ester derivatives of these TP can be readily utilized as a source of carbon by some microbes known to be privileged by the host. This strategy of targeting microbes with nutrients is analogous to the secretion of mucin by the human gut aimed at attracting those microbes with the ability to metabolize the carbohydrate-rich polymer. The ability of plants to synthesize these compounds of vital ecological significance helps these sedentary organisms communicate with their friends and foes alike (**Figure 9.14**).

9.16 Microbes, Methanol and Plant Connection

Methanol is the second most abundant volatile organic compound in the atmosphere after methane (CH_4). The methylation and demethylation of biomolecules are the important modification that enables the intracellular transmission of signals and bestow disparate physico-chemical properties essential for structural integrity. Plants emit methanol as a by-product of demethylation reactions occurring in DNA, proteins and carbohydrate within cell walls. The latter consists of homogalacturonans (pectins) that are heavily methylated. The esterification by methyl groups is a pivotal determinant of the properties of the cell wall and of growth. These modified

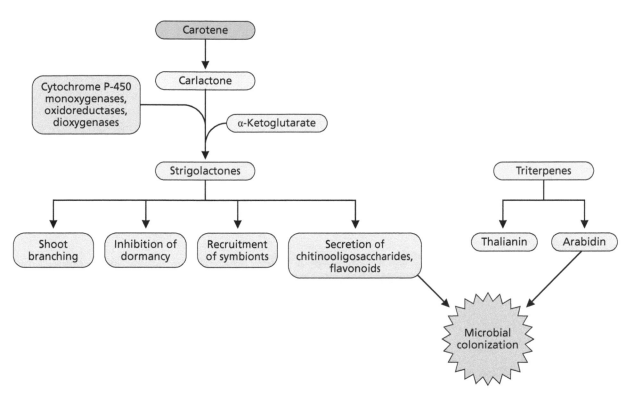

Figure 9.14. Radical signals and biological activity. Plants produce strigolactones and terpenes involved in signalling a variety of functions including the recruitment of microbial partners. Carotene is the source of carlactone that can be transformed into different strigolactones. Acetyl CoA is the precursor of triterpenes, metabolites promoting select microbial colonization.

carbohydrates also specify the positioning of leaves within a plant. Pectin methylesterases are enzymes involved in the reversible modification of the carboxylic moiety with methyl groups. The hydrolysis of the ester liberates a carboxylic group ready for binding with Ca^{2+}, a reaction aimed at fortifying the cell wall. Methanol, the other product of this enzymatic cleavage is released by the plant and usually concentrates on leaves. Furthermore, aggression by herbivores triggers demethylation, a reaction resulting in increased emission of methanol. This molecule activates the methanol-inducible genes in the adjoining plants priming them to respond to any eventual assault. Attack by insect pest such as aphid or any mechanical damage also tends to prompt the liberation of this alcohol that leads to the expression of the defensive genes. Hence, the availability of this alcohol is a potent determinant of microbes residing on the plants. The presence of methanol on leaves attracts very select microbes with the ability to utilize this metabolite as a source of carbon. Methylotrophs possess methanol dehydrogenases, enzymes that convert methanol into formaldehyde with the assistance of such cofactors as NAD^+ and Pyrroloquinoline quinone. The reduced cofactors can be channelled in the synthesis of ATP while the formaldehyde is processed via disparate metabolic pathways. The phyllosphere of methanol-rich plants comprises *Methylophilus* spp., *Methylobacterium* spp. and *Hyphomicrobium* spp. that are known to harbour the genetic information to consume methanol and assist the plants in numerous functions (**Figure 9.15**).

9.17 Plant Microbiome and Environmental Temperature

Temperature changes in the environment have major impact on the plants and the associated microbial communities. A rise in temperature results in an increase in leaf transpiration and an augmentation in respiration rate. The latter is fuelled by an increase in affinity of ribulose 1,5-bisphosphate

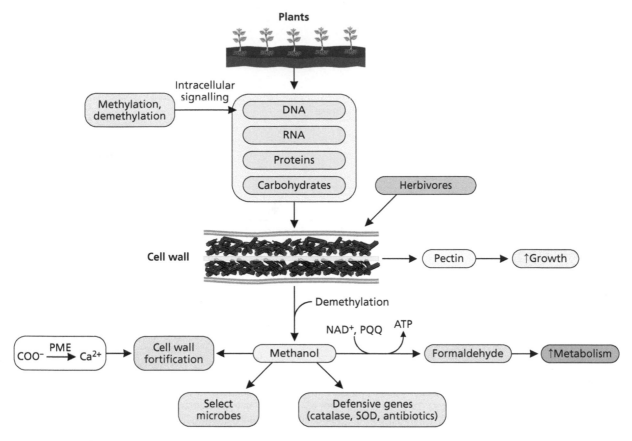

Figure 9.15. Methanol metabolism and microbe–plant interactions. Methanol is an important by-product of demethylation reaction, an important process in all biological systems. The presence of methanol on leaves dictates the assembly of the microbiome and also serve numerous other functions including stimulating genes and dissuading herbivores. Methanol produced from the metabolism of plant cell wall can be processed into ATP and formaldehyde. (PME: pectin methylesterase; SOD: superoxide dismutase.)

carboxylase/oxygenase (Rubisco) for O_2, an event characterized by a decrease in photosynthesis. To prevent evaporation, the stomata are closed leading to a reduced intake of CO_2. Thus, less photosynthates are synthesized and channelled to roots for the microbial community. This temperature-induced decrease in photosynthesis is accompanied by an increase in heat-shock proteins and the stimulation of the anti-oxidative defence mechanisms. Enzymes like catalase, superoxide dismutase and peroxidases are upregulated to combat lipid peroxidation prominent during heat stress. As plants cope with this stress, they are aided in this task by a shift in microbial community assembled to mitigate the dangers associated with a rise in temperature. Microbes with enhanced 1-aminocyclopropane-1-carboxylic acid (ACC) degrading activity are selected. The deamination of ACC mediated by ACC deaminase helps limit the synthesis of C_2H_4, a promoter of senescence and dormancy. The NH_4^+ released from the ACC, the precursor of C_2H_4 is shunted for cellular growth. For instance, *Paraburkholderia phytofirmans* which expresses ACC deaminase activity allows the normal growth of potato plant under heat stress.

A decrease in temperature is characterized by numerous changes including diminished bud development, dehydration of plant tissue and disruption of plasma membrane. To combat this stress, the expression of cold-shock proteins is upregulated and the synthesis of anti-freeze metabolites such as polyols is augmented. Additionally, this biochemical reconfiguration within the host is accompanied by a shift in microbial population dedicated to alleviate the cold-provoked injuries. The ability of microbes to produce EPS is favoured as these moieties regulate the micro-environment and limit the injury associated with cold stress. Microbes like *Pseudomonas syringae* and *Erwinia herbicola* known for their ice-nucleating attribute are selected. These endophytes and epiphytes promote ice formation on their cell wall and help the host adapt to the dangers inherent in ecosystems with cold temperatures. Microbes are important cellular components of the plant holobiont and can readily be assembled to confront any temperature fluctuation in the environment. The limited biochemical specialization that is the hallmark of these bacterial cells is an asset as it lends itself to easy recruitment and adaptation to abiotic stress, a feature key to the survival of the plant host (**Figure 9.16**).

9.18 Conclusions

The sedentary lifestyle of plants has forced them to develop intimate relationship with their microbial communities which are essential cellular components of these photosynthetic hosts. Although modified microbes constituting the chloroplast and the mitochondrion are part of plants, these multicellular organisms also harbour microbes on the surface and within. The microbial community is also an important component of the adaptive tool plants have at their disposal and can be activated in order to survive in a changing environment. For instance, carnivorous plants will be unable to resort to a protein diet without the assistance of their microbial community in trapping and digesting of their prey. Proliferating in a habitat with elevated

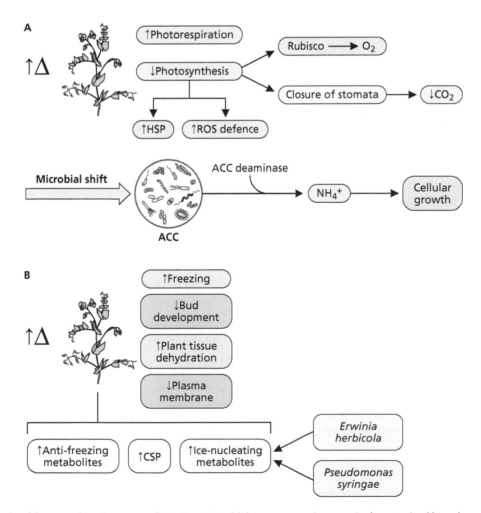

Figure 9.16. Plant microbiome and environmental temperature. (a) A temperature increase is characterized by a decrease in photosynthesis and an increase in photorespiration. Heat-shock proteins (HSPs) and ROS detoxification are enhanced. Microbes with the ability to degrade ACC and liberate NH_4^+ are promoted. (b) A decrease in temperature provokes freezing and inhibits development. Cold-shock proteins (CSP) are increased and microbes with anti-freezing/ice-nucleating metabolites are selected.

BOX 9.2 BIOFERTILIZERS: PLANT GROWTH-PROMOTING MICROBES

Once the significance of microbes as important cellular components of plants is unravelled and the diverse essential functions they fulfilled are deciphered, this realization will lay the foundation for improved agricultural products and will aid in combatting climate change. Currently, microbial inoculations are commonly utilized to fortify a desired function in plants, to provide a more conducive environment for improved productivity, to mitigate adverse conditions and to defend against any pathogenic infestation. Biofertilizers such as *Azospirillum* spp., *Azotobacter* spp. and *Rhizobium* spp. are applied to plants to enhance nitrogen fixation and to increase the protein content of the edible products they produce. To improve phosphate nutrition, plants are treated with *Pseudomonas* spp. with the ability to solubilize this important nutrient. Siderophore-secreting bacteria help plants acquire Fe, a pivotal ingredient in a variety of biological processes. The hydroxamate and catecholate moieties these siderophores possess scavenge Fe and make it easily bioavailable to the host. *Bacillus* spp. is a potent defence against the necrophytic fungus *Botitris cinerea* responsible for the damage caused to numerous plants. These bacteria release such cyclic lipopeptides as surfactin and fengycin that are associated with fungicidal activity.

The climate change that the planet is experiencing is punctuated with erratic weather patterns that result in extreme heat, prolonged drought and frost conditions in regions accustomed to tropical temperatures. These dramatic fluxes in the ecosystem impose tremendous stress on the flora and fauna, a situation bound to impact the biodiversity and agricultural output. Microbes can help limit the negative impact on plants as they participate in numerous functions aimed at enabling their hosts to adapt to environmental stress. For instance, grapevines treated with *Paraburkholderia phytofirmans* tend to tolerate cold conditions as this microbe impedes the synthesis of C_2H_4, while water stress leading to oxidative stress is mitigated by the application of microbes with enhanced glutathione peroxidase and ascorbate peroxidase activity. *Penicillium* spp. also helps during drought situation as

this microbe with its extensive hyphae network allows for improved water conductivity and soil exploration by the roots. Microbes with ability to generate terpenes and other VOCs can promote the assembly of microbes contributing to the synthesis of gibberellin, a phytohormone responsible to the overexpression of aquaporins. These are H_2O transporters mediating the increased uptake of H_2O. Hence, different strategies can be adopted to prevent plants from falling victims to the deteriorating ecosystem triggered by climate change. The inclusion of microbial remedies to improve the health of the coral reef, human health and plant health is becoming a common practice as a better understanding of the intimate interactions of these microbial cells with their hosts is achieved (**Box Figure 2**).

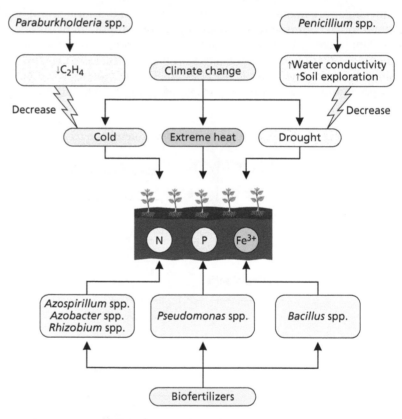

Box Figure 2. **Microbial remedies aimed at food security and climate change.**

concentrations of inorganic and organic pollutants is dependent on the resident microbes with the attribute of detoxifying these contaminants. A rise and a decrease in temperature that can manifest in drastic cellular injury and even death are mitigated with the input of microbial partners. The intricate social network operating amongst the residents of the plant holobiont is central in dissuading predators from infesting the host and in this case the volatile organic messengers produced by the microbes can even trigger adaptive responses in organisms rooted far away.

The appreciation and understanding of the contribution of the microbial community which is part of multicellular photosynthetic organisms will pave the way to improve food security, mitigate climate change and preserve biodiversity. The application of proper microbial inoculants is already yielding better agricultural output and promoting the protection of the environment. Even the nutritional quality of the agricultural products generated is seeing significant improvement with plant-derived foods being rich is such nutrients as vitamins, unsaturated lipids, proteins and anti-oxidants. Proper signalling network triggered by microbial partners may result in diminishing demand for pesticides and other chemicals aimed at maintaining plant health. Introduction of microbial partners with a diverse biochemical repertoire within plants may allow the production of agricultural goods in regions that are arid, are subjected to drastic temperature fluxes and are prone to continual flooding. Plant microbiome may hold the solutions to the numerous challenges the globe is facing as the host needs their microbial components to survive and proliferate.

SUGGESTED READINGS

Araya, J. P., González, M., Cardinale, M., Schnell, S., & Stoll, A. (2020). Microbiome dynamics associated with the atacama flowering desert. *Frontiers in Microbiology, 10*, eCollection2019. doi:10.3389/fmicb.2019.03160

Armitage, D. W. (2016). Bacteria facilitate prey retention by the pitcher plant *Darlingtonia californica. Biology Letters, 12*(11), 1–4. doi:10.1098/rsbl.2016.0577

Barrow, J. R., et al. (2008). Do symbiotic microbes have a role in regulating plant performance and response to stress? *Communicative & Integrative Biology, 1*(1), 69–73. doi:10.4161/cib.1.1.6238

Beans, C. (2021). Drought causes lasting changes to the rice root microbiome. *PNAS.* https://blog.pnas.org/2021/08/drought-causes-lasting-changes-to-the-rice-root-microbiome/

Berg, G. (2014). The plant microbiome and its importance for plant and human health. *Frontiers in Microbiology, 5*, 1–2. doi:10.3389/fmicb.2014.00491

Canarini, A., et al. (2019). Root exudation of primary metabolites: mechanisms and their roles in plant responses to environmental stimuli. *Frontiers in Plant Science, 10*, eCollection2019. doi:10.3389/fpls.2019.00157

Carvalho, S. D., & Castillo, J. A. (2018). Influence of light on plant–phyllosphere interaction. *Frontiers in Plant Science, 9*, eCollection2018. doi:10.3389/fpls.2018.01482

Chen, T. et al., (2020). A plant genetic network for preventing dysbiosis in the phyllosphere. *Nature, 580*(7805), 653–657. doi:10.1038/s41586-020-2185-0

Compant, S., et al. (2010). Climate change effects on beneficial plant-microorganism interactions. *FEMS Microbiology Ecology.* 3(2), 197–214. doi:10.1111/j.1574–6941.2010.00900.x

Del Cueto, J. et al., (2017). Cyanogenic glucosides and derivatives in almond and sweet cherry flower buds from dormancy to flowering. *Frontiers in Plant Science, 8*, eCollection2017. doi:10.3389/fpls.2017.00800

Dorokhov, Y. L., et al. (2018). Methanol in plant life. *Frontiers in Plant Science, 9*, 1–6. doi:10.3389/fpls.2018.01623

Finkel, O. M., et al. (2017). Understanding and exploiting plant beneficial microbes. *Current Opinion in Plant Biology, 38*, 155–163. doi:10.1016/j.pbi.2017.04.018

Fitzpatrick, T. B., & Chapman, L. M. (2020). The importance of thiamine (vitamin B1) in plant health: from crop yield to biofortification. *Journal of Biological Chemistry, 295*(34), 12002–12013. doi:10.1074/jbc.rev120.010918

Frank, A., et al. (2017). Transmission of bacterial endophytes. *Microorganisms, 5*(4), 70. doi:10.3390/microorganisms5040070

Grady, K. L., et al. (2019). Assembly and seasonality of core phyllosphere microbiota on perennial biofuel crops. *Nature Communications, 10*(1), 1–10. doi:10.1038/s41467-019-11974-4

Herrera, H., et al. (2020). Mycorrhizal fungi isolated from native terrestrial orchids from region of La Araucanía, Southern Chile. *Microorganisms, 8*(8), 1120. doi:10.3390/microorganisms8081120

Hsiao, C., et al. (2019). Synthesis and biological evaluation of the novel growth inhibitor streptol glucoside, isolated from an obligate plant symbiont. *Chemistry - A European Journal, 25*(7), 1722–1726. doi:10.1002/chem.201805693

Huang, A. C., et al. (2019). A specialized metabolic network selectively modulates *Arabidopsis* root microbiota. *Science, 364*(6440), 1–10. doi:10.1126/science.aau6389

Huang, A. C., & Osbourn, A. (2019). Plant terpenes that mediate below-ground interactions: Prospects for bioengineering terpenoids for plant protection. *Pest Management Science. 75*(9), 2368–2377. doi:10.1002/ps.5410

Ionescu, I. et al. (2016). Chemical control of flowering time. *Journal of Experimental Botany. 68*(3), 369–382. doi:10.1093/jxb/erw427

Kenny, D. J., & Balskus, E. P. (2018). Engineering chemical interactions in microbial communities. *Chemical Society Reviews, 47*(5), 1705–1729. doi:10.1039/c7cs00664k

Kim, Y., et al. (2020). Root response to drought stress in rice (Oryza sativa L.). *International Journal of Molecular Sciences, 21*(4), 1513. doi:10.3390/ijms21041513

Koo, A. J. (2017). Metabolism of the plant hormone jasmonate: A sentinel for tissue damage and master regulator of stress response. *Phytochemistry Reviews, 17*(1), 51–80. doi:10.1007/s11101-017-9510-8

Kurkjian, H., et al. (2020). The impact of interactions on invasion and colonization resistance in microbial communities. *Plos Computational Biology*, 17 (1), e1008643. doi:10.1101/2020.06.11.146571

Lanfranco, et al. (2017). Strigolactones cross the kingdoms: Plants, fungi, and bacteria in the arbuscular mycorrhizal symbiosis. *Journal of Experimental Botany, 69*(9), 2175–2188. doi:10.1093/jxb/erx432

Leach, J. E., et al. (2017). Communication in the phytobiome. *Cell, 169*(4), 587–596. doi:10.1016/j.cell.2017.04.025

Marasco, R., et al. (2018). Rhizosheath microbial community assembly of sympatric desert speargrasses is independent of the plant host. *Microbiome, 6*(1), 1–19. doi:10.1186/s40168-018-0597-y

Meena, K. K., et al. (2017). Abiotic stress responses and microbe-mediated mitigation in plants: The omics strategies. *Frontiers in Plant Science, 8*, 1–25. doi:10.3389/fpls.2017.00172

Mendes, R., Garbeva, P., & Raaijmakers, J. M. (2013). The rhizosphere microbiome: Significance of plant beneficial, plant pathogenic, and human pathogenic microorganisms. *FEMS Microbiology Reviews, 37*(5), 634–663. doi:10.1111/1574-6976.12028

Mitter, B., et al. (2017). A new approach to modify plant microbiomes and traits by introducing beneficial bacteria at flowering into progeny seeds. *Frontiers in Microbiology, 8*, eCollection2017. doi:10.3389/fmicb.2017.00011

Mishra, S., et al. (2020). The plant microbiome: A missing link for the understanding of community dynamics and multifunctionality in forest ecosystems. *Applied Soil Ecology, 145*, 1–6. doi:10.1016/j.apsoil.2019.08.007

Ortíz-Castro, R., et al. (2009). The role of microbial signals in plant growth and development. *Plant Signaling & Behavior, 4*(8), 701–712. doi:10.4161/psb.4.8.9047

Park, Y., & Ryu, C. (2021). Understanding plant social networking system: Avoiding deleterious microbiota but calling beneficials. *International Journal of Molecular Sciences, 22*(7), 3319. doi:10.3390/ijms22073319

Rebolleda-Gómez, et al. (2019). Gazing into the anthosphere: Considering how microbes influence floral evolution. *New Phytologist, 224*(3), 1012–1020. doi:10.1111/nph.16137

Sangiorgio, D., et al. (2020). Facing climate change: Application of microbial biostimulants to mitigate stress in horticultural crops. *Agronomy, 10*(6), 794. doi:10.3390/agronomy10060794

Sickel, W., et al. (2019). Venus flytrap microbiotas withstand harsh conditions during prey digestion. *FEMS Microbiology Ecology, 95*(3), 1–12. doi:10.1093/femsec/fiz010

Sirová, D., et al. (2018). Hunters or farmers? Microbiome characteristics help elucidate the diet composition in an aquatic carnivorous plant. *Microbiome, 6*(1), 225–239. doi:10.1186/s40168-018-0600-7

Taghinasab, M., & Jabaji, S. (2020). Cannabis microbiome and the role of endophytes in modulating the production of secondary metabolites: An overview. *Microorganisms, 8*(3), 355. doi:10.3390/microorganisms8030355

Toju, H. et al. (2019). Factors influencing leaf- and root-associated communities of bacteria and fungi across 33 plant orders in a grassland. *Frontiers in Microbiology, 10*, eCollection2019. doi:10.3389/fmicb.2019.00241.

Wheatley, R. M., & Poole, P. S. (2018). Mechanisms of bacterial attachment to roots. *FEMS Microbiology Reviews. 42*(4), 448–461. doi:10.1093/femsre/fuy014

CONCLUSIONS AND FUTURE PERSPECTIVES

Since the origin of planet earth, the prevailing random physical and chemical activities have helped guide and shape the emergence of the precise (bio)molecular diversity which has eventually coalesced into chemical entities with the ability to store information, generate energy, and replicate in discrete amphipathic and asymmetrical vesicles. These phenomena gave rise to the first living system referred to as the last universal common ancestor (LUCA), the precursor of eubacteria and archaea that populate our environment today. This long and arduous process punctuated by happenstance but dictated by the physicochemical, geological and biological constraints resulted in the establishment of microbial life in all its diverse manifestations and functional specificity. These organisms with a wide range of adaptive attributes are well-rooted in the habitats where they are found and perform a variety of tasks essential for the survival of the ecosystem. The eukaryotes that possess numerous intracellular organelles where specific biochemical networks are housed arose from eubacteria and archaea. The mitochondrion, the energy powerhouse in all aerobic organisms, and the chloroplast, an organelle dedicated to the conversion of light energy into ATP, can be traced to their bacterial ancestors.

The fingerprint of microbial life is not only evident in the emergence of eukaryotes but is also prominent in the advent of multicellular organisms. In all multicellular systems, microbial components form an integral part of the energy-generating machinery, signalling networks and replication apparatus. In fact, bacteria with their own genetic information are important constituents of these complex cellular organisms where they fulfil vital functions. As these organisms grew in size and the need for specialized cells became pivotal, bacteria provided the cues, resulting in tailored-made cells designed for select functions. The signals the microbes generate is key in modulating the development of function-specific cells. In hydra, microbes expressing lipoxygenases contribute to morphogenic changes, leading to the differentiation of the stem cells into specific cells. Furthermore, antimicrobial peptides secreted by the resident microbes help prime the immune network of this water-dwelling creature. In humans, butyrate plays an important role in the genesis of colonocytes. Gnotobiotic organisms are characterized by numerous developmental and functional abnormalities. Thus, bacteria are essential constituents of most if not all multicellular organisms. This occurrence is not unusual as it is well known that nature utilizes ingredients common in its environment in order to mould any living system whether it is a unicellular or supra-structured organism. Any constituent resulting in a failure or a non-functional entity is quickly discarded in favour of a working concoction; hence, the intimate association of microbes with their hosts.

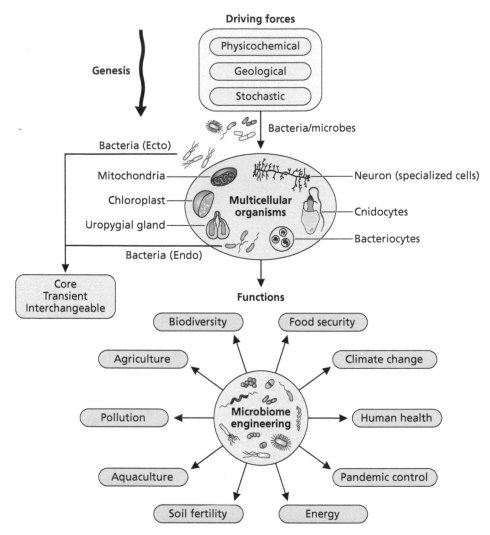

Figure 1. The physicochemical and geological forces leading to the establishment of microbial and multicellular organisms. Microbes are an integral constituent of most multicellular organisms and their functions are linked to the well-being of the host.

Microbes afford a degree of adaptability and interchangeability not provided by stem cells. Stem cells have a set of genetic information that can be switched on to differentiate into specific cell type in order to rejuvenate senescent or non-functional cells. These are part of the cellular package inherent in the host and are usually confined to the information they contain, which is limited to the functions they can carry out. They usually respond to signals generated by the host and the microbes within. The stem cells are usually incapable of Morphing into cell types they do not have the information for. This is in sharp contrast to microbial cells that readily adapt to external situations, a characteristic essential for the survival of the host. They can usually be replaced if they do not contribute to the well-being of the host. For instance, the gut microbes change with the diet consumed or the geographical location of the host. A vegetarian has a different set of microbes in the digestive tract compared to an omnivore. Sea anemones residing on the surface of oceans comprise photosynthetic microbes while those living at the bottom of the ocean where light can barely penetrate and methane is an important carbon source, discard the photosynthetic microbes in favour of methanogens. The same organism can adopt this disparate lifestyle only due to the microbial cells it harbours. The functional plasticity and replaceability inherent in microbes are far-reaching compared to the limited ability of the stem cells to adjust. Thus, microbes encompass a degree of adaptability not bestowed by the stem cells. The adaptive range of the latter is narrow compared to the highly malleable qualities associated with microbes. Thus, the adaptable microbes that can be readily substituted by other microbial systems are more suited to the functional needs of the host. Microbes provide that extra survivability power compared to functionally limited stem cells.

It is also important to note that organisms with limited mobility usually recruit a wider assortment of microbes with specific traits to navigate the intricacies of life. Organisms with a sessile lifestyle tend to have more microbes. For instance, plants depend on their microbial consortia to acquire nutrients and fend against parasites and predatory invasion. The root is replete with microbes that are busy communicating and helping the host accommodate with the sedentary life it is confined to. Although plants have specialized cells to recognize these predatory assaults, they depend on their microbes to elaborate warning signals against the intruders and prompt the other resident microbes to mount intricate defence mechanisms to combat the pathogens. Their supply of nitrogen and essential minerals is readily provisioned with the assistance of the microbial partners. In corals, algae and bacteria work in tandem to allow this sessile animal to lead a highly reproductive lifestyle. The algae supply energy while the bacteria contribute to Ca^{2+} homeostasis, important contributors that fuel the growth of the coral holoorganism. Multicellular organisms with a limited number of specialized cells depend on their microbes for numerous functions and are more adaptable to fluxes in their ecosystem. Living organisms with more specialized cells involved in fulfilling precise functions are somewhat less reliant on their microbes than organisms with fewer specialized cells. However, higher degree of specialization is inherently associated with less adaptability and more prone to extinction following a sudden change in the ecosystem. The inclusion of microbes as an integral part of an organism increases the survivability of the host. Thus, denuding an organism of its microbes is akin to eliminating an organ or a specialized cell, a situation the host cannot tolerate.

In numerous cases, these microbes live as symbionts. Such an arrangement based on the interdependence that one partner cannot survive without the other is less amenable for replacement. The nitrogen-fixing bacteria in leguminous plants, the light-emitting microbes in squids, the scent-producers in hyenas or the bacteria residing in the avian uropygial glands are part of the microbe-containing specialized organs that these organisms have evolved with. Regardless of the degree of intimacy or interdependence in the microbe-host relationship, these two components evolved in tandem with the former shaping a plethora of anatomical features and functions the holobiont possesses. The immune system, the intracellular signalling network and the gut are shaped by microbial constituents. The digestive tract, where most microbes are harboured in numerous organisms, reflects this intimate conversation between the host and the microbial population. Herbivores, in general, tend to have a longer digestive system than omnivores and carnivores. In ruminants, the rumen, part of the foregut is designed to maintain most microbes where cellulolytic materials are broken down. Flying squirrels living on top of trees have a diet rich in seeds, flowers, fruits and leaves. They possess a simple stomach but have an elongated caecum where food processing continues over a longer period and toxins are detoxified. These morphological adaptations guided by the microbial residents enable the hosts to maximize the optimal extraction of nutrients from their diets, a process further enhanced by microbial activities. The partnership between multicellular organisms and their microbial components evolved concomitantly. Thus, it is impossible to imagine a host devoid of its microbes as, in most likelihood, it would not exist and if it did, the organism would be completely different. Such a scenario will be unlikely in the current prevailing conditions on planet earth as microbial life is ubiquitous. It is thus quite obvious that microbes and multicellular organisms are intimately and intricately linked despite having their disparate genetic networks. These information modules work in a cooperative fashion for the betterment of the holobiont in a manner analogous to the mitochondrion. The latter has its own genes. However, numerous proteins it elaborates are synthesized jointly with input from 'host' as do the plant-residing chloroplasts. Numerous products are generated jointly by the genetic input of the microbes and the hosts. Hence, these organisms constituting the holobiont are interdependent.

Ageing is a process that affects most multicellular organisms and eventually results in death. DNA lesions, decrease in protein quality control, organelle malfunction, diminished cellular rejuvenation aberrant signalling network, oxidative stress and change in the microbiome are some of the factors that promote ageing and diminish longevity. Amongst vertebrates, the pygmy goby lives up to 8 weeks while the Greenland sharks can survive for 400 years. Herbivores usually have a longer lifespan than omnivores or carnivores. Carnivores tend to have the shortest longevity. For instance, tigers live for 20 years and giant tortoises with herbivorous diet can live up to 150 years. The significance of microbes in the ageing process and in modulating longevity is becoming increasingly evident. The decrease in microbial diversity is linked to a diminished life span. Humans usually have a life expectancy of 80 years. Population living fewer years is associated with a diminished diversity in microbial constituents while centenarians harbour an increased diverse microbial community. Microbial transplant of gut microbes from young killifish to middle-aged counterpart results in an increase in the lifespan of the latter. These fishes are short-lived vertebrates. The older fish has more *Proteobacteria* compared to the young one that is enriched in *Bacteroidetes* and *Firmicutes*. Humans having increased amount of *Bifidobacteria* and *Lactobacillus* tend to live longer than those where these microbes are diminished. Humans with a shorter lifespan tend to have an increased amount of *Clostridia* and *Enterobacteria*. Germ-free mice are short-lived and die fast. Microbe-generated metabolites such as nicotinic acid promote stem-cell proliferation and modulate sirtuins, features that aid in extending longevity. The former can rejuvenate different cell types while the latter responsible for epigenetic modification can program biochemical processes aimed at mitigating the accumulation of age-related molecular damage. Thus, age-mediated decline is restricted resulting in longer lifespan. It is not surprising then that centenarians possess a more diverse microbial population.

The incorporation of microbes in multicellular organisms as organelles, as symbionts and as integral microbial cells is central to the biodiversity we experience on planet earth. These microbial constituents make multicellular organisms the way they are and contribute to numerous functions in a manner analogous to any organ. This burgeoning discipline of microbiome has already revealed some of the essential roles microbial communities play in the well-being of their hosts. In fact, these microbes are so well integrated with the holoorganisms that the latter cannot exist without their microbes. Once the assembly of the microbial partners is fully delineated and molecular cross-talk amongst the protagonists is deciphered, a new era of biological engineering will emerge. This will lay the foundation to generate new therapies dedicated to diseases triggered by dysbiosis, to fortify the immune response and to extend longevity. Food security mediated by microbiome engineering of plants having unique features to yield more nutritional products and to grow virtually in any environment will become a reality. Climate change can be managed by promoting more CO_2 sequestration and more O_2 release. Additionally, the emission of CH_4 can be quelled by programming the metabolic networks in ruminants to utilize this carbon and limit its discharge. Biodiversity can be maintained and promoted by managing the microbial residents in various hosts. For instance, the bleaching of coral reefs induced by the elimination of microbial life can be reversed by microbiome engineering. Seeding these reefs with beneficial microbes and with prebiotics favouring the proliferation of select microbes can bring back life to these bleached $CaCO_3$- rich habitats. Pest control assisted by microbes will be beneficial to farmers and the environment. In this instance, the desired insects can be targeted by microbes with the information to render them infertile, a situation that can effectively arrest their reproduction. On the other hand, the dwindling population of insect pollinators can be resuscitated by an effusion of a dose of microbes promoting their reproductive capacity. Prebiotics stimulate the proliferation of these insect reproduction - enhancing microbes can also be contemplated. Almost all multicellular organisms need their microbes and these partners have evolved in tandem to generate the flora and fauna in the earthly ecosystem. The molecular understanding of this relationship is poised to reveal immense opportunities that will be beneficial to the planet.

INDEX

Milton Keynes UK
Ingram Content Group UK Ltd.
UKHW050441111024
449327UK00050B/438

9 780367 749897